PHYSICAL CHEMISTRY
PRACTICAL

PHYSICAL CHEMISTRY PRACTICAL

[for Honours, Postgraduate, and Chemical Engineering students]

Dr Saroj Kumar Maity PhD
Associate Professor, and Head of the Department of Chemistry
Chakdaha College, Nadia
Formerly, Lecturer in Chemistry
Nabadwip Vidyasagar College, Nadia
and Senior Lecturer in Chemistry
Vidyanagar College, South 24 Parganas

Dr Naba Kumar Ghosh PhD
Assistant Professor, Jangipur College, Murshidabad
Ex-Technical Assistant, Grade I
Physical Chemistry Section
Department of Chemistry, University of Kalyani

New Central Book Agency (P) Ltd

Co-Published by

Bijay Publications

E-mail : bijaypublications@gmail.com

Exclusively Distributed by

Bijay Pustak Bhandar (P) Ltd

90/6A M. G. Road, Bhawani Dutta Lane, Kolkata-700007

✆ : (033) 4605-7856, 8777898576, 8274916104

E-mail : bijaypustakbhandar@gmail.com

NCBA

REGD OFFICE
8/1 Chintamoni Das Lane, Kolkata 700 009, India
email: ncbakolkata@gmail.com

OVERSEAS
NCBA (UK) Ltd, 149 Park Avenue North, Northampton, NN3 2HY, UK
email: ncbauk@yahoo.com

CORPORATE OFFICE
212 Shahpur Jat, New Delhi 110 049
email: delhincba@gmail.com

BIJAY PUBLICATIONS
90/6A M. G. Road, Bhawani Dutta Lane, Kolkata-7
E-mail : bijaypublications@gmail.com

BIJAY PUSTAK BHANDAR (P) LTD
90/6A M. G. Road, Bhawani Dutta Lane, Kolkata-7
E-mail : bijaypustakbhandar@gmail.com

PHYSICAL CHEMISTRY PRACTICAL • Saroj Kumar Maity and Naba Kumar Ghosh

First Published: July 2012
Reprinted: January 2015, 2024

PUBLISHER • TYPESETTER • PRINTER
New Central Book Agency (P) Ltd
8/1 Chintamoni Das Lane, Kolkata 700 009

CO-PUBLISHED BY
Bijay Publications
90/6A M G Road, Bhawani Dutta Lane, Kolkata 700007

TECHNICAL EDITOR
Dipan Roy

PROJECT TEAM
Swapan Sarkar, Koushik Samanta and Prabhat Jas

ISBN: 978 81 7381 698 7

Price ₹ 495.00

Dedicated to

My Late Parents and Late Father-in-law

—Saroj Kumar Maity

and my Late Father

—Naba Kumar Ghosh

Contents

Contents

Preface

ıysical Chemistry concerns itself mainly with theoretical discussions based on Mathematics—both mple and difficult problems. Those theories provide physico-chemical data, needed for all branches of ience, through precise experimental techniques. Students at their first entry in the gateway of science not find any scope to be acquainted with so-called Physical Chemistry Practical till their footstep in e B.Sc. courses, and, to be honest, in the Honours courses. They have had some idea about Inorganic actical and, though faint, about Organic Practical too. In their very first practical class in Physical hemistry, they remain far from the essence of this subject. And, as we see, they are advised to prepare lot of solutions, arrange bottles, take notes and so on. They follow the instructions as non-technical llowers. The reason is their lack of theoretical background about the experiments to be done and struments or equipments to be used by them.

During classes, we have always felt the dire need of a student-friendly guide-book for Physical hemistry Practical which may largely motivate and instruct the students towards their desired works. ith this aim in view this book, **Physical Chemistry Practical**, has been written.

In our first attempt as many as seventy-seven experiments—along with suggestions for some further periments—have been discussed in nine chapters. Each chapter begins with a theoretical background ıder the name 'Introductory Words' and ends with a relevant Appendix. An instrumental background s been provided wherever necessary. We have observed that the students feel relaxed when they go rough the 'Procedure' first and then write the Laboratory Notebook.

The book covers almost all types of experiments in Physical Chemistry **spread over the syllabi of fferent Indian universities and abroad for Undergraduate, Postgraduate**, and **Chemical ngineering Courses**. We are committed to include many more experiments in the next edition. A short mmunication has been dedicated exclusively to the students at the beginning of the book. Cleanliness, actness, and economy (i.e., use of as small amount of chemicals as possible) is the goal of the book. mention precisely, all the experiments have been done by the authors themselves.

Some shortcomings may have crept in despite our sincere efforts. We earnestly invite constructive ıd intuitive suggestions from our fellow-teachers, demonstrators, laboratory instructors and, surely, ır beloved students, for the betterment of the book.

We attribute our sincere and respectful thanks to our teacher, Prof. Sujit Chandra Lahiri, Professor meritus, ex-teacher of the University of Kalyani—who still works in the laboratory—for his continuous ggestions. We believe that his insistence on accuracy has been reflected in this book.

We do remember, with a deep sense of gratitude, Late Prof. Dilip Kumar Majumder of University of alyani, whose peevishness and apparent vexation to students are important and inspiring elements for riting this book meticulously—as far as possible.

We pay our heartfelt respects to teachers as honourable as Prof. S. C. Rakshit, Prof. R. S. Banerje Prof. D. Pramanik, Prof. N. G. Mukherjee, Prof. A. Guha and a good many respected teachers wi whom we had been lucky to be directly or indirectly in contact with regard to Physical Chemist Practical.

We thank Dr. Ashok Prasun Chatterjee and Prof. Bidyut Guhaniyogi of the University of Kalya Dr. Swapan Kumar Chakraborty, WBES, and Prof. Ashok Mukherjee of Nabadwip Vidyasagar Colleg Nadia, for their constant enthusiasm towards our arduous effort in the preparation of this book.

It will be unfair if we fail to mention about our students and acknowledge them, without who continuous queries during practical classes we would not have been able to prepare the manuscript this book according to their needs.

Last but not the least, we thank Mr. Amitabha Sen, Director, Mr. R. M. Nath, Mr. Prabhat Jas a Mr. Koushik Samanta of New Central Book Agency (P) Ltd for their tireless effort in bringing out th title expeditely in its present form.

Saraswati Puja
28 January 2012
Kolkata

Maity and Ghos

Introductory Notes for Students

First read the 'theory' vividly from this book and, if necessary, from a standard Textbook on Physical Chemistry.

Go through the experiment only when you are well-conversant with the theory and the procedures mentioned against each experiment.

For this purpose you draw a scheme on your **'Hand-note Diary'** for Physical Chemistry Experiments with the following points:

(a) Write the name of the experiment, date, names of your partners (as generally happens in a lab—with 2-3 workers performing an experiment jointly).

(b) List the equipments and materials necessary for the experiment.

(c) Prepare a brief but useful working procedure. This will help you in your future works (even for research purposes) and during the Practical Examination.

(d) Insert all data neatly and stepwise.

Remember, this hand-note diary by no means is less important than your main Laboratory Note Book (LNB).

For an instrument or any glass equipment like Viscometer, Stalagmometer, etc., be confident that you are in a position to smoothly handle it.

For electrical and electronic instruments, be well-conversant with the electric poles, switch, and connections before handling the instruments. **A rough or wrong handling may often damage a precious instrument.**

In case of any confusion regarding working with an instrument or at any stage of the experiment, immediately consult your teacher or lab-instructor.

Label all the solutions quoting the name, strength, bottle no., your name, and date of preparation of each and every solution.

For drying of glass goods after washing with distilled water, a few drops of acetone or absolute alcohol may be used to rinse, followed by introducing in an oven for a short period.

Be economical with your lab-reagents and both tap-water and distilled water.

Apart from the instructions in this book, before going into the experiment, make an assessment of the volume of a particular solution needed.

For Physical Chemistry Practical, in particular, volumetric flasks of 25 ml and 50 ml capacities, pipettes of good quality glass of 2 ml, 5 ml and 10 ml capacities are of great use. Though these cost proportionately a bit high, they save the institution economically as well as help a student to save time and achieve good results.

8. Since different pipettes and burettes may differ slightly in volume, avoid using a number c such equipments as far as possible.
9. Strong alkali solutions corrode glass. Since these are secondary standard solutions, us polythene bottles or common reagent bottles for preparation of NaOH and KOH solutions. is advisable not to use volumetric flasks. If that be inevitable, as in burette also, immediatel drain out the solution after work is over and wash well with water. Polythene burettes shoul also be used.
10. NaOH contains many impurities. Always take a bit larger weight than what is instructed c calculated by you.
11. Keep your working table/rack/basin/sink clean and well-arranged after completion of worl
12. **Recording of laboratory/room temperature**

    Place sufficient water in a 250 ml beaker. After half an hour, insert a thermometer in th water and note the temperature. Again take the temperature after the experiment is ove Note the mean of the two temperatures.
13. **Preparation and use of chromic acid for cleansing of glass apparatus**

    Preparation of chromic acid is done very cautiously. Dissolve about 250 g $K_2Cr_2O_7$ in 50 ml water in a one-litre beaker. **Warm gently** during dissolution but **neither boil nor eve heat strongly**. Slowly add conc. H_2SO_4 and stir with a glass rod until the solution become dark red and crystals of chromium trioxide begin to settle. Stop addition of H_2SO_4. Th chromic acid so formed is ready for cleaning of glasswares.

    An uncleaned pipette, burette, viscometer or stalagmometer is filled up with slightl warm chromic acid and kept for a few hours. Pour back the acid in the container for reus The apparatus is washed with water several times. For drying, it is rinsed with a few ml c acetone and dried in an oven or by passing dry air.
14. **Recording in LNB and setting of Graphs**

    Most of the experiments in Physical Chemistry are characterised by graphs. The quality an preciseness of your work is immediately reflected in the graph. So you will have to be wel acquainted with the nature of the graph of a particular experiment.

    On the other hand, there is every scope to manipulate the result with a very small variatio in the slope. In fact, much can be said about graphs. Which points are to be accepted an which are to be rejected depend on a number of factors.

    The important thing is that you are to perform your work sincerely with due attentivenes In most cases you are supposed to be the best judge to accept or cancel a point on the grapl But, after all, **you must consult your teacher after drawing the graph with th experimental data**. He will guide you properly. Do not insist on a pre-thought result or book-value. Your lab-conditions, quality of materials and apparatus all play important role behind your ultimate result. So, never try to manipulate—it will harm you in many respect Just note precisely what you achieve from the experiment.

    Each of the experiments is to be done 2-3 times to be well-conversant with the procedure and use of equipments—as you do in Organic or Inorganic Practicals. In those cases yo have to complete the analysis of at least 5/6 unknown samples.

Take the dependent variable as ordinate (Y-axis). For example, in kinetic graphs, time is represented by the X-axis and concentration by the Y-axis, because concentration changes with time. Similarly, in potentiometric curves, emf is placed at the Y-axis and drops of oxidant or reductant at the X-axis, because emf changes with the addition of the oxidant or reductant.

Before setting the graph in LNB, show it to your teacher. Obtaining his green signal, paste it in the appropriate place.

The items to be incorporated in your LNB are:

i. Name of the experiment in the Index page with page number.

ii. Name of the experiment as heading in the appropriate page with date and experiment number.

iii. Theory/Principle in your own words, and, as far as practicable, in brief. Mind that you are to recollect this theory while writing in the examination script.

iv. A short procedure of the experiment. This is not essential, yet, for quick revision, it is very handy, and so prepare one in your LNB.

v. Laboratory temperature.

vi. All data in tabular form.

vii. Calculations and results with proper units wherever necessary. Calculations may be shown in the left hand white pages of the LNB also.

viii. Conclusion.

CHAPTER 1

Experiments Related to Viscosity of Liquids

CHAPTER 1

Experiments Related to Viscosity of Liquids

I. Introductory Words

During the flow of a liquid through a tube, the liquid is assumed to be made up of a number of concentric cylindrical layers with a small distance dr in between two consecutive layers. The layer in contact with either of the wall is stationary due to adhesion. The velocity of the subsequent layers approaching the centre of the tube increases with a velocity gradient $\frac{du}{dr}$, the central layer having the highest velocity. Afterwards, again the velocity decreases and at the opposite wall it becomes zero.

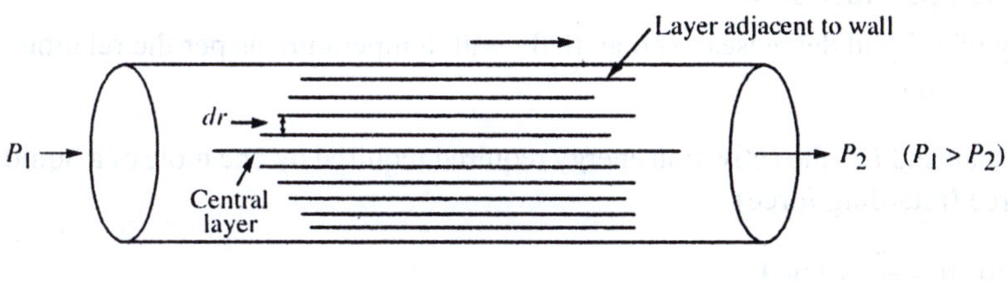

The difference in velocities may be attributed to some kind of backward force acting on a layer in motion. A relatively faster layer tries to accelerate the relatively slower layer adjacent to it and a slower layer tries to retard the relatively faster layer in contact with it. This resistance to relative motion between two layers of a liquid is termed as 'viscosity'.

The existence of viscous forces comes into play only when the liquid is in motion. And this motion developes when a force is applied to the liquid at one end of the tube. Gases (also fluids) exhibit the viscous property too.

If the applied force be F, the retarding force or the viscous force will be $-F$. This force is proportional to the area of contact (A) between two moving layers and to the velocity gradient.

$$\therefore \quad F \propto A\frac{du}{dr}$$

$$\text{or,} \quad F = \eta A\frac{du}{dr}, \tag{1}$$

where η is called the viscosity coefficient, a property of the particular liquid, **which is defined as the force required to maintain unit velocity gradient between two layers of unit area of contact**.

It is to be mentioned that equation (1) is valid for slow motion of any fluid, more appropriately, for laminar flow or streamline motion.

Unit of η

From (1), $$\eta = \frac{F}{A} \cdot \frac{1}{du/dr} \qquad (2)$$

$\therefore$ in C.G.S. units, $\eta = \frac{\text{dyne}}{\text{cm}^2} \cdot \frac{1}{\frac{\text{cm/s}}{\text{cm}}} = \frac{\text{g} \cdot \text{cm/s}^2}{\text{cm}^2} \cdot \text{s} = \mathbf{g \cdot cm^{-1} s^{-1} = poise}$.

Poiseuille equation

If V be the volume of a liquid flowing in time t through a capillary tube of length l, radius r due to difference in pressures at the two ends, then according to Poiseuille,

$$\eta = \frac{\pi r^4 t (P_1 - P_2)}{8lV}. \qquad (3)$$

Equation (3) is known as Poiseuille equation.

Writing $P_1 - P_2 = P$, the pressure head,

$$\eta = \frac{\pi r^4 t P}{8lV}. \qquad (4)$$

Equation (4) is used to determine the absolute viscosity of a liquid.

Influence of temperature on η

The viscosity of a liquid decreases exponentially with temperature as per the relation

$$\eta = Ae^{E/RT}, \qquad (5)$$

where A is a constant and E is the activation energy required required by one mole of a liquid to overcome the frictional force (retarding force).

Taking log, $$\log \eta = \frac{E}{RT} + \log A. \qquad (6)$$

The dependence of η on absolute temperature are shown graphically following equations (5) and (6) as under.

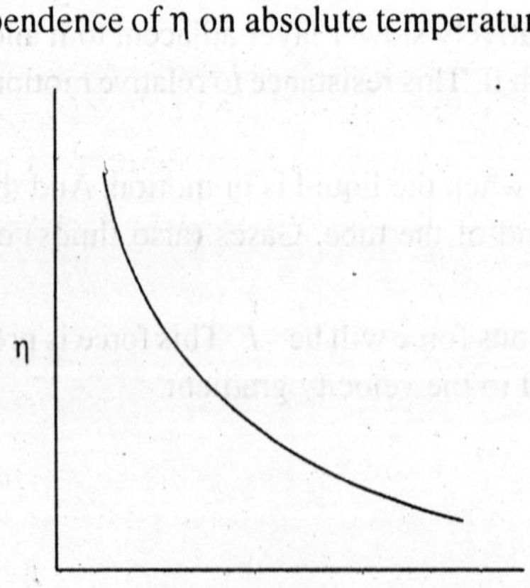

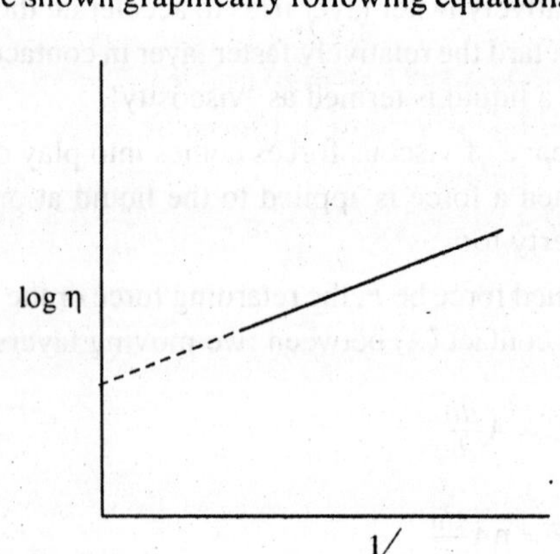

η vs. density or concentration of liquid or solution

From equation (4), for a definite height of liquid, $P = h\rho g$.

$$\therefore \eta \propto \rho,$$

i.e., viscosity increases with the density of the solution. Again, V is the volume of the liquid flowing in time t. With increase in concentration, V will decrease. Hence, as per equation (4),

$\eta \propto c$,

i.e., viscosity of same kind of solution will increase with the concentration of the solution.

η and Molecular Weight

For compounds falling under the same class (say, halogen acids, alcohols, paraffins, etc), viscosity increases with the molecular weight.

II. The Apparatus

The most widely used instrument to determine the viscosity coefficient of a liquid or a solution is **Ostwald Viscometer**. As direct measurement of η using equation (4) is very difficult, this instrument provides a simple means to determine the relative viscosity of a liquid (or a solution) in comparison with another liquid, say, water.

The Ostwald viscometer, made of glass, has two limbs joined to give a U-like shape. One of the limbs comprises a small bulb *C*, below which is a 8-10 cm long capillary *D*. There are two definite marks (*A*, *B*) above and below *C*. On the other limb, there is a bigger bulb (*E*) with wider mouth through which the experimental solution is poured down. On the month of the other limb, an arrangement of rubber tubing is kept to suck the solution. During experiment the instrument is clamped vertically and time is recorded against the passage of the solution from *A* to *B*.

A definite amount of the experimental solution is pipetted out and poured in *E*. The solution is sucked through a rubber tube and allowed to fall. When the surface of the solution reaches at *A*, the stopwatch is started on, and stopped when it reaches *B*. The time interval is noted.

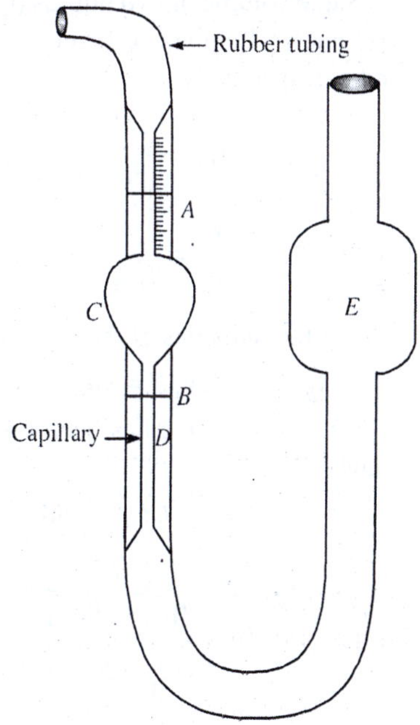

Ostwald's Viscometer

It is important to take the proper amount of the experimental solution. An excess amount either takes much time or may not reach at all at *B*. This adjustment is to be done by the worker depending upon the volume of the particular viscometer. Generally, a 15 ml solution will be helpful.

Another viscometer, known as **Ubbelhode Viscometer**, is also very popular. However, it is not discussed here.

Experiment 1

To determine the viscosity coefficient of a given solution (or liquid) by Ostwald Viscometer.

Theory: When a liquid is allowed to flow slowly through a narrow tube, the relative motions of the adjacent layers, by which the liquid is supposed to be composed of, are retarded due to an inherent property of the liquid, called viscosity. This is some kind of internal friction within the liquid.

As per Newton's law regarding viscous force,

$$F = \eta A \frac{du}{dr}, \tag{1}$$

where F is the force required to maintain a velocity gradient $\frac{du}{dr}$ between two adjacent layers with area of contact A. η is called the coefficient of viscosity. It is defined as the force required to maintain unit velocity gradient between two consecutive layers with unit area of contact.

η is determined by using the Poiseuille equation

$$\eta = \frac{\pi r^4 t P}{8lV}, \qquad (2)$$

where V is the volume of a liquid flowing in time t through a capillary tube of length l and radius r and P is the applied average pressure. The instrument used for determination of viscosity coefficient is known as Ostwald Viscometer. By this viscometer the viscosity of a liquid is compared to another solution with known viscosity.

Same volume of two liquids (or solutions) are allowed to flow through the same length in a particular viscometer. Let the coefficients of viscosity of two solutions be η_1 and η_2, and time taken for flow is t_1 and t_2 respectively.

$$\therefore \quad \eta_1 = \frac{\pi r^4 t_1 P_1}{8lV} \quad \text{and} \quad \eta_2 = \frac{\pi r^4 t_2 P_2}{8lV}$$

$$\therefore \quad \frac{\eta_1}{\eta_2} = \frac{t_1 P_1}{t_2 P_2} \qquad (3)$$

Let the densities of the two solutions be d_1 and d_2.

Now $P = hdg$.

As same volume has been taken, hence the difference in height of two levels in two arms will be same.

$$\therefore \quad P_1 = hd_1 g \quad \text{and} \quad P_2 = hd_2 g.$$

$$\therefore \text{ from (3), } \frac{\eta_1}{\eta_2} = \frac{t_1 h d_1 g}{t_2 h d_2 g} = \frac{t_1 d_1}{t_2 d_2}.$$

$$\therefore \quad \eta_1 = \eta_2 \frac{t_1 d_1}{t_2 d_2}. \qquad (4)$$

If η_1, be the viscosity coefficient of the solution and η_2 be that of water, using suffixes that s for 1 and w for 2, equation (4) becomes

$$\eta_s = \eta_w \frac{t_s}{t_w} \cdot \frac{d_s}{d_w}$$

$$\text{or,} \quad \boxed{\eta_s = \eta_w . s \frac{t_s}{t_w}} \qquad (5)$$

where s is specific gravity of the solution, $s = \frac{d_s}{d_w}$. Equation (5) is the working formula for determination of the coefficient of viscosity of a given solution. η_w is computed from the Table at the experimental temperature.

Apparatus and Equipments

Viscometer fitted with rubber tubing, burette stand with clamp, 10 ml graduated pipette, 2 ml pipette, weighing bottle, beakers, thermometer, stopwatch.

Materials: Given liquid or solution, distilled water.

Procedure

1. Note the laboratory temperature before and after the experiment. Take the mean.
2. Clean the viscometer with chromic acid and then wash 3-4 times with distilled water.
3. Clamp the viscometer vertically with the burette-stand (clamp the wider arm).
4. Pour down 12-15 ml distilled water by the graduated pipette in the wider arm (*E*), suck the solution through rubber tubing and check whether the solution smoothly comes down and crosses the marking zone (*AB*). Adjust the exact volume (you will have to take the same volume for the solution).

   After sucking water, allow it to fall. Start stopwatch when the meniscus touches the upper mark of the capillary arm and stop the watch as it just touches the lower mark. Note the time. Repeat the process twice and take the mean of time required (t_w).
5. Rinse the viscometer with the supplied solution or the liquid.
6. Perform the process 4 with the solution or the liquid and note the time required (t_s).
7. Determine the specific gravity of the given solution.

   Proceed as follows:

   a. Weigh a clean perfectly dried weighing bottle (w_1).

   b. Take 2 ml of distilled water in the bottle with the help of the 2 ml pipette and weigh (w_2).

   c. Add 2 ml of the supplied solution with the same 2 ml pipette (after rinsing it with the solution) in (b) and weigh (w_3).

   d. Specific gravity of the solution

$$s = \frac{\text{Weight of 2 ml of the solution}}{\text{Weight of 2 ml of water}} = \frac{w_3 - w_2}{w_2 - w_1}.$$

8. Collect η_w from literature at the working temperature.
9. Put the value of η_w, t_w, t_s and s in equation (5) and calculate η_s.

Results and Calculations

Table 1: Recording of room temperature in usual tabular form

Table 2: Noting of time of flow

No. of obs. with water	Time of flow (sec)	Mean time of flow (sec) (t_w)	No. of obs. with soln/liq.	Time of flow (sec)	Mean time of flow (sec) (t_s)
1	...		1	...	
2	...	...	2	...	...
3	...		3	...	

Table 3: Determination of specific gravity of the solution / liquid

Weight of the empty bottle (w_1)g	w_1 + weight of 2 ml water (w_2) g	w_2 + weight of 2 ml of soln/liq. (w_3)g	Specific Gravity $s = \frac{w_3 - w_2}{w_2 - w_1}$
...	...	...	...

Results: At ... °C temperature, $\eta_w = ...$ centipoise

$$\therefore \quad \eta_s = \eta_w s \frac{t_s}{t_w} = ... \text{ centipoise.}$$

Conclusion: Hence the coefficient of viscosity of the given solution or liquid is ... centipoise at ... °C temperature.

Experiment 2

To determine the concentration of a given solution of glycerol (or sucrose) by measuring viscosity of the solution.

Theory: Write the theory of experiment 1 and then add:

To determine the concentration (%) of a given solution of glycerol (or sucrose), a number of solutions with different concentrations are prepared and viscosity coefficients of all the solutions are measured with Ostwald's Viscometer. η of the given solution is also measured ($\eta_{unknown}$).

Now a plot of η vs. C (in %) is drawn, which will be a straight line with positive slope. From the graph, the percentage concentration of the given solution is estimated against the corresponding $\eta_{unknown}$ value.

[For practice in laboratory, a solution of known strength of the same solute is prepared, which may be regarded as the supplied solution. Having obtained the experimental value of concentration, the result may be compared with the prepared strength.]

The nature of the graph will be as follows:

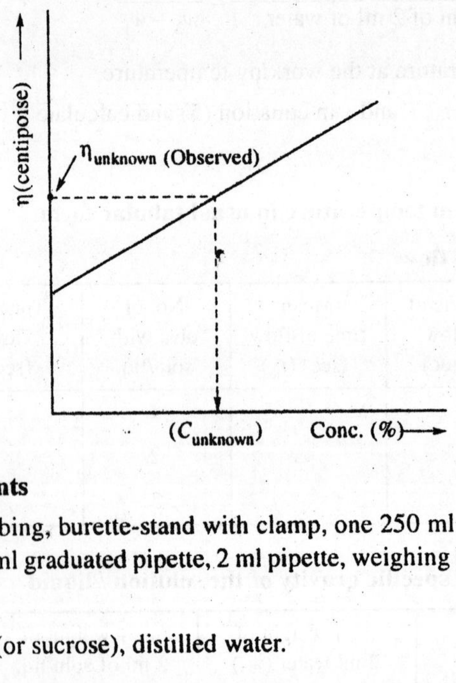

Apparatus and Equipments

Viscometer with rubber tubing, burette-stand with clamp, one 250 ml volumetric flask, four 50 ml volumetric flasks, burette, 10 ml graduated pipette, 2 ml pipette, weighing bottle, beakers, thermometer, stopwatch.

Materials: Pure glycerol (or sucrose), distilled water.

Procedure

1. Note the laboratory temperature at the beginning and at the end of the experiment.
2. Prepare 250 ml 12% glycerol solution (*v/v*) in water. (For sucrose, 12% *w/v* solution).

3. From the 12% stock solution prepare 50 ml each of 9%, 6% and 3% solution by proper dilution and label the flasks. (*A* 10% stock solution and therefrom 8%, 6%, 4%, 2% solution may be prepared alternatively).

   Prepare also a 50 ml 7.2% solution, which may be used as the unknown solution of which percentage strength is to be determined.
4. Clean the viscometer with chromic acid, wash 2-3 times with distilled water and pour a definite volume of distilled water (say, 15 ml) by the graduated pipette in the wider arm. Clamp the wider arm with burette-stand. Suck at the capillary end through rubber tube and note the time of passing of water over the marked zone. Repeat twice and take the mean of the falling time (t_w).
5. Clamp off the viscometer. Wash well the pipette and the viscometer with water.

   Take exactly the previous volume (that for water) of 3% solution, pour in the wider arm and clamp again with the burette-stand. Allow the solution to come down and note the time of falling. Repeat twice or at least once more. Take mean ($t_{3\%}$).
6. Exactly repeat procedure 5 for 6%, 7.2%, 9% and 12% solution (from lower conc. to higher) and time of falling ($t_{x\%}$) for each solution is noted.
7. For determination of specific gravity of each solution, proceed as follows:
   a. Weigh a perfectly clean and dried weighing bottle (w_1).
   b. Add 2 ml distilled water by pipette in the bottle, cover and weigh (w_2).
   c. Add further 2 ml 3% solution using the same 2 ml pipette in (b), cover and take weight (w_3).
   d. Continue the process of addition of 2 ml solution and weighing with 6%, 7.2%, 9% and 12% solution successively, never emptying the bottle. Let the weights be w_4, w_5, w_6 and w_7 respectively.

      [**N.B.** A total of 12 ml volume of liquid has been added so far. If it appears too much for the bottle, 1 ml of water and each solution may be taken using a single 1 ml pipette.]
   e. Calculate the specific gravity of each solution. For example,

$$s_{6\%} = \frac{w_4 - w_3}{w_2 - w_1}.$$

8. Calculate η of each solution using equation (5).
9. Plot η vs. *C* (%) and from the plot, find out the percentage strength of the given solution (here 7.2% solution).

Results and Calculations

Table 1: Recording of room temperature

Preparation of 250 ml 12% solution of glycerol

Take 30 ml of pure glycerol in a clean 250 ml volumetric flask from burette, make up the volume with distilled water and mix well.

(For 12% sucrose solution, weigh out exactly 30g sucrose, in a 250 ml volumetric flask, dissolve in water and make up the volume).

Table 2: Preparation of glycerol solutions with different strengths from the stock 12% solution

Vol. of 12% solution to be added from burette (ml)	Final volume after make-up with water (ml)	conc. of the solution (%)
37.5	50	9.0
30.0	50	7.2
25.0	50	6.0
12.5	50	3.0

Table 3: Determination of the specific gravities of the solutions

Weight of the empty weighing bottle = w_1g

Conc. of the solution (%)	Weight of (empty bottle + all previous solution + present solution) (g)	Weight of the solution (present wt. – just previous wt. (g)	Specific gravity of the solution $= \frac{\text{wt. of the soln.}}{\text{wt. of water}}$
0.0 (i.e. H_2O)	w_2	$(w_2 - w_1) = w$	1
3.0	w_3	$w_3 - w_2$	$\frac{w_3 - w_2}{w}$
6.0	w_4	$w_4 - w_3$	$\frac{w_4 - w_3}{w}$
7.2	w_5	$w_5 - w_4$	$\frac{w_5 - w_4}{w}$
9.0	w_6	$w_6 - w_5$	$\frac{w_6 - w_5}{w}$
12.0	w_7	$w_7 - w_6$	$\frac{w_7 - w_6}{w}$

Table 4: Noting of time flow and calculation of η of different solutions

η_w at ··· °C = ··· poise t_w = ··· sec

Conc. of Soln. (%)	Time of flow (sec)	Mean time of flow (t_s) (sec)	Sp. Gravity $s_{x\%}$	$\eta_s = \eta_w s \frac{t_s}{t_w}$
3.0	··· ···	···	···	···
6.0				
7.2				
9.0				
12.0				

Results

From the graph, it is found that the concentration of the given solution is ··· %.

[**N.B.** η at 0% means η_w, which falls on the η-axis. This value is the literature-value. Hence, during drawing the graph, weightage is to be given to this value. The line should start as far as close to this point].

Conclusion

The concentration of the supplied solution (observed) = ··· %

The concentration of the supplied solution (calculated) = ··· %.

Experiment 3

To determine the Isoelectric point of gelatin and to find out the Intrinsic Viscosity at isoelectric point. [For P.G. courses]

Theory

Certain lyophilic colloids like gelatin, albumin, etc. exhibit peculiar electrical property. In acid medium, they become positively charged and suffer electrophoretic migration towards cathode; in alkaline medium, they are negatively charged and move towards anode. Due to this dependence of charging of lyophilic protein sols on pH, a definite protein colloid at a definite pH carries no charge and thereby does not move in any direction in an electric field. This pH-value is called the isoelectric point of the particular colloid.

The isoelectric point of a protein is defined as a definite pH-value for that protein colloid, at which it does not suffer any electrophoresis.

Different colloids have different isoelectric point, due to which they can be isolated from their mixture.

At the isoelectric point, the colloidal particles are least solvated and hence exhibit minimum viscosity.

Intrinsic viscosity: The viscosity of the solutions of proteins or other polymers increases from solvent to solution due to presence of solute particles, their size and shape and molecular weights.

The relative increase in viscosity in going from solvent to solution is called specific viscosity, η_{sp}, i.e.,

$$\eta_{sp} = \frac{\eta_{\text{solution}} - \eta_{\text{solvent}}}{\eta_{\text{solvent}}}$$

or, $$\eta_{sp} = \frac{\eta_{\text{solution}}}{\eta_{\text{solvent}}} - 1 = \frac{t_1 d_1}{t_2 d_2} - 1$$ [from equation (4) of experiment 1]

or, $$\eta_{sp} = \frac{t_1 d_1 - t_2 d_2}{t_2 d_2}, \qquad (2)$$

where d_1, d_2 are densities of solution and solvent and t_1 and t_2 are time of flow of solution and solvent respectively.

A plot of η_{sp} vs. concentration (c) of the solution gives a straight line, which when extrapolated to zero concentration ($c = 0$), gives the intrinsic viscosity, $[\eta]$.

$$\therefore \quad [\eta] = \lim_{c \to 0}\left(\frac{\eta_{sp}}{c}\right). \qquad (3)$$

The equation for the straight line is

$$\frac{\eta_{sp}}{c} = A + Bc \, . \tag{4}$$

The intercept, A, is the intrinsic viscosity.

The nature of the graph will be as under:

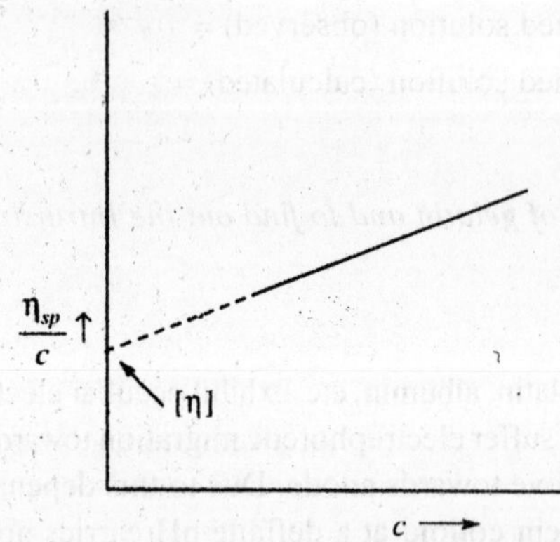

Intrinsic viscosity is also known as **limiting viscosity number** or **Staudinger index**.

Equipments: Viscometer with rubber tubing, stopwatch, one 50 ml volumetric flask, ten 100 ml clean stoppered reagent bottles, two 500 ml clean stoppered reagent bottles, 6-8 conical flasks of 100 ml capacity, 10 ml pipette, burette, two 250 ml volumetric flasks, three 250 ml reagent bottles with stopper, beaker and thermometers.

Materials: Gelatin, oxalic acid, NaOH, acetic acid, phenolphthalein, distilled water.

Procedure

1. Prepare 100 ml 2% gelatin solution (generally this is supplied). Take 2g gelatin in a clean beaker, add 25 ml water, shake and keep for hours (preferably overnight). Shake well. Slightly warm, if necessary . Add rest amount of water (75 ml), stir.
2. Prepare 50 ml standard 0.4 N Oxalic acid solution, weight required = 1.26 g.
3. Prepare about 400 ml 0.5 N NaOH solution. Dissolve not less than 8 g NaOH in 400 ml distilled water.
4. Prepare about 400 ml 0.5 N CH_3COOH solution. Assuming the strength of CH_3COOH in the lab bottle be 17(N), dissolve 12 ml CH_3COOH in water to make 400 ml solution.
5. Titrate NaOH solution (in burette) against oxalic acid with phenolphthalein indicator and calculate the strength of the alkali solution. From the standard solution, prepare exactly 250 ml 0.4 N NaOH solution by quantitative dilution in a volumetric flask.
6. Standardise acetic acid solution against standard NaOH solution using phenolphthalein indicator and from the known strength, prepare exactly 250 ml 0.4N acetic acid solution by proper dilution in a volumetric flask.
7. Take 100 ml 0.4 N acetic acid by burette in a 250 ml reagent bottle. Add 100 ml water in it and mix gently.

   Thus 200 ml 0.2 N acetic acid solution is prepared.
8. Take 100 ml 0.4 N acetic acid by burette in a 250 ml reagent bottle. Add 100 ml 0.4 N NaOH solution in it and mix gently.

   Thus 200 ml 0.2 N sodium acetate solution is prepared.

9. Prepare the buffer solutions as per Table 5. Add 5 ml of 2% gelatin solution in each buffer.
10. Measure the time of flow of each buffer including that of water by Ostwald Viscometer.
11. Plot pH vs. time of flow. From the minimum time of flow, find the isoelectric point as the corresponding pH.

Intrinsic viscosity

12. Prepare 100 ml buffer solution, for which the isoelectric point is determined, by mixing proportionately 0.2 N acetic acid and 0.2 N NaOAc solution.
    [Suppose, the isoelectric point corresponds to bottle no. 4, i.e. pH = 4.45. Hence, you are to mix 60 ml 0.2 N AcOH and 40 ml 0.2N NaOAc solution, etc.]
13. Add 2 g gelatin in this buffer and dissolve to prepare 2% solution.
14. From the 2% gelatin solution in the buffer, prepare a number of solutions in 100 ml conical flasks as per Table 7.
15. Measure the density of each solution. For this purpose, weigh a clean and perfectly dried weighing bottle. Add 2 ml buffer solution (without gelatin) and take weight.

$$d_2 = d_{\text{buffer}} = \frac{\text{Weight of the buffer}}{2} \text{ g/ml}.$$

    Do not throw away the solution. To this, add 2 ml of the lowest concentrated solution (with gelatin). Take weight. The difference is the weight of the gelatinous buffer solution. Divide by 2 and get the density (d_1). Continue with other solution without rejecting the solution.
16. Study the time of flow of each solution (t_1).
17. Find η_{sp} using equation (2) and then $\frac{\eta_{sp}}{c}$.
18. Plot $\frac{\eta_{sp}}{c}$ vs. c, extrapolate to $c = 0$. The intercept gives the intrinsic viscosity.

Results and Calculations

Table 1: Recording of temperature (in usual tabular form)

Table 2: Preparation of 50 ml 0.4 N Oxalic acid solution

1st wt. of Oxalic acid (w_1) g	2nd wt. of Oxalic acid (w_2) g	wt. taken ($w_1 - w_2$)	wt. required	Strength of the solution $\frac{w_1 - w_2}{1.26}(0.4\text{N})$
...	...	...	1.26 g	...

Table 3: Standardisation of NaOH solution by standard Oxalic acid solution

No. of Titrations	Vol. of Oxalic (ml)	Initial burette reading (ml)	Final burette reading (ml)	Volume of NaOH required (ml)	Mean Vol. of NaOH (ml)	Conc. of NaOH solution (N)
1	10.0	0.0	...	...		
					...	...
2	10.0	0.0	...	...		

From this standardised NaOH solution (> 0.4 N), prepare 250 ml 0.4 N solution.

Table 4: Standardisation of AcOH solution by standard NaOH solution (take the original alkali solution)

Indicator : Phenolphthalein

No. of Titrations	Vol. of AcOH (ml)	Initial burette reading (ml)	Final burette reading (ml)	Volume of NaOH required (ml)	Mean Vol. of NaOH solution (ml)	Conc. of AcOH soln. (N)
1	10.0	…			…	…
2	10.0	…	…	…		

Table 5: Preparation of the buffer solutions (in 100 ml reagent bottles)

No. of buffer solution (Bottle No.)	Vol. of 0.2 N AcOH solution (ml)	Vol. of 0.2 N NaOAc solution (ml)	pH	Vol. of 2% gelatin solution added (ml)
1	27.0	3.0	3.72	5.0
2	24.0	6.0	4.05	5.0
3	21.0	9.0	4.27	5.0
4	18.0	12.0	4.45	5.0
5	15.0	15.0	4.65	5.0
6	12.0	18.0	4.80	5.0
7	9.0	21.0	4.99	5.0
8	6.0	24.0	5.23	5.0
9	3.0	27.0	5.57	5.0

Table 6: Reading of time of flow through viscometer

Bottle No.	pH	Time of Flow (sec)
Water	…	…
1	3.72	…
2	4.05	
3	4.27	
4	4.45	
5	4.65	
6	4.80	
7	4.99	
8	5.23	
9	5.57	

[**N.B.** For gelatin, it will be enough to take the bottles 2–7 and study the time of flow]

Draw time of flow vs. pH curve and find the isoelectric point [approximate value : 4.4 - 4.7].

The isoelectric point is at pH ···, i.e. solution no. ···

100 ml buffer solution corresponding to pH ··· is prepared and a 2% gelatin solution with this buffer is prepared as per procedures 12 and 13. **Before addition of gelatin, time of flow of this buffer solution is to be studied first.**

Table 7: Preparation of different solution of gelatin with buffer solution no. ... and study of time of flow

Conical flask No.	Vol. of 2% gelatin solution (ml) (use burette)	Vol. of water (ml) (use burette)	Total Volume (ml)	Conc. of the solution (%)	Time of flow (s)
1	1.5	28.5	30	0.1	...
2	3.0	27.0	30	0.2	...
3	6.0	24.0	30	0.4	...
4	9.0	21.0	30	0.6	...
5	15.0	15.0	30	1.0	...
6	30.0	—	30	2.0	...
Buffer soln. (without gelatin)	—	—	—	—	...

Table 8: Determination of the densities of the solutions (d_1) and of the buffer (d_2)

Weight of the empty bottle = w_1 g

Add 2 ml solution from a 2 ml pipette in each time.

Conc. of the soln. c (%)	Wt. of (empty bottle + all previous solns. (g)	Wt. of the solution (present wt. – just previous wt.) (g)	density (d_1) $d = \frac{\text{weight}}{2}$ (g/ml)
Buffer solution (without gelatin)	w_2	$w_2 - w_1$	$d_2 = \frac{w_2 - w_1}{2}$
0.1	w_3	$w_3 - w_2$	$d_1 = \frac{w_3 - w_2}{2}$
0.2	w_4	$w_4 - w_3$	$d_1 = \frac{w_4 - w_3}{2}$
0.4	w_5	$w_5 - w_4$	$d_1 = \frac{w_5 - w_4}{2}$
0.6	w_6	$w_6 - w_5$	$d_1 = \frac{w_6 - w_5}{2}$
1.0	w_7	$w_7 - w_6$	$d_1 = \frac{w_7 - w_6}{2}$
2.0	w_8	$w_8 - w_7$	$d_1 = \frac{w_8 - w_7}{2}$

Table 9: Calculation of η_{sp} of different solutions

$d_2 = \cdots$ g/ml; $t_2 = \cdots$ s

Conc. of the solution c (%)	Density (d_1) (g/ml)	Time of flow (t_1) (s)	$\eta_{sp} = \frac{d_1 t_1}{d_2 t_2} - 1$	$\frac{\eta_{sp}}{c}$ (g^{-1} cm^3)
0.1	...	...	...	...
0.2	...	...	...	...
0.4				
0.6				
1.0				
2.0				

Results: From the graph of $\frac{\eta_{sp}}{c}$ vs. *c*, the intercept is the intrinsic viscosity at isoelectric point, i.e.,

$$[\eta] = \lim_{c \to 0}\left(\frac{\eta_{sp}}{c}\right) = \cdots \text{ g}^{-1}\text{cm}^3 .$$

Experiment 4

To determine the Molecular Weight of a Polymer from Viscosity Measurements

Theory: Polymers are giant molecules whose molecular weights range from several thousands to even millions. They get dissolved in suitable organic solvents like benzene, toluene, etc. Such solutions are highly viscous even at very low concentrations. Measuring the viscosities of the solutions of polymers at low concentration, the average molecular weights of several polymers may be determined.

The molecular weight *M* (more correctly, the viscosity average molar mass) of a polymer is empirically related to its intrinsic viscosity, $[\eta]$, by the modified Staudinger equation,

$$[\eta] = KM^{\alpha} , \tag{1}$$

where *K* is a constant for a definite type of polymer and depends upon solvent and temperature and α, also a constant, is function of shape or geometry of the polymer molecule. The equation is also known as Mark-Houwink equation.

The instrinsic viscosity is represented as

$$[\eta] = \lim_{c \to 0}\left(\frac{\eta_{sp}}{c}\right), \tag{2}$$

where η_{sp}, the specific viscosity, is defined as

$$\eta_{sp} = \frac{\eta_{\text{solution}}}{\eta_{\text{solvent}}} - 1 . \tag{3}$$

c is the concentration of the polymer solution which is the number of grams of polymer in 100 ml of the solution.

The fundamental equation for determining the relative viscosity of a solution is

$$\frac{\eta_{\text{solution}}}{\eta_{\text{solvent}}} = \frac{\eta}{\eta_0} = \frac{dt}{d_0 t_0}, \tag{4}$$

where d and d_0 are the densities of solution and solvent respectively and t and t_0 are the corresponding time flow. For polymer solution with low concentrations, d and d_0 are almost equal, so that equation (4) reduces to

$$\frac{\eta}{\eta_0} = \frac{t}{t_0}. \tag{5}$$

From (3) and (5),

$$\eta_{sp} = \frac{t}{t_0} - 1. \tag{6}$$

Hence, a plot of η_{sp} vs. c, when extrapolated to zero concentration, gives $[\eta]$.

Knowing K and α for a particular polymer-solvent combination, molecular weight of the polymer can be calculated from equation (1).

Sometimes $[\eta]$ is determined from the relationship

$$[\eta] = \lim_{c \to 0}\left[\frac{1}{c}\ln\frac{\eta}{\eta_0}\right], \tag{7}$$

which results from the expansion of the logarithmic function $\ln\frac{\eta}{\eta_0}$ and neglecting the higher terms as $c \to 0$.

In that case, $[\eta]$ will be intercept on the graph of $\frac{1}{c}\ln\frac{\eta}{\eta_0}$ vs. c.

To have a reliable value, $[\eta]$ is calculated from both the graphs.

Nature of graph: As shown in experiment 3.

Equipments: Ostwald viscometer, thermostat, stopwatch, one 100 ml volumetric flask, clean and dry 100 ml beakers or small volumetric flasks (25 ml / 50 ml), graduated pipette.

Material: Polystyrene (preferably dried) and toluene.

Procedure

1. Prepare about 2% solution of polystyrene by accurately weighing ~2 g of the polymer, allowing it to dissolve in small volume of toluene and then making up volume up to mark (i.e., 100 ml). If the polymer is not dried, it may take overnight to dissolve.
2. Prepare different solutions from the original 2% solution as per Table 3.
3. Perfectly dry the cleaned viscometer, fill up with a definite volume of toluene (12-15 ml, depending upon the smooth flowing in the particular viscometer) and note the flow time (t_0), taking mean of 2-3 observations.
4. Rinse the viscometer with the polystyrene solution of lowest concentration. Fill up the viscometer with the same volume of the solution as above and note the flow time (t).

5. Repeat process 4 with the remaining solutions with ascending concentrations.
6. For processes 3, 4 and 5, keeping the viscometer in a thermostat, after filling up with the solution, for 10-15 minutes, will yield a better result.
7. Calculate η_{sp} using equation (6) and then $[\eta]$ as per Table 4.
8. Calculate the molecular weight of polystyrene using equation (1).

Table 1: Recording of temperature (in usual tabular form)

Table 2: Preparation of 100 ml standard ~ 2% solution of polystyrene in toluene

1st wt. of Polystyrene w_1 (g)	2nd wt of Polystyrene w_2 (g)	Weight taken $w_1 - w_2$ (g)	Conc. of the solution after dissolving in toluene (%)
...	...	...	$(w_1 - w_2)\% = x\%$

Table 3: Preparation of polystyrene solutions of different concentrations

Bottle No.	Vol. of 2% Polystyrene (ml)	Vol. of toluene added (ml)	Total Vol. of the solution (ml)	Conc. of the solution (%) (To be calculated from x)
1	4.0	16.0	20.0	...
2	8.0	12.0	20.0	...
3	12.0	8.0	20.0	...
4	15.0	5.0	20.0	...
5	20.0 (The remaining soln. in the original flask)	0.0	20.0	x

N.B. If the viscometer needs more volume, prepare 25 ml solution instead of 20 ml by proper calculation and dilution thereon.

Table 4: Noting of flow time and calculation of η_{sp}

Flow of toluene, t_0 =... sec

Bottle No.	Conc. of the solution C(g / 100 ml) (i.e. %)	Flow time (s)	Mean flow time t (s)	$\eta_{sp} = \frac{t}{t_0} - 1$	η_{sp} / c (100 ml/g)
1	...	... ...	...	...	...
2.	...	... ...	...	...	...
3	...	... ...	...	...	...
4	...	... ...	...	...	...
5	x	... ...	...	—	—

Results: From the graph of $\frac{\eta_{sp}}{c}$ vs. c,

$[\eta]$ = intercept on y-axis

= ··· $(100\ cm^3\ g^{-1})$

$\therefore$ from the equation, $[\eta] = KM^{\alpha}, M = \left(\frac{[\eta]}{K}\right)^{1/\alpha}$.

Putting, $K = 3.7 \times 10^{-4}\ cm^3\ g^{-1}$ and $\alpha = 0.62$ (all at 25 °C),

the molar mass, M, of polystyrene = ··· [Adjust K by dividing 100 to match unit of η]

[**N.B.** For Benzene - Polystyrene system : $K = 9.52 \times 10^{-3}$, $\alpha = 0.744$]

APPENDIX

Table 1: Viscosity of some common liquids at different temperatures

η in centipoises

Liquids	20°C	25°C	30°C
Acetone	0.3290	0.3160	0.2950
Aniline	4.40	3.71	3.16
Benzene	0.652	0.601	0.564
Carbon Tetrachloride	0.9690	0.8980	0.8430
Chloroform	0.5630	0.542	0.514
Ethyl alcohol	1.20	1.094	1.003
n-Hexane	0.326	0.294	—
Methyl alcohol	0.597	0.547	0.510
Toluene	0.590	—	0.526
Water	1.005	0.8937	0.8007

Table 2: Viscosity (in centipoise) of water at different temperatures

Temp. (°C)	η (centipoise)	Temp. (°C)	η (centipoise)	Temp. (°C)	η (centipoise)
15	1.1404	22	0.9579	29	0.8180
16	1.1110	23	0.9358	30	0.8007
17	1.0828	24	0.9142	31	0.7840
18	1.0559	25	0.8937	32	0.7679
19	1.0299	26	0.8737	33	0.7523
20	1.0050	27	0.8545	34	0.7371
21	0.9810	28	0.8360	35	0.7225

CHAPTER 2

Experiments Related to Surface Tension of Liquids

I. **Introductory Words**

II. **The Apparatus**

III. **Experiment 1**

To determine the surface tension of a supplied liquid or a solution by drop-weight (or drop-counting) method using stalagmometer.

IV. **Experiment 2**

To determine the concentration of a given solution of acetic acid by measuring surface tension of the solution

V. **Experiment 3**

To determine the Critical Micelle Concentration (CMC) of sodium lauryl sulphate from the measurement of surface tension of different concentrations of SLS by Tensiometer or Torsion balance method (for P.G. Courses)

VI. **Experiment 4**

Determination of surface excess of amyl alcohol (for P.G. courses)

VII. **Experiment 5**

To determine the amount of acetic acid adsorbed at its different concentrations by charcoal and hence to verify the Freundlich adsorption isotherm

VIII. **Appendix**

Table 1. Surface tension of some common liquids at different temperatures

Table 2. Some physical constants of water (including surface tension) at different temperatures.

CHAPTER 2

Experiments Related to Surface Tension of Liquids

I. Introductory Words

Surface chemistry constitutes an important part of physical chemistry exploring some of the most important characteristics of liquids, solutions of electrolytes, nonelectrolytes and surface active agents. Measurement of surface tension of a liquid or a solution, of course, is the primary physical parameter, through which a number of other important physical aspects are achieved. To mention are the adsorption, surface film, surface area, surface excess, etc. which have immense practical as well as industrial importance.

Surface tension

In liquid, the molecules experience intermolecular force of attraction from all directions. If we concentrate our attention on two molecules, one in the bulk and the other on the surface of the liquid, the resultant intermolecular force will be different for the two.

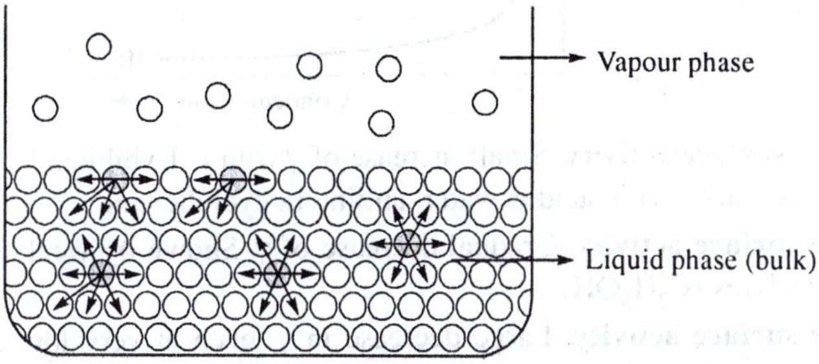

The molecule inside the bulk suffers no resultant force of attraction, being equally attracted from all directions. But that on the surface is pulled towards the bulk as fewer molecules are present on the vapour phase. Due to this unbalanced forces acting on all the molecules on the surface, the surface of a liquid behaves as a stretched membrane having tendency to contract to minimum area. Such property of a liquid is known as *surface tension* of the liquid, which is a measure of inward pulls resulting from the unbalanced force on the surface-molecules.

Definition of surface tension

The force acting along the surface of a liquid at right angle to any imaginary line of unit length drawn on the surface at a definite temperature is called the surface tension of the liquid. It is denoted by γ.

S.I. unit of surface tension is $N\text{-}m^{-1}$ and C.G.S. unit is $dyne\text{-}cm^{-1}$.

Surface energy

To increase the surface area more molecules will have to be brought to the surface. This requires some work to be done. So, *surface energy is defined as the work done in increasing the surface area by unity.* This may be treated as an alternative definition of surface tension. Both have the same unit.

Surface tension and Temperature

Surface tension of a liquid decreases with increasing temperature. Ramsay-Shields-Eötvös equation explains the temperature-dependence of surface tension, which is

$$\gamma (Mv)^{2/3} = k(t_c - t - 6),$$

where Mv is the molar volume of liquid, t_c, the critical temperature and t, the temperature in discussion in Celsius scale and k is a constant.

Surface Tension and Concentration of Solutions

Additions of solid or liquid to the liquid solvent changes the surface tension of the liquid. The nature of changes depends on the nature of the solute (liquid or solid), which is shown in the following graph.

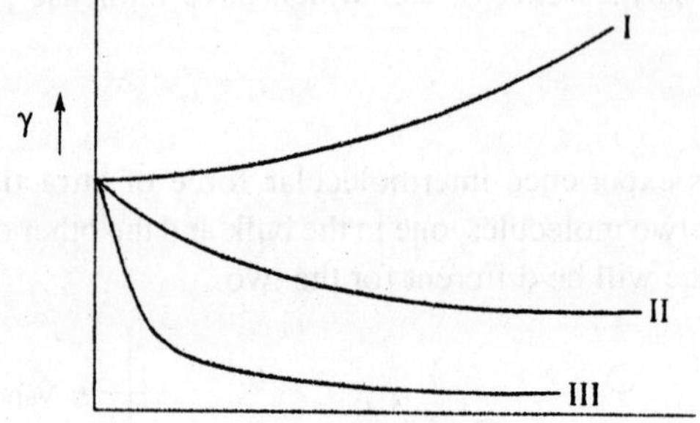

I. **Negative surface activity.** Small increase of γ with c. Exhibited by strong electrolytes, sucrose in water, aminobenzoic acid is water, aniline in cyclohexane.

II. **Positive surface activity.** Gradual decrease of γ. Shown by weak electrolytes (CH_3COOH) or nonelectrolytes (C_2H_5OH).

III. **Positive surface activity.** Large decrease in γ even at very low concentration of the solute. Exhibited by soaps, detergents, etc. These are surface-active agents.

Surface Tension and Adsorption

Like liquids, solid-surfaces also possess a state of unbalanced or unsaturated force. As a result of this unsaturation, solid and liquid surfaces tend to satisfy their residual forces by attracting gases or dissolved substances with which they come in contact retaining them on their surfaces. This phenomenon of concentration of substance on the surface of a solid or a liquid is called *adsorption.* Thus, it is the unbalanced force of a liquid-surface which causes surface tension as well as adsorption of solute or gas by liquid or solid.

• The Apparatus

Surface tension measuring apparatus—The Stalagmometer.

A different varieties of stalagmometer is available in the market. It consists of a pipette-like glass tube with a bulb in the middle and a fine capillary with sharp edge. At the upper end, a rubber tube is

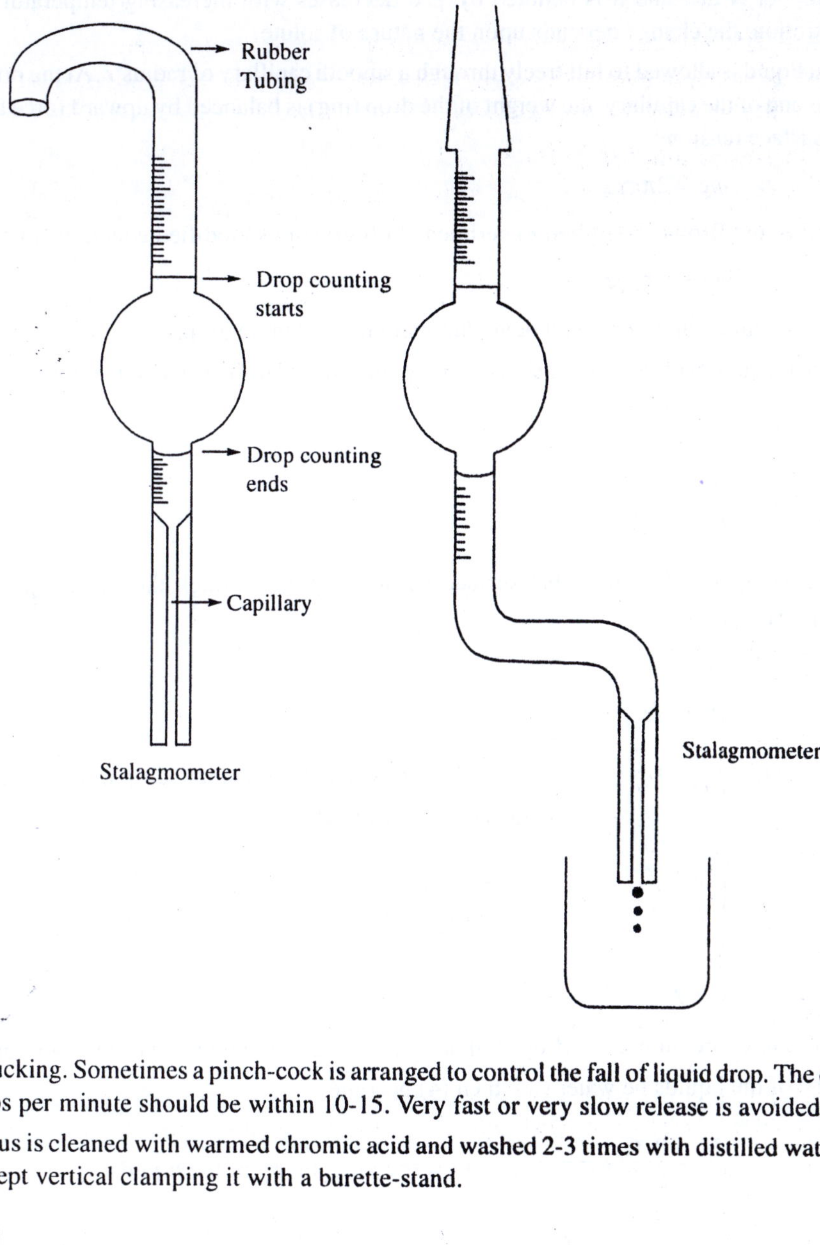

connected for sucking. Sometimes a pinch-cock is arranged to control the fall of liquid drop. The desirable number of drops per minute should be within 10-15. Very fast or very slow release is avoided.

The apparatus is cleaned with warmed chromic acid and washed 2-3 times with distilled water before working. It is kept vertical clamping it with a burette-stand.

Experiment 1

To determine the surface tension of a supplied liquid or a solution by drop-weight (or drop-counting) method using stalagmometer.

Theory: A molecule inside the bulk of a liquid is in a uniform field of attraction from all directions around it, thereby experiencing zero resultant force of attraction on it. But a molecule on the surface of the liquid suffers a net resultant inward pull as fewer number of molecules attract it on the vapour phase above it. In effect, the surface of any liquid (or its solution) behaves as a stretched membrane.

The force acting along the surface of a liquid at right angle to any imaginary line of unit length drawn on the surface at a particular temperature is called the surface tension of the liquid. Its unit is dyne-cm^{-1} or N m^{-1} and it is denoted by γ. It decreases with increasing temperature and varies with concentration, the change depends upon the nature of solute.

Let a liquid is allowed to fall freely through a smooth capillary of radius r. At the point of detachment from the end of the capillary, the weight of the drop (mg) is balanced by upward force acting on the drop due to surface tension.

$$\therefore \quad mg = 2\pi r\gamma. \tag{1}$$

Harkins and Brown from their experimental observations modified equation (1) as

$$mg = 2\pi r\gamma\phi, \tag{2}$$

where ϕ is a function of $r / v^{1/3}$, v being the volume of a single drop.

Let two liquids with surface tensions γ_1 and γ_2 are allowed to fall separately through the same caprillary.

Then $m_1 g = 2\pi r\gamma_1\phi_1$ and $m_2 g = 2\pi r\gamma_2\phi_2$.

$$\text{Dividing,} \quad \frac{m_1}{m_2} = \frac{\gamma_1\phi_1}{\gamma_2\phi_2}. \tag{3}$$

For two liquids of comparable surface tensions falling through the same capillary, $\phi_1 \sim \phi_2$ and equation (3) reduces to

$$\frac{m_1}{m_2} = \frac{\gamma_1}{\gamma_2}$$

$$\text{or,} \quad \frac{v_1 d_1}{v_2 d_2} = \frac{\gamma_1}{\gamma_2}$$

[v_1 and v_2 are volumes of a single drop of two liquids and d_1, d_2 are respective densities]

$$\text{or,} \quad \frac{\left(\frac{v}{n_1}\right) d_1}{\left(\frac{v}{n_2}\right) d_2} = \frac{\gamma_1}{\gamma_2} \qquad \text{or,} \quad \frac{n_2 d_1}{n_1 d_2} = \frac{\gamma_1}{\gamma_2}, \tag{4}$$

where n_1 and n_2 are number of drops fallen against a definite volume v of the two liquids.

If one of the liquids be water (w), then (4) become

$$\frac{\gamma_l}{\gamma_w} = \frac{n_w d_l}{n_l d_w} = \frac{n_w}{n_l} \cdot s_l$$

[s_l = specific gravity of the liquid (l)]

$$\text{or,} \quad \boxed{\gamma_l = \gamma_w \cdot \frac{n_w}{n_l} \cdot s_l} \tag{5}$$

Equation (5) is the working formula for calculation of the surface tension of a liquid or a solution.

Stalagmometer provides the definite volume v and the capillary of a radius r with which the equation (5) is arrived at. Thus, counting the number of drops and determining the specific gravity of liquid, surface tension of the liquid is measured.

Apparatus and equipments

Stalagmometer fitted with rubber tubing (pinch-cock, if required), burette-stand with clamp, 5 or 10 ml graduated pipette, weighing bottle, beakers, thermometer.

Materials : Distilled water, given liquid or solution.

Procedure

1. Note the laboratory temperature before and after experiment.
2. Clean the stalagmometer by chromic acid, wash 3-4 times with distilled water, each time sucking through rubber tube and allowing the washings to be released through capillary.
3. Clamp the stalagmometer vertically with the burette-stand.
4. Take distilled water in a clean 100 ml beaker.
5. Immerse the end of stalagmometer in water, suck, adjust rate of fall of water drop (10-15 drops per minute). Count the number of drops while passing through two specified graduation of the stalagmometer. Repeat thrice, take mean (n_w).
6. Repeat the work with the supplied liquid after rinsing the apparatus with the liquid. Find n_l.
7. **Specific gravity** of the liquid can be measured in a simple way as follows:
   a. Weigh (w_1) a clean, perfectly dried weighing bottle.
   b. Take 2 ml distilled water in the bottle by graduated pipette (or a 2 ml clean pipette) and weigh (w_2).
   c. Add 2 ml of the supplied liquid using the same pipette in (b) and weigh (w_3).
   d. $s = \dfrac{\text{weight of a definite volume of the liquid}}{\text{weight of the same volume of water}} = \dfrac{w_3 - w_2}{w_2 - w_1}$.
8. Collect γ_{H_2O} from literature at the working temperature.
9. Put the values in equation (5) to calculate γ_{liquid}.

Results and Calculations

Table 1: Recording of room temperature

Table 2: Counting of drops

No. of observation	No. of drops of water	Mean (n_w)	No. of observation	No. of drops of liquid	Mean (n_l)
1	...		1	...	
2	...	...	2	...	...
3	...		3	...	

Table 3: Determination of specific gravity of the liquid

Weight of empty bottle (w_1)g	w_1 + weight of 2 ml water (w_2)g	w_2 + weight of 2 ml liquid (w_3)g	Specific Gravity $s = \dfrac{w_3 - w_2}{w_2 - w_1}$
...	...	...	...

Results: At ... °C temperature, γ_{H_2O} = ... dyne/cm

$$\therefore \quad \gamma_l = \gamma_w \cdot \frac{n_w}{n_l} \cdot s_l = \ldots \text{ dyne/cm}$$

Conclusion: Hence, the surface tension of the given liquid or solution is ...dyne/cm.

The intermediate equation $\frac{m_1}{m_2} = \frac{\gamma_1}{\gamma_2}$ may be used to determine the surface tension of a solution by **drop-weight method.** If one of the liquid be water, then

$$\frac{m_l}{m_w} = \frac{\gamma_l}{\gamma_w}, \qquad \therefore \quad \boxed{\gamma_l = \frac{m_l}{m_w} \cdot \gamma_w}$$

Here equal number of drops (way, 10 or 15) of water and the given liquid are allowed to fall through the same stalagmometer and their weights (m_l and m_w) are taken. Thus γ_l is calculated.

Experiment 2

To determine the concentration of a given solution of acetic acid by measuring surface tension of the solution.

Theory: Write theory of experiment 1 in toto and then add: Acetic acid is a weak electrolyte, hence shows positive surface activity. Its surface tension gradually decreases with increasing concentration. To determine the concentration (%) of a given acetic acid solution, a number of solutions with different concentrations are prepared and their surface tensions are measured at laboratory temperature using stalagmometer. Surface tension of the given solution is also measured.

A plot of γ vs. *c* (in %) is drawn. From the graph, the percentage composition of the given solution is estimated against its corresponding observed surface tension.

[For practice in laboratory, a solution of known composition of the same electrolyte is prepared, which may be regarded as the supplied solution. Compare the strength obtained from graph to that of your known strength.]

The nature of the graph will be as follows:

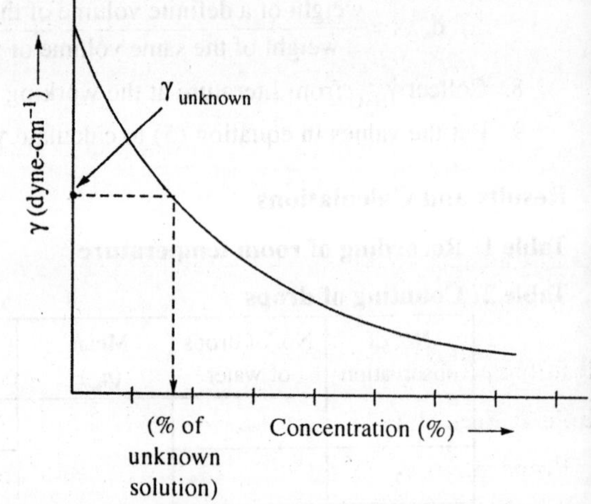

Apparatus and equipments

Stalagmometer with rubber tubing (pinch-cock, if required), burettte-stand with clamp, one 250 ml volumetric flask, four 50 ml volumetric flasks, burette, weighing bottle, graduated pipette or a 2 ml pipette, beakers, thermometer.

Materials: Acetic acid, distilled water.

Procedure

1. Note the laboratory temperature before and after the experiment.
2. Prepare 250 ml 12% (*v/v*) acetic acid solution.

3. Prepare from the 12% solution 50 ml each of 9%, 6% and 3% solutions by proper dilution and label the flasks. Prepare also a 50 ml 7.2% solution, which may be used as the unknown solution, the strength of which is to be determined.
4. Clean the stalagmometer by chromic acid, wash with distilled water and clamp it to the burette-stand.
5. Take distilled water in a clean beaker. Fill up the stalagmometer with water by sucking through the rubber tube. Count the number of drops. Repeat twice more. Pour out water from both stalagmometer and beaker.
6. Rinse twice both the empty beaker and the stalagmometer by the 3% solution. Take 10-15 ml solution in beaker, suck in stalagmometer and count number of drops thrice.
7. Perform step 6 above successively with 6%, 7.2%, 9% and 12% solution (from lower to higher concentration).
8. For determination of specific gravity of each solution, step 7 of experiment 1 is to be followed, which is discussed again.
   a. Weigh a perfectly clean and dried weighing bottle (w_1).
   b. Add 2 ml distilled water in the bottle, cover and weigh (w_2).
   c. Add further 2 ml 3% solution using the same pipette in (b), weigh (w_3).
   d. Continue the process of addition of 2 ml solution and weighing with 6%, 7.2%, 9% and 12% solutions successively, never emptying the bottle. Let the weights be w_4, w_5, w_6 and w_7 respectively.

      [**N.B.** A total of 12 ml volume has been added so far. If it appears too much for the weighing bottle, 1 ml of water and other solutions may be taken using a same one ml pipette or a graduated pipette].
   e. Calculate the specific gravity of each solution. For example,

$$s_{3\%} = \frac{w_3 - w_2}{w_2 - w_1} \text{ and } s_{6\%} = \frac{w_4 - w_3}{w_2 - w_1}, \text{ etc.}$$

9. Calculate surface tension of each solution using equations (5).
10. Plot γ vs. c and from the plot, find out the percentage strength of the given solution.

Results and Calculations

Table 1: Recording of room temperature

Preparation of 250 ml 12% solution of acetic acid

Add 30 ml pure acetic acid from burette in a clean 250 ml volumetric flask and make up the volume with distilled water.

Table 2: Preparation of acetic acid solutions of different strengths from the 12% solution

Conc. of the solution (%)	Vol. of 12% solution to be added from burette (ml)	Final volume after make up with water (ml)
9.0	37.5	50
7.2	30.0	50
6.0	25.0	50
3.0	12.5	50

Table 3: Determination of the specific gravities of the solutions

Weight of the empty weighing bottle = (...) + (...) + (...) + (...) = w_1 g.

Conc. of the solution (%)	Weight of the bottle + all previous solutions + present solution (g)	Weight of the soln. (present wt. – just previous wt.) (g)	Sp. gravity of the solution $\left(\frac{\text{wt. of the soln.}}{\text{wt. of water}}\right)$
0.0 (i.e. H_2O)	... + + ... = w_2	$(w_2 - w_1) = w$	1
3.0	w_3	$w_3 - w_2$	$\frac{w_3 - w_2}{w}$
6.0	w_4	$w_4 - w_3$	$\frac{w_4 - w_3}{w}$
7.2	w_5	$w_5 - w_4$	$\frac{w_5 - w_4}{w}$
9.0	w_6	$w_6 - w_5$	$\frac{w_6 - w_5}{w}$
12.0	w_7	$w_7 - w_6$	$\frac{w_7 - w_6}{w}$

Table 4: Counting of drops and calculation of γ

γ_w at ... °C = ... dyne-cm^{-1}

Conc. of Soln. (%)	No. of Drops (Counted thrice)	Mean of no. of drops	Sp. Gravity $s_{x\%}$	$\gamma_{x\%} = \gamma_w \times \frac{n_w}{n_{x\%}} s_{x\%}$
3.0	... ...	...	...	...
6.0				
7.2 (unknown)				
9.0				
12.0				

Result: From the graph, it is found that the conc. of the given solution is ... %

Conclusion: The conc. of the solution (calculated) = ... %

The conc. of the solution (observed) = ... %

Experiment 3

To determine the Critical Micelle Concentration (CMC) of sodium lauryl sulphate from the measurement of surface tension of different concentration of SLS by Tensiometer or Torsion Balance method. [For P.G. courses]

Theory: Write the theory of **experiment 11 of conductance measurement** section (page 104) regarding CMC beyond the last paragraph. Then proceed as follows.

Soaps, detergents etc. are positive surface active agents, surface tensions of their solutions fall rapidly with increase in concentrations. After attaining a minimum value, γ remains almost constant for further concentration rise. The concentration corresponding to the minimum value of γ is the critical micelle concentration.

Tensiometer provides simple, rapid and much accurate results of surface tensions for solutions, especially those of detergents like cetyltrimethyl ammonium bromide (CTAB), sodium lauryl sulphate (SLS), cetyl pyrridinium chloride (CPC), etc.

A number of dilute solutions of SLS (or other surface active agents which form association colloids) are prepared and their surface tensions are measured by a tensiometer or torsion balance method. Then γ's are plotted against concentration. The concentration corresponding to the minima of the curve is the CMC of SLS.

The nature of the graph will be as below:

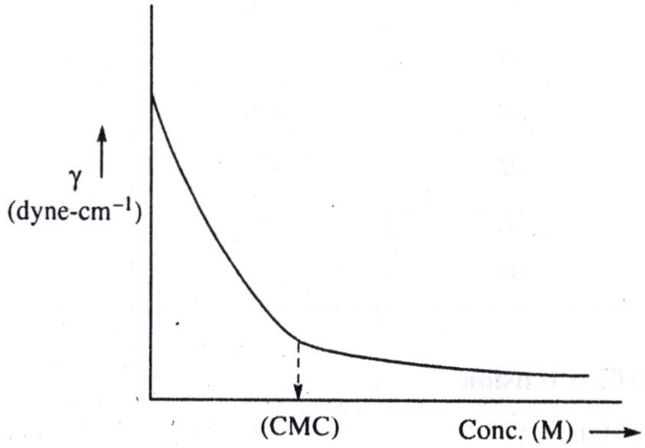

Apparatus and equipments

Tensiometer (or, Torsion balance; then a platinum ring is required), 50 ml volumetric flasks (7-8 nos.), burette.

Materials : Sodium Lauryl Sulphate (AR), conductivity water.

Procedure

1. Prepare 500 ml 10^{-2} M standard solution of sodium lauryl sulphate. Weight required = 1.4419 g (M.W. of SLS = 288.38)
2. Prepare 9×10^{-3}M, 8×10^{-3} M, 7×10^{-3} M, 6×10^{-3} M, 5×10^{-3} M, 4×10^{-3} M, 3×10^{-3} M and 2×10^{-3} M solutions each of 50 ml from the 10^{-2} M solution by proper dilution.
3. Measure the surface tension of the above solutions along with that of 10^{-2} M solution.
4. Note the laboratory temperature and collect γ_{H_2O} from literature at this temperature.
5. Plot γ against c (including water).
6. Find out the CMC from the graph.

Results and Calculations

Table 1: Recording of room temperature

Table 2 : Preparation of 500 ml 10^{-2} M solution of SLS
Weight required = 1.4419 g

1st weight w_1 (g)	2nd weight w_2 (g)	Weight taken $w_1 - w_2$ (g)	Conc. of SLS Solution (M)
...	...	...	$\frac{w_1 - w_2}{1.4419}(10^{-2})$

Table 3 : Preparation of solutions of various concentrations of SLS from 10^{-2} M solution

[The stock solution may not be exactly 10^{-2} M conc. The exact strength after dilution should be calculated properly from the original strength.]

Vol. of flask No.	Vol. of 10^{-2} M SLS solution (ml)	Final vol. after make up with water (ml)	Conc. of the SLS solution [$c \times 10^3$ (M)]
1	45	50	e.g., 9.0, if the original conc. be 10^{-2}M
2	40	50	...
3	35	50	...
4	30	50	...
5	25	50	...
6	20	50	...
7	15	50	...
8	10	50	...

Table 4 : Measurement of Surface Tension

γ_{H_2O} at ... °C = ... dyne-cm^{-1}

Conc. of the SLS solution [$c \times 10^3$ (M)]	Surface Tension (dyne-cm^{-1})
10 (if 10^{-2} M)	...
9 (if 10^{-2} M)	...
...	...
...	...
...	...
...	...
...	...
...	...

Results : From the graph of γ vs. c, the C.M.C. of sodium lauryl sulphate is found to be ... (M) at ... °C temperature.

[**N.B.** Compare the value with that from conductance measurement]

Experiment 4

Determination of surface excess of amyl alcohol [for P.G. Courses]

Theory: For a solution, the bulk concentration of the solute is different from the surface concentration or more precisely, the concentration of the solute at the interphase. This difference in concentration was thermodynamically established by Gibbs by the **adsorption isotherm** equation

$$\Gamma = -\frac{c}{RT}\left(\frac{d\gamma}{dc}\right), \tag{1}$$

where c = concentration of the solute in the bulk,

γ = surface tension of the solution at temperature T,

and Γ = surface excess of the solute.

Surface excess is defined as the excess concentration in terms of number of moles of solute present per unit area for a given volume of the liquid on the surface over the bulk.

If for a volume v of the solution having area A, the number of moles of solute present on the surface be n_2 and that in the bulk be n_1, then the expression for surface area will be

$$\Gamma = \frac{n_2 - n_1}{A}. \tag{2}$$

$\frac{d\gamma}{dc}$ in equation (1) may be positive, zero or negative. For substances like soaps, detergents, weak or non-electrolytes, $\frac{d\gamma}{dc}$ is negative, i.e., surface tension of the solutions of those substances decreases with concentration. Hence, Γ will be positive, i.e., a positive value of surface excess will be obtained. In other words, those substances are said to have positive adsorption. Amyl alcohol, acetic acid, butyric acids, various surface active agents have positive surface excess.

Equation (1) may be written as

$$\Gamma = -\frac{1}{2.303RT}\left(\frac{d\gamma}{d\log_{10} c}\right). \tag{3}$$

Measuring surface tensions of a number of solutions of amyl alcohol of different concentrations, a plot of γ vs. $\log_{10}c$ is drawn, the slope is nothing but $\frac{d\gamma}{d\log_{10} c}$.

Since values of c are less than 1, take $-\log_{10}c$ instead of $\log_{10}c$.

The nature of the graph will be as below:

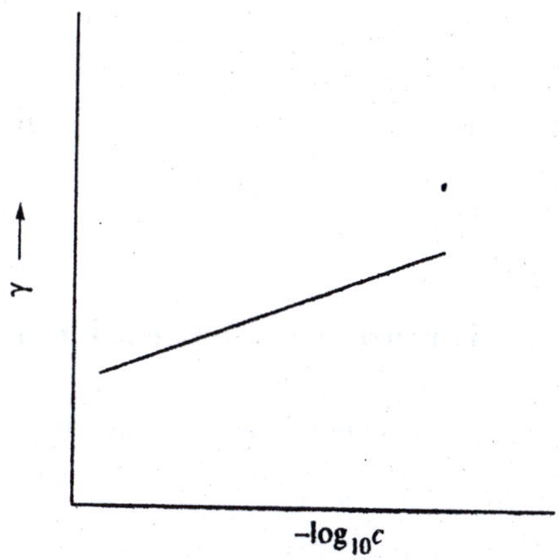

Hence, the surface excess following equation (3) will be

$$\Gamma = \frac{1}{2.303RT} \times \text{slope} \tag{4}$$

Surface tension may be measured either by

i. Drop-weight (or drop-volume) method or by

ii. Tensiometric method.

Equipments: Stalagmometer / Tensiometer, 8-10 nos. of 50 ml volumetric flasks, 250 ml volumetric flask, burette, weighing bottle.

Materials: Amyl alcohol, double distilled water.

Procedure

1. Prepare 250 ml 0.1 M amyl alcohol solution.
2. Prepare a number of solutions from this solution as per Table 2.
3. Note laboratory temperature.
4. (For Stalagmometer) Count number of drops of water and all other solutions against a definite volume of liquid taken in the stalagmometer.
5. Weigh a definite volume of each of water and the solutions.
6. Either calculate specific gravity of each solution following **Table 3 of Experiment 2** or calculate density of each solution as

$$\text{density} = \frac{\text{weight of the solution}}{\text{volume of the solution}}.$$

7. Calculate surface tension of each solution from either of relation (5) and (6)

$$\gamma_l = \frac{n_w \rho_l}{n_l \rho_w} \cdot \gamma_w, \tag{5}$$

$$\gamma_l = \frac{n_w \cdot s_l}{n_l} \cdot \gamma_w, \tag{6}$$

where l and w stand for liquid and water respectively, ρ = density, and s_l = specific gravity of the liquid.

$$s_l = \frac{\rho_l}{\rho_w}$$

$$= \frac{\text{weight of a definite volume of the solution}}{\text{weight of same volume of water}}.$$

Results and Calculations

Table 1: Recording of room temperature

Table 2 : Preparation of 250 ml 0.1 M amyl alcohol solution and from this preparation of other solutions

Calculate molarity of the original amyl alcohol from m.w. and density given on the label of the bottle.

No. of Vol. flask	Vol. of 0.1 M Amyl alcohol (ml)	Final vol. after make up with water (ml)	Conc. of the solution (M)
1	5.0	50.0	0.01
2	10.0	50.0	0.02
3	15.0	50.0	0.03
4	20.0	50.0	0.04
5	25.0	50.0	0.05
6	30.0	50.0	0.06
7	35.0	50.0	0.07
8	40.0	50.0	0.08

Table 3: Determination of specific gravity (s_l) or density (ρ_l) of the solution (write anyone)

For specific gravity, follow table 3 of experiment 2, only changing the conc. (M) column as 0.01—0.1 (9 flasks). For density, do the same work and then divide the weight of the solution by the volume (either 1 ml or 2 ml).

Table 4: Counting of drops and calculation of γ

γ_{H_2O} at ... °C temperature = ...dyne-cm^{-1}

[ρ_{H_2O} at ... °C temperature = ... g/cm^3, if needed]

Follow Table 4 of experiment 2. Write conc. (M) column as 0.01, 0.02, 0.03, 0.04, 0.05, 0.06, 0.07, 0.08 and 0.1.

Table 5: Results

Vol. flask No.	Conc. of the soln. (M) c	$-\log_{10}c$	γ of the soln. (dyne/cm)
1	0.01	2.0	...
2	0.02	1.7	...
3	0.03	1.52	...
4	0.04	1.4	...
5	0.05	1.3	...
6	0.06	1.22	...
7	0.07	1.15	...
8	0.08	1.1	...
9	0.10	1.0	...

From the plot of γ vs. $-\log_{10}c$, the slope is obtained. Hence, surface excess of amyl alcohol,

$$\Gamma = \text{slope} \times \frac{1}{2.303\,RT} \text{ moles/cm}^2$$

Conclusion: The literature value is 5×10^{-10} moles/cm^2.

Experiment 5

To determine the amount of acetic acid adsorbed at its different concentrations by charcoal and hence, to verify the Freundlich adsorption isotherm.

Theory: Liquid and solid surfaces always remain in a state of unsaturated force (in contrast to that in the bulk). As a result, solid or liquid surfaces have tendency to satisfy their residual forces by attracting solute from solution or gaseous mixture with which they come in contact and thus retain them on surfaces. **This phenomenon of concentration of solute (liquid or gas) on the surface of a solid is called adsorption**. Liquid surface may adsorb gas molecules also.

When powdered charcoal having a large surface area is shaken with an aqueous solution of acetic acid, some of the solute particles (here acetic acid molecules) are adsorbed by charcoal dust on their surfaces. The result is the decrease in the concentration of the acid solution. The extent of adsorption is a function of the concentration of acetic acid.

The relation between the amount of substances adsorbed per unit mass of the adsorbent (here charcoal) and the equilibrium concentration of the adsorbate (here acetic acid) at constant temperature is called **adsorption isotherm**.

Freundlich adsorption isotherm represents adsorption from solution by solid surface, which is

$$\frac{x}{m} = kc_e^n, \tag{1}$$

where x is the number of moles of adsorbate (acetic acid) adsorbed by m g of adsorbent (charcoal),

c_e = equilibrium concentration of the solution (i.e., conc. of acetic acid after adsorption)

k, n = empirical constants, which depend on the nature of solid and solute in solution; n is less than unity.

Taking logarithm of equation (1),

$$\log\left(\frac{x}{m}\right) = \log k + n \log c_e. \tag{2}$$

Plotting $\log\left(\frac{x}{m}\right)$ against $\log c_e$, a straight line will be obtained as per equation (2). Experimental values support the prediction and the nature of the plot will be as under:

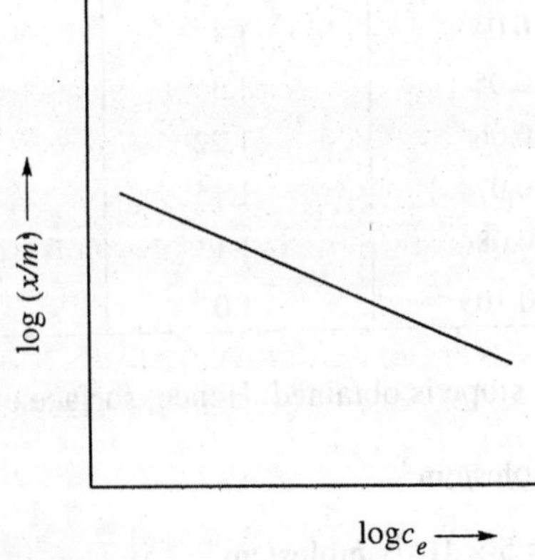

N.B. All values in x-and y- coordinates will be negative.

The intercept gives log k, from which k can be calculated. The slope gives the value for n, which will be less than one.

Apparatus and equipments

Six 100 ml stoppered reagent / Jena bottles, volumetric flasks one each of 50 ml, 250 ml and 500 ml, beakers, burette, pipette and 3-4 small conical flasks.

Materials : Oxalic acid, acetic acid, NaOH, phenolphthalein.

Procedure

1. Prepare 50 ml standard oxalic acid solution of the order of 0.1 N strength. Weight required = 0.315 g.
2. Prepare about 600 ml 0.1 N NaOH solution in a one litre beaker. Weight required = 2.4 g. Dissolve about 3 g NaOH in 600 ml water.
3. Prepare 250 ml 0.5 N acetic acid solution. First calculate the normality of acetic acid from the specific gravity (or density) and wt% written on the label of the acid bottle. Then find out the required volume and add water to a little more of the required volume of acid taken to make up volume to 250 ml.

   [Generally the conc. of acetic acid is 17 (N). If that be, take 8.5 - 9.0 ml acid and dilute to 250 ml. The actual volume needed is 7.35 ml.]
4. Standardise NaOH solution against the standard oxalic acid solution using phenolphthalein indicator. The concentration will be more than 0.1 N. Dilute properly to prepare **exactly 500 ml 0.1 N solution** in the 500 ml volumetric flask.
5. Standardise acetic acid against standardised NaOH solution using phenolphthalein indicator. Take 5 ml of acetic acid solution.

   [This titration should be done without pouring down the alkali from burette after titration against oxalic acid solution. Thereafter, 0.1 N alkali solution should be prepared.]

   Prepare **exactly 250 ml 0.5 N acetic acid solution** by proper dilution from the stronger solution.
6. Take the clean and dry 100 ml reagent bottles and number them from 1-6.
7. Weigh **accurately** about 1.5-1.7 g powdered animal charcoal by a properly weighed cellophane paper and take in each bottle.
8. Add 0.5 N acetic acid and water in each bottle as per **Table 5**, stopper well, shake for 30-40 minutes in mechanical shaker or by hand occasionally. Allow to stand for 15 minutes.
9. Filter the solutions separately starting from bottle no. 1 by a small filter paper in a clean and dry beaker or conical flask. For each solution, a new filter paper is to be used and first 5-10 ml filtrate will have to be rejected.
10. Titrate 10 ml of each filtrate against 0.1 N NaOH solution from burette using phenolphthalein indicator. The burette readings in ml of NaOH solution correspond to c_e for each solution.
11. Titrate 10 ml of 0.5 (N) acetic acid solution by the same 0.1 N NaOH solution (theoretically 50 ml NaOH will be required). This is blank titration and the burette reading in ml of NaOH solution corresponds to c_0.
12. Calculate $x(= c_o - c_e)$, then $\log\frac{x}{m}$ and $\log c_e$. Plot $\log\left(\frac{x}{m}\right)$ against $\log c_e$.

Results and Calculations

Table 1: Recording of laboratory temperature

Table 2 : Preparation of 50 ml 0.1 N oxalic acid solution

1st weight w_1 (g)	2nd weight w_2 (g)	Weight taken $w_1 - w_2$ (g)	Strength of Oxalic acid solution
...	...	...	$\frac{w_1 - w_2}{0.315}\left(\frac{N}{10}\right)$

Table 3 : Standardisation of NaOH solution by oxalic acid solution

No. of titration	Vol. of oxalic acid soln. (ml)	Initial burette reading (ml)	Final burette reading (ml)	Vol. of NaOH required (ml)	Mean vol. (ml)	Strength of NaOH soln.
1	10	...	...	...	...	...
2	10	...	...	...		

Table 4 : Standardisation of acetic acid solution by NaOH solution

As Table 3. Take 10 ml acetic acid each time. From the calculated strength of NaOH, prepare 500 ml 0.1 N NaOH solution and from the strength of AcOH obtained, prepare 0.5 N AcOH solution.

[You may not dilute NaOH to exactly 0.1 N or AcOH to 0.5 N, rather use the stock standardised solution as usual; but then x and c_e should be calculated properly.]

Table 5 : Preparation of different concentration of AcOH solution

Bottle No.	Vol. of 0.5 N AcOH soln. (ml) (To add through burette)	Vol. of water (ml) (To add through burette)
1	50	0
2	40	10
3	30	20
4	20	30
5	15	35
6	10	40

Table 6 : Standardisation of Filtrate by NaOH solution

Volume of 0.1 N NaOH for blank titration = v ml

Bottle No.	No. of titrations	Vol. of the AcOH soln. (ml)	Burette reading Initial (ml)	Burette reading Final (ml)	Vol. of NaOH required (ml)	Mean Vol. in ml. ≡ eqbm conc. (c_e)
1	1	10	...	...	...	...
	2	10	...	...	...	

Bottle No.	No. of titrations	Vol, of the AcOH soln. (ml)	Burette reading Initial (ml)	Burette reading Final (ml)	Vol. of NaOH required (ml)	Mean vol. in ml. ≡ eqbm conc. (c_e)
2	1	10	...	...	...	...
	2	10	...	...	...	
3	1	10	...	...	...	...
	2	10	...	...	...	
4	1	10	...	...	...	...
	2	10	...	...	...	
5	1	10	...	...	...	...
	2	10	...	...	...	
6	1	10	...	...	...	...
	2	10	...	...	...	

Table 7 : Results for Adsorption isotherm

Bottle No.	Wt. of charcoal taken, m (g)	Initial conc. of AcOH, c_o (in ml of NaOH)	Eqbm. conc. c_e (in ml of NaOH) (From Table 6)	Amount of AcOH adsorbed $x = c_o - c_e$ (in ml of NaOH)	$\frac{x}{m}$	$\log\left(\frac{x}{m}\right)$	$\log c_e$
1	...	v	...	...	...	...	...
2	...	$\frac{4}{5}v$	...	...	...	...	...
3	...	$\frac{3}{5}v$	...	...	...	...	...
4	...	$\frac{2}{5}v$	...	...	...	...	...
5	...	$\frac{3}{10}v$	...	...	...	...	...
6	...	$\frac{1}{5}v$	...	...	...	...	...

Results: From the plot of $\log\left(\frac{x}{m}\right)$ vs. $\log c_e$, k and n are calculated.

$\log k$ = intercept ; n = slope.

Conclusion: Since a straight line is obtained, hence Freundlich adsorption isotherm is obeyed.

APPENDIX

Table 1

Surface Tensions (dyne cm^{-1}) of some common liquids at different temperatures

Liquid	10 °C	20 °C	30 °C
Accetic acid	28.8	27.42	—
Benzene	30.2	28.87	27.56
Carbon Tetrachloride	28.0	26.75	25.33
Chloroform	28.5	27.2	—
Ethyl alcohol	23.15	22.3	21.43
Methyl alcohol	—	22.55	—
Toluene	—	28.53	27.32
Water	—	72.75	71.18

Table 2

Some physical constants of water (including surface tension) at different temperatures

Temp. (°C)	Density (g/ml)	Vapour Pressure (mm Hg or Torr)	Surface Tension (dyne / cm)
15	0.99913	12.8	73.49
16	0.99894	13.63	73.34
17	0.99880	14.53	73.19
18	0.99869	15.50	73.05
19	0.99843	16.48	72.90
20	0.99820	17.54	72.75
21	0.99802	18.65	72.59
22	0.99780	19.83	72.44
23	0.99757	21.07	72.28
24	0.99730	22.38	72.13
25	0.99707	23.76	71.97
26	0.99681	25.21	71.82
27	0.99654	26.74	71.66
28	0.99626	28.35	71.50
29	0.99597	30.04	71.35
30	0.99567	31.80	71.18
31	0.99537	33.69	—
32	0.99505	35.66	—
33	0.99473	37.63	70.73
34	0.99440	39.90	—
35	0.99406	42.18	70.38

CHAPTER 3

Ionic Equilibria

(Solubility, Solubility Product, Heat of Solution and pH by Colour Matching)

I. **Introductory Words**

II. **Experiment 1**

Determination of the solubility of benzoic acid (or salicylic acid) at different temperatures and

(a) construction of solubility curve of the solute and

(b) calculation of heat of solution (for U.G. / P.G.)

III. **Experiment 2**

Determination of the heat of solution of oxalic acid at its saturation from the solubility measurements at different temperatures (for U.G. / P.G.)

IV. **Experiment 3**

Determination of the solubility and solubility product of potassium hydrogen tartarate (KHTa) in water and in presence of different concentrations of a common ion (e.g., KCl) at room temperature (for P.G.)

V. **Experiment 4**

To determine the activity solubility product (or thermodynamic solubility product) of potassium hydrogen tartarate (for P.G.)

VI. **Experiment 5**

Determination of activity solubility product of $CaSO_4$ by complexometric titration (for P.G.)

VII. **Experiment 6**

To determine the solubility product of silver acetate in water and to study the effect of a common cation (e.g., $AgNO_3$) and the effect of a common anion (e.g., AcONa) on its solubility (for U.G. / P.G.)

VIII. **Experiment 7**

To determine the pH of a given solution through matching its colour developed by an indicator (for U.G.)

IX. **Appendix**

CHAPTER 3

Ionic Equilibria

(Solubility, Solubility Product, Heat of Solution and pH by Colour Matching)

I. Introductory Words

Electrolytes dissociate in suitable ionising solvents, mainly water, into its constituent ions and an equilibrium is set up between the ions and the undissociated molecules of the electrolyte. A strong electrolyte dissociates almost completly, but the dissociation of weak electrolytes vary from very low to some low but in appreciable degree.

With the extent of dissociation, the concentration of the solute in solution plays an important role in providing valuable physico-chemical data in solution. Very dilute solutions, generally below 10^{-4} M, may be regarded as ideal solutions, where molal concentrations be well accepted as molar concentrations and the activity coefficient may be taken as unity. But stronger than this concentration, solutions lack ideality and activity coefficient, f or γ, needs considerations.

The relation between activity and concentration (molar or molal) is given by

$$\left.\begin{aligned} a &= c \times f \\ \text{and}\quad a &= m \times \gamma \end{aligned}\right\} \tag{1}$$

where c is the molar concentration and m is the molal concentration, f and γ are activity coefficients.

For a salt A_xB_y, which dissociates as

$$A_xB_y \rightleftharpoons xA^{q+} + yB^{q-},$$

the **mean ionic activity** coefficient,

$$\gamma_{\pm} = \left(\gamma_{+}^{x} \cdot \gamma_{-}^{y}\right)^{\frac{1}{x+y}}, \tag{2}$$

where γ_+ and γ_- are the activity coefficients of the cation and anion respectively.

They Debye-Hückel Limiting law for dilute solution correlating activity coefficient and ionic strength of the solution is given by

$$\log \gamma_{\pm} = -A\left|z_+z_-\right|\sqrt{\mu}, \tag{3}$$

where z_+ and z_- are charges of cation and anion respectivly and A is a constant whose value is 0.509 at 25 °C. μ is the **ionic strength** of the solution and is expressed as

$$\mu = \frac{1}{2}\sum m_i z_i^2, \tag{4}$$

where m_i and z_i are the molality and charge of the i-th ion in solution.

Solubility

Solubility is defined as the maximum number of grams of solute required to prepare a saturated solution of 100 gram of a solvent at a particular temperature.

Thus, if S be the solubility of a compound in a particular solvent at a definite temperature, the concentration of the solution will be $\frac{10S}{M}$ **molal**, M being the molecular weight of the solute.

For very low soluble compounds, the molality may be taken as molarity. For example, in case of low solubility, the above concentration may be written as $\frac{10S}{M}$(M).

The solubility of solid solute in a liquid depends on temperature, pressure, nature of the solute and solvent. For electrolytes, the solubility, in addition, depends on the concentration of the other ions—common or uncommon to either ions of the solute, present in the solution.

As per basic definition, solubility is a unitless quantity. But as disscused above, its unit may be molal, molar or even it may be expressed in percentage strength.

Solubility Curve

Solubility of a solute measured at different temperatures, when plotted against temperature (x-axis), give what is known as solubility curve of the solute in a particular solvent.

The curve may or may not be a straight line. The slope is generally positive, but examples are not very a few where some part of curve gives negative slope. Hydrated salts like Na_2SO_4, $10H_2O$, show such nature. On the other hand, NaCl shows a very small increase in solubility with rise in temperature.

Solubility Product

When a simple (1 : 1) sparingly soluble salt is dissolved in water, the portion of the salt dissolved is completely dissociated (salt being a strong electrolyte) and an equilibrium is set up between the undissolved solid and the ions, as

$$\text{MA (solid)} \rightleftharpoons \text{MA}\ (aq) \rightleftharpoons \text{M}^+(aq) + \text{A}^-(aq)$$

The equilibrium constant,

$$K_a = \frac{a_{\text{M}^+} \cdot a_{\text{A}^-}}{a_{\text{MA}(s)}} = a_{\text{M}^+} \cdot a_{\text{A}^-} \qquad [\because \text{ activity of solid} = 1]$$

$$\text{or,} \quad K_a = (c_{\text{M}^+} \cdot \gamma_{\text{M}^+})(c_{\text{A}^-} \cdot \gamma_{\text{A}^-}) \qquad [c = \text{molar conc.,} \quad \gamma = \text{activity coefficient}]$$

$$\text{or,} \quad K_a = (c_{\text{M}^+} \cdot c_{\text{A}^-}) \times (\gamma_{\text{M}^+} \cdot \gamma_{\text{A}^-})$$

$$\text{or,} \quad K_a = K_{sp} \cdot \gamma_{\pm}^2 \tag{5}$$

K_a is the **true thermodynamic solubility product** or **activity solubility product** and K_{sp} is the **concentration solubility product** or **simply solubility product** of the sparingly soluble salt at a particular temperature.

For very dilute solution, as in the case of a sparingly soluble salt, $\gamma_{\pm} = 1$; hence,

$$K_a = K_{sp}.$$

For a sparingly soluble salt of the type A_xB_y,

$$K_{sp} = (xc)^x (yc)^y$$

or, $K_{sp} = x^x y^y c^{x+y}$ [c = molar concentration] (6)

and $K_a = K_{sp} \gamma_{\pm}^{x+y}$ (7)

For, $Ba_3(PO_4)_2$,

$$K_{sp} = (3c)^3 \times (2c)^2 = 108c^5$$

and $K_a = 108c^5 \gamma_{\pm}^5$.

Effect of external ions on Solubility Product

K_a, the thermodynamic solubility product is constant at a constant temperature. But at the same temperature, K_{sp} varies when some external ions, may be common or not to either of the ions produced from the electrolyte, added to the solution.

i. **An inert ion is added:**

$K_a = K_{sp} \gamma_{\pm}^2$ (for 1 : 1 electrolyte)

K_a is constant. According to equation (3), $\gamma_{\pm}$ decreases with increase in μ. When an inert solute is added (like KNO_3 to a solution of AgCl), μ increases, So $\gamma_{\pm}$ decreases. Hence, K_{sp} will increase to keep the constancy of K_a.

ii. **A common ion is added:**

Say, KCl is added to a saturated solution of AgCl.

$\therefore \quad K_{sp} = c'(c' + x) = c^2$, (8)

where c is the previous solubility, c' is the new solubility and x is the molarity of Cl^- from added KCl solution.

Obviously, $c' < c$, i.e., solubility will decrease.

Heat of solution, ΔH :

Solubility largely depends upon temperature. Dissolution of a solute in water is mostly endothermic. It may be exothermic also.

When the dissolution process is **endothermic**, the **solubility increases with temperature**.

If it be **exothermic, solubility will decrease with temperature**.

Measuring the solubilities of a solute at different temperatures, the enthalpy of the process can be determined from van't Hoff isochore, which is

$$\log_{10} S = -\frac{\Delta H}{2.303RT} + \text{constant}, \quad (9)$$

where S is the solubility.

Instruments

Solubility-related experiments have wide scope to use a number of instruments starting from **pipette** and **burette**. Use of **potentiometer** and **conductometer** for this purpose has been discussed. **Spectrophotometer**, **pH-meter** also find their uses.

However, for precise results as well as for some specific experiments, a **thermostat** with about 40-50 °C interval is an inevitable equipment. Facility of lower temperatures like 10-15 °C will provide useful data even for research works. However, that will cost a little higher.

Experiment 1

Determination of the solubility of benzoic acid (or salicylic acid) at different temperatures and

(a) construction of a solubility curve of the solute

and (b) calculation of heat of solution.

Theory: Solubility is defined as the maximum number of grams of a solute required to saturate 100 grams of a solvent at a definite temperature.

By definition, solubility is a unitless physical property of a solute. However, it is also expressed in terms of molality (number of moles per 1000 g of solvent) or molarity (number of moles present in 1000 cm^3 of solution).

If x be the solubility at a particular temperature, then it may be easily expressed as $\frac{x \times 10}{M}$ molal (i.e. no. of moles per 1000 g of solvent), M being the molecular weight of the solute.

However this value cannot be easily converted to molarity except the solution is very dilute.

Since benzoic acid (or salicylic acid) is not too much soluble in water, the solubility can be expressed as $\frac{x \times 10}{M}$ molar (i.e. moles per 1000 cm^3 of solution).

Solubility is largely influenced by temperature and obeys Le Chatelier principle. If the dissolution is endothermic, solubility will increase with temperature-rise. If it is exothermic, solubility decreases with increase in temperature.

a. For construction of a solubility curve of the given solute, its solubilities at different suitable temperatures are measured and plotted aganist temperature.

b. For determination of the heat of solution, van't Hoff isochore is used, which is

$$\left(\frac{\delta \ln S}{\delta T}\right)_p = \frac{\Delta H}{RT^2}, \text{ where } S = \text{molarity of the solution ie, } S = \frac{x \times 10}{M}(\text{M}).$$

$$\text{or, } \int d \ln S = \frac{\Delta H}{R} \int \frac{dT}{T^2} \quad [\text{Assuming } \Delta H \text{ is independence of temperature}]$$

$$\text{or, } \ln S = -\frac{\Delta H}{RT} + c$$

$$\text{or, } \log_{10} S = -\frac{\Delta H}{2.303R} \cdot \frac{1}{T} + c'. \qquad \left[c' = \text{constant} = \frac{c}{2.303}\right] \qquad (1)$$

$\log_{10} S$ is plotted against $\frac{1}{T}$. The slope $= -\frac{\Delta H}{2.303R}$.

$\therefore \; \Delta H = -\text{slope} \times 2.303R$.

Thus, ΔH is evaluated.

The solubility in case of dilute saturated solution is, in fact, the equilibrium constant of solid $\rightleftharpoons$ solution system, as an equilibrium is set up between the undissolved solid and its solution. If the solute be A, then

$$A \text{ (solid)} \rightleftharpoons A \text{ (soln)}$$

$$\therefore \quad K = \frac{a_{A(\text{soln})}}{a_{A(\text{solid})}} = a_{A(\text{soln})} = c_{A(\text{soln})} \qquad [\because \; a_{A(\text{solid})} = 1].$$

Equipments: Thermostat, 10 ml pipettes (2 nos.), thermometer (with $\frac{1}{10}$th division), a small drawn-out glass-tube fitted with rubber tubing and partly packed with cotton/glass wool, four 100 ml reagent bottles or conical flasks, one 250 ml conical flask, burette.

Materials: Pure benzoic acid (or salicylic acid), NaOH, oxalic acid, phenolphthalein.

Procedure

1. Prepare about 250 ml 0.05 N NaOH solution is a reagent bottle. Weight required = 0.5 g.
2. Prepare 100 ml 0.05 N standard oxalic acid (or succinic acid) solution. Weight required for oxalic acid = 0.315 g.
3. Dissolve 3-4 g benzoic acid in about 300 ml water in a 500 ml reagent bottle. Shake well time to time for about half an hour and place in the thermostat at 50 °C temperature. Keep it for 20-30 minutes.
4. Meanwhile standardise NaOH solution against 10.0 ml of standard oxalic acid solution using phenolphthalein indicator. Do not evacuate the burettle.
5. Insert the end of a 10 ml pipette in the rubber tubing fitted with a drawn-out tube and packed with glass or cotton wool. This is done to avoid creeping of any solute particle during withdrawal. Withdraw 10 ml saturated solution of benzoic acid with this arrangement, remove the rubber tubing and immediately release in the conical flask already having some 60-70 ml water in it. Titrate against standard NaOH solution from the burette using phenolphthalein indicator. Wash off the conical flask.
6. Label the reagent bottles with 40 °C, 35 °C and 25 °C. Transfer about 50 ml original saturated solutions of benzoic acid in these three bottles, no matter if some solid particles pass through.
7. Set the thermostat at 40 °C. Immerse the respective bottle and keep for 20-30 minutes. Pipette out 10 ml solution as done in (5) and titrate against NaOH solution.
8. Repeat (7) for 35 °C and 25 °C temperatures. (Always check the temperature by immersing the thermometer in solution).
9. Calculate S for benzoic acid solution at each temperature using $v_1 s_1 = v_2 s_2$ equation [For benzoic acid, normality = molarity].
10. Plot (a) S vs. T for solubility curve and (b) $\log_{10} S$ vs. $\frac{1}{T}$ for claculation of heat of solution.

Results and Calculation

Table 1: Preparation of 100 ml standard oxalic acid solution (~ 0.05 N).

Required weight = 0.315 g

1st weight w_1 (g)	2nd weight w_2 (g)	Weight of oxalic acid transferred $w_1 - w_2$ (g)	Strength of oxalic acid solution
...	...	...	$\frac{w_1 - w_2}{0.315}(0.05)$N

Table 2: Standardisation of NaOH solution by standard oxalic acid solution

No. of titrations	Vol. of oxalic acid solution	Burette reading		Vol. of NaOH required (ml)	Mean vol. of NaOH soln. (ml)	Strength of NaOH Soln. (N)
		Initial (ml)	Final (ml)			
1	10.0	0.0	...	...		
2	10.0	...	...	...	...	...
3	10.0	...	...	...		

Table 3: Titration of saturated benzoic acid solutions at different temperatures against standard NaOH solution

Temp. (°C)	Vol of Satd. benzoic acid solution	Burette reading		Vol. of NaOH soln. required (ml)	Strength of benzoic acid soln. S (mol/l)
		Initial (ml)	Final (ml)		
50	10.0	0.0	...	...	...
40	10.0	...	...	...	...
35	10.0	...	...	...	...
25	10.0	...	...	...	...

Table 4: Results for solubility curve and ΔH of the solution

Temp. (°C)	Solubility S (mol/l)	$\text{Log}_{10}S$	T (K)	$\frac{1}{T}$
50	...	...	323	...
40	...	...	313	...
35	...	...	308	...
25	...	...	298	...

Results: From the graph of $\log S$ vs. $\frac{1}{T}$,

$$\Delta H = -\text{Slope} \times 2.303R$$
$$= \ldots \text{ kJ/mol.}$$

Conclusion

Solubility curve: From the nature of graph, it is seen that solubility of benzoic acid in water (increases/decreases) with temperature.

ΔH: Since the value of ΔH is (*+ve/–ve*), hence the process of dissolution of benzoic acid in water is (endothermic/exothermic).

Experiment 2

Determination of the heat of solution of oxalic acid at its saturation from the solubility measurements at different temperatures.

Theory: Solubility of a substance is defined as the number of grams of the substance required to saturate 100 grams of a solvent through dissolution at a particular temperature. This definition leaves solubility a unit less quantity.

However, solubility may also be expressed in molality unit, which comes from the above definition. If x be solubility at a particular temperature, then the solubility will be $\frac{10x}{M}$ molal, M being the molar mass.

Determination of solubility in molal unit is a little bit labourious work. For determination of heat of solution, it may be calculated in molar unit, i.e., number of moles of the solute present in 1000 ml of saturated solution of the substance. However, in contrast to solutions of low soluble compounds like benzoic acid, salicylic acid, etc, molality cannot be regarded as molarity for solution of oxalic acid, as it is sufficiently soluble in water.

For saturated solution of any compound with any degree of solubility, the existence of solid and solution together actually establishes an equilibrium between the two phases. i.e.,

$$A \text{ (solid solute)} \rightleftharpoons A \text{ (saturated solution)}.$$

∴ the equilibrium constant

$$K = \frac{a_{A(\text{soln.})}}{a_{A(\text{solid})}} = a_{A(\text{soln.})} \qquad [\because\ a_{A(\text{solid})} = 1].$$

$$\therefore \quad K = a_{A(\text{soln.})} \tag{1}$$

Using molar concentration, S, for activity,

$$K = S_{A(\text{soln.})} \tag{2}$$

Using equation (2), the heat of solution (ΔH) of a solute in its saturated solution can be computed. The temperature dependence of equilibrium constant has been mathematically expressed by van't Hoff isochore as

$$\frac{d \ln K}{dT} = \frac{\Delta H}{RT^2}. \tag{3}$$

From (2) and (3), we have

$$\frac{d \ln S}{dT} = \frac{\Delta H}{RT^2} \tag{4}$$

On integration, (4) becomes

$$\ln S = -\frac{\Delta H}{RT}$$

$$\text{or,} \quad \log_{10} S = -\frac{\Delta H}{2.303R} \cdot \frac{1}{T} \tag{5}$$

(assuming ΔH be independent of T, at least for the present purpose, which is not a complicated dissolution process).

Now, plot of $\log_{10} S$ vs. $\frac{1}{T}$ will give a straight line, from the slope of which ΔH can be calculated.

$$\Delta H = -2.303R \times \text{slope}. \tag{6}$$

Equipments: Thermostat, 5 ml and 10 ml pipettes, burette, 250 ml reagent bottle, 100 ml conical flask, thermometer with $\frac{1}{10}$th graduation.

Materials: Oxalic acid, NaOH, phenolphthalein.

Procedure

1. Prepare 250 ml ~ 0.5 (N) NaOH solution. Required weight = 5 g (a small larger weight is preferred).
2. Prepare 50 ml standard oxalic acid solution of ~ 0.5 N strength. Weight required = 1.575 g.
3. Standardise NaOH solution with the standard oxalic acid solution.
4. Prepare about 100 ml saturated oxalic acid solution in the 250 ml reagent bottle and keep it in the temperature bath set at 35 °C.

   [Because of high solubility of oxalic acid, it is advisable not to exceed the temperature beyond 40 °C].
5. Carefully pipette out 5 ml oxalic acid solution with the rubber-tubing-cum-withdrawal tube arrangement as in the previous experiment and keep in the 100 ml conical flask. Immediately add 25-30 ml water and titrate against standard NaOH solution from the burette.

   [The addition of excess water is to avoid setting down any oxalic acid crystal due to temperature fall. **It is better to keep water first and then to add the solution.**]
6. Set the bath at 30 °C and allow the original solution to avail of this temperature. Check the temperature with thermometer. Pipette out 5 ml solution into conical flask with some water and titrate against standard NaOH solution.
7. Perform the procedure (6) at 25 °C and 20 °C temperatures.
8. Calculate the solubility (S) of oxalic acid in water in **mol/l** unit, at the above-said temperatures, viz. 35 °C, 30 °C, 25 °C and 20 °C using $v_1 s_1 = v_2 s_2$ equation. Divide the normalities so obtained by 2 to get the results in mol/l unit.
9. Plot $\log_{10} S$ vs. $\frac{1}{T}$ and calculate ΔH from equation (6).

Results and Calculations

Table 1: Preparation of 50 ml standard ~ 0.5 N oxalic acid solution

Required weight = 1.575 g

Use Table 1 in experiment 1.

Table 2: Standardisation of NaOH solution by standard oxalic acid solution (use 10 ml acid solution).

Use Table 2 in experiment 1.

Table 3: Titration of saturated oxalic acid solutions at different temperatures against standard NaOH solution.

Temp. (°C)	Vol. of saturated Oxalic acid Soln. (ml)	Burette reading		Vol. of NaOH soln. required (ml)	Strength of oxalic acid soln. S_1 (N)	Solubility of oxalic acid (mol/l) $S = \frac{S_1}{2}$
		Initial (ml)	Final (ml)			
35	5.0	...	...	...	...	...
30	5.0	...	...	...	...	...
25	5.0	...	...	...	...	...
20	5.0	...	...	...	...	...

Table 4: Results for calculation of ΔH

Temp. (°C)	T (K)	$\frac{1}{T}$	Solubility, S (mol/l)	$\log_{10} S$
35	308	...	...	...
30	...	...	...	...
25	...	...	...	...

Results: From the plot of $\log_{10} S$ vs. $\frac{1}{T}$,

$$\Delta H = -\text{slope} \times 2.303\, R \text{ kJ/mol.}$$

Conclusion: Since the value of ΔH is (+ve/–ve), hence the process of dissolution of oxalic acid in water is (endothermic/exothermic) at its saturation point.

Experiment 3

Determination of solubility and solubility product of potassium hydrogen tartarate (KHTa) in water and in presence of different concentrations of a common ion (e.g., KCl) at room temperature.

Theory: [Write definition of solubility, its units from theories of previous experiments or from introductory words].

Potassium hydrogen tartarate is a sparingly soluble salt in water. In its saturated solution an equilibrium is established between the solid phase and the ions.

$$\text{KHTa (solid)} \rightleftharpoons \text{KHTa (solution)} \rightleftharpoons K^+(aq) + HTa^-(aq)$$

ie, $$\text{KHTa (solid)} \rightleftharpoons K^+(aq) + HTa^-(aq) \qquad (1)$$

Due to the low concentration, the activities may be regarded as molar concentrations and that of the undissolved solid is unity. Hence, the equilibrium constant of the solid-ion system [eqn. (1)].

$$K_a = \frac{a_{K^+} \cdot a_{HTa^-}}{a_{KHTa}}$$

or, $$K_{sp} = [K^+]\,[HTa^-] \qquad (2)$$

where K_{sp} is the solubility product of KHTa and K_a is its activity solubility product.

In general, solubility product of a compound (generally a low soluble or spaningly soluble electrolyte) is defined as the product of the maximum concentrations of its constituent ions, raised to appropriate stoichiometric powers, at a particular temperature.

Presence of an ion, common to either of the ions produced from the electrolyte, lowers the solubility of the electrolyte. Thus, in presence of KCl, the solubility of KHTa decreases.

Let S be the solubility of KHTa in mol/l in presence of KCl and c be concentration of KCl in mol/l.

Then from eqn. (2),

$$K_{sp} = [K^+]\,[HTa^-]$$

or $$K_{sp} = (S + c)S \qquad (3)$$

Hence adding KCl solution of known concentration and standardising the saturated solution of KHTa against a standard NaOH solution, the changed solubility of KHTa can be estimated. That the K_{sp} remains almost constant can also be verified.

Equipments and Materials

NaOH, oxalic acid, potassium hydrogen tartarate, KCl, water-trough, thermometer, 5 ml pipette, burette, phenolphthalein, five 100 ml reagent bottles, five 100 ml beakers, 100 ml conical flasks, filter paper, funnels, vol. flasks— 50 ml, 250 ml.

Procedure

1. Prepare 250 ml ~0.05 N NaOH solution.
2. Prepare 50 ml standard oxalic acid solution of the order 0.1 N. Weight required = 0.315 g.
3. Prepare 250 ml 0.1 M (exact) KCl solution. Weight required = 1.864 g.
4. Standardise NaOH solution against 10 ml standard oxalic acid solution.
5. Arrange 5 reagent bottles as in Table 5.
6. Place the bottles in a mechanical shaker and allow to be shaken for 30-45 minutes. Place them in water-trough for 15 minutes. Note the temperature of water.
7. Take 5 clean dry beakers and label 1-5.
8. Filter solution no. 1 in beaker 1, rejecting first 5-6 ml of the filtrate. Take 5 ml of the filtrate in a conical flask and titrate against standard NaOH solution using phenolphthalein indicator. Titrate twice.
9. Perform 8 for solutions 2-5, **each time using a new filter paper and dry funnel.**
10. Calculate strength of KHTa (= $[HTa^-]$) in each bottle.
11. Calculate K_{sp} for each solution.

Results and Calculations

Table 1: Recording of temperatue: Take the mean of temperatures of water in the trough during filtering of the first and last bottle. Put in tabular form.

Table 2: Preparation of 50 ml 0.1 N standard oxalic acid solution

Required weight = 0.315 g.

Use Table 1 of experiment 1.

Table 3: Preparation of 250 ml exact 0.1 M KCl solution

Weight required = 1.864 g.

Record the weight in tabular form. Take exact weight or a little more. Adjust strength by proper dilution.

Table 4: Standardisation of NaOH solution by standard oxalic acid solution

Use Table 2 of experiment 1.

Table 5: Preparation of saturated KHTa solution in water and in different KCl concentrations.

Bottle No.	Vol. of 0.1 M KCl solution (ml)	Vol. of water (ml)	Total vol. of Solution (ml)	Conc. of KCl solution c (mol/l)	Approximate wt. of KHTa (g)
1	0	50	50	—	~ 0.5
2	10	40	50	0.02	~ 0.5
3	20	30	50	0.04	~ 0.5
4	30	20	50	0.06	~ 0.5
5	40	10	50	0.08	~ 0.5

Table 6: Determination of solubility and solubility product

Bottle No.	Conc. of KCl soln. (c) mol/l	Vol. of filtrate (ml)	Burette reading		Vol. of NaOH required (ml)	Mean vol. of NaOH (ml)	Solubility mol/l (S)	Solubility Product $K_{sp} = (S + c)S$
			Initial (ml)	Final (ml)				
1	0	5	···	···	···	···	···	···
		5	···	···	···			
2	0.02	··· ···						
3	0.04							
4	0.06							
5	0.08							

Conclusion

1. The solubility product of KHTa at ··· °C temperature is ···.
2. From the results it is seen that the solubility (S) decreases with the increase of the strength (c) of the common ion and that K_{sp} remains almost constant.

N.B. *The activity solubility product, K_a, can also be determined from this experiment. See experiment 4.*

Experiment 4

To determine the activity solubility product (or thermodynamic solubility product) of potassium hydrogen tartarate.

Theory: When a saturated solution of a sparingly soluble salt, like potassium hydrogen tartarate (KHTa), is in contact with its solid phase, the following equilibram is set up.

$$\text{KHTa (solid)} \rightleftharpoons \text{KHTa (solution)} \rightleftharpoons \text{K}^+\text{(soln.)} + \text{HTa}^-\text{(soln.)}$$

$$\text{i.e., KHTa(solid)} \rightleftharpoons \text{K}^+\text{(soln.)} + \text{HTa}^-\text{(soln.)}$$

$$\text{or,}\quad K_a = \frac{a_{\text{K}^+} \cdot a_{\text{HTa}^-}}{a_{\text{KHTa(solid)}}}$$

or, $K_a = a_{K^+} \cdot a_{HTa^-}$ $[\because\ a_{KHTa(solid)} = 1]$ (1)

K_a is known as activity solubility product.

Since $a = c \cdot \gamma_{\pm}$, (2)

where c = molar concentration and $\gamma_{\pm}$ = mean ionic activity coefficient of ions,

$$\therefore\ K_a = [K^+][HTa^-] \cdot \gamma_{\pm}^2$$

or, $K_a = K_{sp}\gamma_{\pm}^2$ $[K_{sp}$ = solubility product] (3)

Taking logarithm on both sides,

$$\log K_a = \log K_{sp} + 2\log \gamma_{\pm}$$

or, $\log K_a = \log K_{sp} - 2A|z_+ z_-|\sqrt{\mu}$, [From Debye-Hückel law, $\log \gamma_{\pm} = -A|z_+ z_-|\sqrt{\mu}$]

where z_+, z_- are charges of cations and anions and μ = ionic strength of the solution.

or, $-\log K_{sp} = -\log K_a - 2A|z_+ z_-|\sqrt{\mu}$. (4)

A plot of ($-\log K_{sp}$) vs. $\sqrt{\mu}$ thus gives a straight line with negative slope. From the intercept (i.e., $-\log K_a$), K_a, the activity solubility product is obtained.

The nature of the graph is

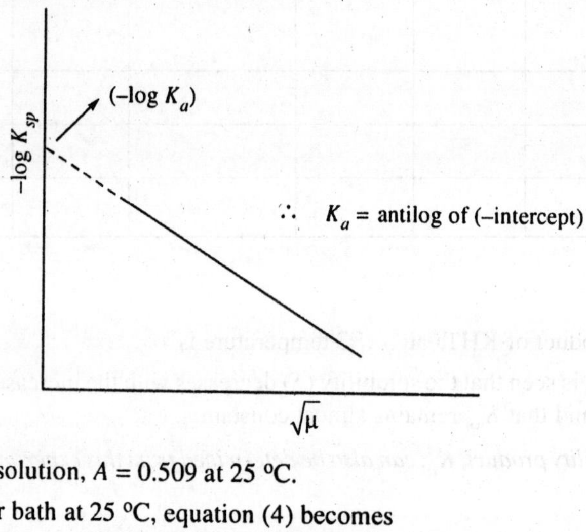

For an aqueous solution, A = 0.509 at 25 °C.

Setting the water bath at 25 °C, equation (4) becomes

$$-\log K_{sp} = -\log K_a - 0.118|z_+ z_-|\sqrt{\mu}$$ (5)

However, equation (5) is not required for the present experiment.

[Equation (4) may also be written as

$$\log K_{sp} = \log K_a + 2A|z_+ z_-|\sqrt{\mu}$$

A plot of $\log K_{sp}$ vs. $\sqrt{\mu}$ will also give a straight line, but with positive slope. From the intercept (i.e., $\log K_a$), K_a can be calculated]

For change of ionic strengths, $BaCl_2$ solutions of different concentration are used, which enhances the solubility of KHTa.

Materials required

NaOH, oxalic acid, $BaCl_2$, potassium hydrogen tartarate, phenolphthalein.

Five each of 100 ml reagent bottles and beakers, volumetric flasks— one each of 250 ml, 50 ml; one 250 ml reagent bottle, conical flasks, filter papers, 5 ml pipette, burette.

Procedure

1. Prepare 250 ml ~0.05 (N) NaOH solution. Dissolve 3-4 beads (~0.5 g) in 250 ml water in reagent bottle.
2. Prepare 50 ml 0.1 (N) standard oxalic acid solution.
3. Prepare 250 ml exact 0.1 (M) $BaCl_2$ solution.
4. Standardise NaOH solution with standard oxalic acid solution using phenolphthalien indicator.
5. Arrange 5 reagent bottles as per Table 5. Add ~0.5 g KHTa in each bottle.
6. Shake all the bottles in a shaker for about 30 minutes, allow to settle for another 15-20 minutes preferably in a trough of water and filter in dry beakers, numbered 1-5, using separate filter paper in each case.
7. Titrate 5 ml of each filtrate against standard NaOH solution using phenolphthalein indicator. Titrate twice.
8. Calculate K_{sp} ($= S^2$), μ and thereby $\sqrt{\mu}$ as below:

$$\mu = \frac{1}{2}\Sigma m_i z_i^2.$$

9. Draw graph of ($-\log K_{sp}$) vs. $\sqrt{\mu}$, extrapolate and calculate K_a.

Results and Calculations

Table 1: Recording of room temperature in tabular form.

Table 2: Preparation of 50 ml 0.1 (N) oxalic acid solution—see experiment 1.

Table 3: Preparation of 250 ml exact 0.1 (M) $BaCl_2$ solution. Put weighing data in tabular form. Required weight = 6.11g. If larger weight is taken, dilute to 0.1 (M).

Table 4: Standardisation of NaOH solution by oxalic acid solution. See experiment 1.

Table 5: Arrangement of different bottles with $BaCl_2$ solutions and ~0.5 g KHTa in each bottle

Bottle No.	Vol. of 0.1 M $BaCl_2$ soln. (ml)	Vol. of water (ml)	Total vol. (ml)	Strength of $BaCl_2$ soln. mol/l
1	10	40	50	0.02
2	20	30	50	0.04
3	30	20	50	0.06
4	40	10	50	0.08
5	50	0	50	0.10

Table 6: Determination of solubility and solubility product

Take 5 ml of the filtrate for each titration

Bottle No.	Burette reading Initial (ml)	Final (ml)	Vol. of NaOH required (ml)	Mean Vol. of NaOH (ml)	Solubility S (mol/l)	Solubility Product $K_{sp} = S^2$
1	0	...	...	...	...	...
	...	...	...			
2	...	...	...	...	...	...
3	...	...	...	...	...	...
4	...	...	...	...	...	...
5	...	...	...	...	...	...

Table 7: Data for graph plotting

Bottle No.	S (mol/l)	$K_{sp} = S^2$	$-\log K_{sp}$	Conc. of $BaCl_2$ (c) (mol/l)	$m = S + 3c$ (see below)	$\sqrt{\mu}$
1	...	...	...	0.02	...	...
2	...	...	...	0.04	...	...
3	...	...	...	0.06	...	...
4	...	...	...	0.08	...	...
5	...	...	...	0.10	...	...

[**Calculation of μ:**

$$\mu = \frac{1}{2}\sum_i \left[m_i\, z_i^2 \right]$$

$$= \frac{1}{2}\left\{[K^+]\cdot 1^2 + [HTa^-]\cdot 1^2 + [Ba^{2+}]\cdot 2^2 + [Cl^-]\cdot 1^2\right\}$$

$$= \frac{1}{2}(S + S + c\cdot 4 + 2c) = S + 3c]$$

From the graph,

Intercept = $-\log K_a$ = ...

∴ K_a = antilog of (– instercept) = ...

Hence, activity solubility product of KHTa = ... at °C.

Conclusion: From the results it is seen that, the solubility and solubility product of KHTa increase with the increasing conc. of $BaCl_2$, but activity solubility product remains constant at the experimental temperature.

Experiment 5

Determination of activity solubility product of calcium sulphate ($CaSO_4$) *by complexometric titration.*

Theory: When a saturated solution of a sparingly soluble salt, say calcium sulphate ($CaSO_4$), is in contact with the solid phase, the equilibrium is

$$CaSO_4\,(s) \rightleftharpoons CaSO_4\,(soln.) \rightleftharpoons Ca^{2+} + SO_4^{2-}$$

At equilibrium $\Delta G = 0$, hence

$$\mu_{Ca^{2+}} + \mu_{SO_4^{2-}} - \mu_{CaSO_4(S)} = \Delta G = 0 \quad (\mu\text{-terms represent chemical potentials})$$

$$\therefore \quad \mu_{Ca^{2+}} + \mu_{SO_4^{2-}} = \mu_{CaSO_4(S)} = \mu^{o}{}_{CaSO_4(S)} + RT\ln a_{CaSO_4(S)} = K_1 \quad \text{(constant)}$$

Hence,

$$\mu^0_{Ca^{2+}} + RT\ln a_{Ca^{2+}} + \mu^0_{SO_4^{2-}} + RT\ln a_{SO_4^{2-}} = K_1 \quad \text{(constant)}$$

$$\text{or,} \quad \ln(a_{Ca^{2+}} \cdot a_{SO_4^{2-}}) = (K_1 - \mu^0_{Ca^{2+}} - \mu^0_{SO_4^{2-}})/RT = K_2 \quad \text{(constant)}$$

At a constant temperature

$$a_{ca^{2+}} \cdot a_{SO_4^{2-}} = K_a \quad \text{(constant).}$$

K_a is called the activity solubility product of the salt at a given temperature. It is the product of the activities of the ions in a saturated solution.

Expressing K_a with molar concentrations and activity co-efficients,

$$K_a = [Ca^{2+}][SO_4^{2-}] \cdot f_\pm^2$$

$$\text{or,} \quad K_a = S^2 \cdot f_\pm^2$$

$$\therefore \quad K_a = K_{sp} \cdot f_\pm^2$$

[where S is the solubility in moles/l and K_{sp} is the solubility product of $CaSO_4$ and $f_\pm$ is the mean activity coefficient]

$$\text{or,} \quad \log K_a = \log K_{sp} + 2\log f_\pm$$

$$\text{or,} \quad \log K_a = \log K_{sp} - 2A|z_+||z_-|\sqrt{\mu}$$

[A = constant, μ = ionic strength = $\frac{1}{2}\sum_i m_i z_i^2$]

$$\text{or.} \quad \log K_{sp} = \log K_a + 2A|z_+||z_1|\sqrt{\mu}\,.$$

[N.B. $\mu_{Ca^{2+}}, \mu_{SO_4^{2-}}$ etc. represent chemical potentials, whereas μ is ionic strength]

If we plot log K_{sp} vs. $\sqrt{\mu}$, a straight line with positive slope will be obtained. From the intercept (logK_a) at $\mu = 0$, the activity solubility product (K_a) can be calculated.

Materials

1. Ammonium hydroxide soln. ($\approx$ 0.5M) 250 ml.
2. Perchloric acid ($\approx$ 0.5M) 250 ml
3. Sodium hydroxide soln. ($\approx$ 0.5M) 100 ml
4. Oxalic acid solution 0.5 N 100 ml
5. Calcium sulphate

6. Ammonium hydroxide—ammonium chloride buffer solution of pH ~ 10.
7. EDTA (disodium salt) solution 0.1 M 100 ml
8. Magnesium sulphate solution (0.1 M) 100 ml
9. EBT indicator, phenolphthalein and methyl orange indicator solution
10. Stoppered bottles (125 ml capacity, 6 Nos.), 125 ml conical flasks (2-3 Nos.), burette, pipettes (10 ml, 5 ml), volumetric flasks (100 ml, 250 ml), beakers (100 ml, 250 ml).

Procedure

1. Prepare 250 ml ≈ 0.5M NH_4OH solution (concentrated NH_4OH is ≈ 14M), 250 ml ≈ 0.5M $HClO_4$ (concentrated $HClO_4$ is ≈ 11.3M), 100 ml ≈ 0.5M NaOH solution (≈ 2 g solid NaOH is dissolved in 100 ml water).
2. Prepare 100 ml 0.5 (N) oxalic acid solution by accurate weighing. Standardise NaOH solution against standard oxalic acid solution and perchloric acid against standardised NaOH solution. Then standardise NH_4OH solution against standardised $HClO_4$ soln.

   Prepare exact 0.4 M $HClO_4$ and 0.4 M NH_4OH (100 ml each) by quantitative dilution and mix them in a stoppered bottle to prepare 200 ml 0.2 M ammonium perchlorate (NH_4ClO_4) solution. Then the following composition of sets are made:

   Composition of sets

Set Nos. (125 ml stoppered bottles)	Vol. of 0.2 M NH_4ClO_4 solution (ml)	Vol. of distilled water (ml)	Amount of $CaSO_4$ added (g)
1	0	50	≈ 1
2	10	40	≈ 1
3	20	30	≈ 1
4	30	20	≈ 1
5	40	10	≈ 1
6	50	0	≈ 1

   [Note that the strengths of NH_4ClO_4 in the above sets are in the range 0.02 M to 0.2 M]

   Shake the above sets for an hour and then allow them to settle for about 15 minutes.
3. Prepare 100 ml each of 0.1 (M) $MgSO_4$ and EDTA disodium salt and prepare 50 ml each of exact 0.01 M $MgSO_4$ solution and EDTA solution by quantitative dilution. Prepare 50 ml Mg-EDTA complex by mixing 25 ml of 0.01 M $MgSO_4$ solution and 25 ml 0.01 (M) (exact) EDTA solution. Again prepare 250 ml 0.01 M EDTA solution by proper dilution.
4. Preparation of $NH_4OH + NH_4Cl$ buffer solution of pH ≈ 10: Dissolve ≈ 17 g of NH_4Cl in 142 ml concentrated NH_4OH (≈ 14 M) and the volume is made to 250 ml. To prepare solution less than 250 ml, lower the amount of NH_4Cl and NH_4OH by appropriate proportions.
5. Filter the solution of set no. 1 (Whatman No.1 filter paper may be used). Pipette 10 ml of the filtrate in a 125 ml conical flask— add 2 ml $NH_4OH + NH_4Cl$ buffer solution and 1 ml of Mg-EDTA complex. Then add pinch of EBT indicator so that a pink colour appears. Titrate it against 0.01 (M) EDTA solution. The end point is noted when just a sky blue colour is appeared. Repeat it twice or thrice for set no. 1. Repeat the same for set nos. 2 to 6. Calculate the solubility and solubility product of $CaSO_4$ for different sets. Also calculate the ionic strength (μ) of each set. Plot $\log K_{sp}$ vs. $\sqrt{\mu}$ and determine K_a from the intercept.

Results and Calculation

Table 1: Recording of laboratory temperature

	Before expt.	After expt.	Mean
Temperature	… °C	… °C	… °C

Table 2: Preparation of 100 ml 0.5 N oxalic acid solution

Initial mass (g)	Final mass (g)	Mass of oxalic acid transferred (g)	Strength of Oxalic acid solution
w_1	w_2	(w_1-w_2)	$\frac{w_1 - w_2}{3.15}\left(\frac{N}{2}\right)$

Table 3: Preparation of 100 ml 0.1 (M) EDTA solution

[Relative molar mass of EDTA (disodium salt), $Na_2H_2C_{10}H_{12}O_8N_2$, $2H_2O$, = 372.24]

Initial mass (g)	Final mass (g)	Mass of EDTA transferred (g)	Strength of EDTA solution
w_1	w_2	(w_1-w_2)	$\frac{(w_1 - w_2)}{3.722}\left(\frac{M}{10}\right)$

From this solution exact 0.01 (M) 250 ml EDTA solution is prepared by diluting … ml to 250 ml.

Table 4: Preparation of 100 ml 0.1 (M) $MgSO_4$ solution

Relative molar mass of $MgSO_4 . 7H_2O$ = 246.3.

Initial mass (g)	Final mass (g)	Mass of $MgSO_4$ transferred (g)	Strength of $MgSO_4$ solution
w_1	w_2	(w_1-w_2)	$\frac{(w_1 - w_2)}{2.463}\left(\frac{M}{10}\right)$

[Exact 0.01 (M) $MgSO_4$ solution is prepared by diluting … ml to 50 ml]

Preparation of Mg-EDTA complex (0.005 M).

25 ml exact 0.01 (M) EDTA solution is mixed with 25 ml exact 0.01 (M) $MgSO_4$ solution to prepare 50 ml 0.005 M Mg-EDTA.

Table 5: Standardisation of ≈ 0.5 (M) NaOH solution against standard oxalic acid solution (nearly 0.5 N)

Observation Nos.	Vol. of oxalic acid solution taken (ml)	Vol. of NaOH solution required (ml)	Mean vol. of NaOH solution (ml)	Strength (M)
1	10	…	…	…
2	10	…		

Table 6: Standardisation of perchloric acid solution (≈ 0.5 M) against standardised NaOH solution (≈ 0.5 M)

Observation Nos.	Vol. of $HClO_4$ soln. taken (ml)	Vol. of NaOH solution required (ml)	Mean vol. of NaOH solution required (ml)	Strength of $HClO_4$ solution
1	10	...	...	... (M)
2	10	...		

Exact 0.4 (M) 100 ml $HClO_4$ solution is prepared by diluting ... ml to 100 ml with distilled water.

Table 7: Standardisation of NH_4OH solution (≈ 0.5 M) against standardised $HClO_4$ solution (≈ 0.5 M) [use methyl orange indicator]

Observation Nos.	Vol. of $HClO_4$ soln. taken (ml)	Vol. of NH_4OH solution required (ml)	Mean vol. of NH_4OH solution required (ml)	Strength of NH_4OH solution
1	10	...	...	... (M)
2	10	...		

100 ml exact 0.4 (M) NH_4OH is prepared by diluting ... ml to 100 ml with distilled water.

Preparation of 200 ml 0.2 (M) NH_4ClO_4 solution: 100 ml 0.4 M (exact) $HClO_4$ is mixed with 100 ml 0.4 (M) (exact) NH_4OH solution to prepare 200 ml 0.2 (M) NH_4ClO_4 solution.

Table 8: Titration results of different sets

Set Nos.	Vol. of the filtrate taken (ml)	Vol. of 0.01 (M) EDTA solution required (ml)	Mean vol. of EDTA solution required (ml)	Molar Concentration of Ca^{2+}, i.e., solubility S (moles/litre)
1	(a) 10 (b) 10	(a) ... (b) ...	...	...
2	(a) 10 (b) 10	(a) ... (b) ...	...	...
3.	(a) 10 (b) 10	(a) ... (b) ...	...	...
4.	(a) 10 (b) 10	(a) ... (b) ...	...	...
5.	(a) 10 (b) 10	(a) ... (b) ...	...	...
6	(a) 10 (b) 10	(a) ... (b) ...	...	...

[*Note* that solubility will increase from set 1 to 6, as ionic strength increases from set 1 to 6.]

Table 9: Table of graph plotting

Set Nos.	Solubility, S (Moles/litre)	$K_{sp} = S^2$	$\log K_{sp}$	Molar concentration of NH_4ClO_4 Solution	μ	$\sqrt{\mu}$
1	...	...	...	0.00		
2	...	...	...	0.04		
3	...	...	...	0.08		
4	...	...	...	0.12		
5	...	...	...	0.16		
6	...	...	...	0.20		

Results from the graph

From the plot of $\log K_{sp}$ vs. $\sqrt{\mu}$, the intercept at $\mu = 0$ is ... = $\log K_a$.

Hence, K_a = ... at ... °C.

Calculation of ionic strenghts of different sets

μ-values can be calculated in the following way:

For set 1: Let the solubility be S_1 moles/litre

Here $[NH_4ClO_4] = 0$

$$\therefore \quad \mu = \frac{1}{2}\left[S_1 \cdot 2^2 + S_1 \cdot 2^2\right] \quad \text{[For } CaSO_4 \text{ only]}$$

$$= 4S_1.$$

For Set 2: Let the solubility be S_2 moles/litre

Here $[NH_4ClO_4] = 0.04\ (M) = C_2$ (Let)

$$\therefore \quad \mu = \frac{1}{2}\left[S_2 \times 2^2 + S_2 \times 2^2 + 0.04 \times 1^2 + 0.04 \times 1^2\right]$$

$$= \frac{1}{2}\left[8S_2 + 2C_2\right] = 4S_2 + C_2$$

Calculation of μ for sets 3, 4, 5 and 6 can be done similarly. In general, if S_n be the solubility and C_n be the concentration of NH_4ClO_4 solution (both in moles/litre), then for a set, $\mu_n = (4S_n + C_n)$.

Experiment 6

To determine the solubility product of silver acetate in water and to study the effect of a common cation (e.g., $AgNO_3$) and the effect of a common anion (e.g., AcONa) on its solubility.

Theory: Silver acetate (CH_3COOAg or AcOAg) is a sparingly soluble salt in water. In its aqueous solution, the following equilibirum is established.

$$\text{AcOAg (solid)} \rightleftharpoons Ag^+(aq) + OAc^-(aq) \qquad (1)$$

The solution being very dilute, the activity of the ions may be taken as molar concentration, so that

$$K_{sp} = [Ag^+][OAc^-] = x^2, \quad [\because \ a_{AcOAg(s)} = 1] \tag{2}$$

where K_{sp} is the solubility product of the salt, which is constant at a constant temperature and x is the ionic concentration of each ion.

A common ion, either a cation or an anion, largely affects the solubility of the compound. In both cases, the solubility decreases.

Presence of common cation

When a saturated solution of AcOAg is prepared in presence of $AgNO_3$ solution, the solubility product, K_{sp} becomes,

$$K_{sp} = [Ag^+][OAc^-]$$

$$\text{or,} \quad K_{sp} = (S + c)S, \tag{3}$$

S being the new solubility of AcOAg in mol/litre and c is the concentration of $AgNO_3$ solution added in mol/litre.

Presence of common anion

When a saturated solution of AcOAg is prepared in presence of AcONa solution, the solubility product, K_{sp} becomes

$$K_{sp} = [Ag^+][OAc^-]$$

$$\text{or,} \quad K_{sp} = S'(S' + c') \tag{4}$$

S' being the solubility of silver acetate in sodium acetate solution in mol/litre and c' is the molar concentration of AcONa added.

Hence titration of definite amount of saturated solutions of silver acetate in water, $AgNO_3$ solution and AcONa solution by a standard KCNS or NH_4CNS solution will give the K_{sp} value of silver acetate in water as well as will help to study the effect of a common ion on its solubility.

Apparatus

Four 100 ml volumetric flasks, three 100 ml well-stoppered reagent bottles (or Jena bottles), two 250 ml stoppered reagent bottles, burette, pipettes, 100 ml conical flasks, beakers, 50 ml volumetric flask.

Materials: Pure siliver acetate, oxalic acid, NaOH, $AgNO_3$, acetic acid, ferric ammonium sulphate (ferric alum), KCNS/NH_4CNS, conc. nitric acid, phenolphthalein.

Procedure

1. Prepare 50 ml 0.4 N standard oxalic acid solution.
   Weight required = 1.26 g.
2. Prepare about 200 ml ~ 0.5 N NaOH solution.
   Weight required ~ 4g.
3. Prepare about 150 ml ~ 0.5 N acetic acid solution. Take about 4.5 ml acetic acid from the original jar (~ 17N).
4. Prepare 100 ml 0.1 N standard $AgNO_3$ solution.
   Weight required = 1.6988 g.

5. 250 ml ~ 0.05 N NH_4CNS or KCNS solution.
6. Prepare 50 ml saturated ferric alum indicator solution. Dissolve about 20 g ferric ammonium sulphate in 50 ml distilled water, add 5 ml conc. HNO_3 and shake gently to mix the ingredients.
7. Standardise NaOH solution against standard oxalic acid solution using phenolphthalein indicator (It will be > 0.4 N).
8. Standardise acetic acid solution against standard NaOH solution using phenolphthalein indicator (It will be > 0.4 N).
9. Prepare exactly 100 ml 0.4 N NaOH and 100 ml 0.4 N acetic acid solution by proper dilution in volumetric flasks.
10. Take 50 ml each of NaOH and AcOH solution (0.4 N) and mix in a 100 ml volumetric flask. Thus 100 ml 0.2 N sodium acetate solution is prepared.
11. Standardise NH_4CNS solution against standard silver nitrate solution. Pipette out 10 ml $AgNO_3$ solution into a conical flask, add 5-6 ml water and 1 ml ferric alum indicator solution. Add NH_4CNS solution from burette. The equivalence point will show permanent **faint pink colour**.
12. Take three reagent bottles. Mark them I, II, III. Place in each bottle ~ 1.0 g siliver acetate and then 50 ml water, 50 ml 0.1 N $AgNO_3$ solution and 50 ml 0.2 N sodium acetate solution (item no. 10) in bottle no. I, II and III respectively. Shake 20-30 minutes in a mechanical shaker (or manually for 45 minutes with pause at times). Place in a water bath at room temperature for another 15-20 minutes. (Make note that the solutions are saturated with AcOAg).
13. Mark three 100 ml dry beakers I, II, III. Filter solution I into beaker I, reject first 5-6 ml solution and then collect.
14. Titrate 10 ml of the filtrate against standard NH_4CNS solution using ferric alum indicator.
15. Perform 13 and 14 for other two bottles, each time using a new filter paper.
16. From the titre values, calculate $[Ag^+]$ in each bottle.
17. Equation (2) gives the solubility product of AcOAg in water at the laboratory temperature. Analysis of II and III gives S and S' as in equations (3) and (4) from K_{sp}.

Results and Calculations

Table 1: Recording of temperature from water bath.

Table 2 : Preparation of 50 ml 0.4 N standard oxalic acid solution.

Weight required = 1.26 g

Note the weights in tabular form.

Table 3 : Preparation of 100 ml 0.1 N standard $AgNO_3$ solution.

Weight required = 1.6988 g

Note in tabular form.

Table 4 : Standardisation of NaOH solution by standard oxalic acid solution.

Take 10 ml oxalic acid solution.

Use Table 2 of experiment 1.

Table 5 : Standardisation of acetic acid solution by standard NaOH solution.

Note data in usual tabular form.

Table 6 : Standardisation of NH_4CNS (or KCNS) solution by standard $AgNO_3$ solution
Indicator: Saturated ferric ammonium sulphate in conc. HNO_3 medium.

No. of Titrations	Vol. of 0.1N $AgNO_3$ soln. (ml)	Burette Reading Initial (ml)	Final (ml)	Vol. of NH_4CNS (ml)	Mean vol. of NH_4CNS (ml)	Strength of NH_4CNS soln.
1	10	...	...	...		
					...	...
2	10	...	...	...		

Table 7 : Preparation of saturated solutions of AcOAg

Bottle No.	Solution/water (50 ml)	Amount of silver acetate (g)
I	water	~ 1.0
II	0.1 N $AgNO_3$	~ 1.0
III	0.2 N AcONa	~ 1.0

Table 8 : Standardisation of different saturated solutions of AcOAg against NH_4CNS solution.
Indicator: Saturated ferric alum soln. in conc. HNO_3 medium.

Bottle No.	Medium	Vol. of satd. AcOAg soln. (ml)	Burette Reading Initial	Final	Vol. of NH_4CNS required (ml)	Mean vol. of NH_4CNS (ml)	Conc. of Ag^+ (mol/litre) (i,e., solubility)
I	Water	10.0	...	...	...		
						...	x
		10.0	...	...	...		
II	0.1 N $AgNO_3$	10.0	...	...	...		
						...	S
		10.0	...	...	...		
III	0.2 N AcONa	10.0	...	...	...		
						...	S'
		10.0	...	...	...		

Results

1. The solubility of silver acetate in water is x mol/l.
   $\therefore$ solubility product of silver acetate, $K_{sp} = x^2$
2. The solubility of silver acetate in 0.1N $AgNO_3 = S$ mol/l.
3. The solubility of silver acetate in 0.2 N sodium acetate = S' mol/l.

Conclusion: (comment on the effect of common ions, its strength on solubility, which will be $x > S > S'$)

N.B. *The data obtained in this experiment may be utilized to calculate activity solubility product, K_a, as done in experiment 4. You are to calculate μ and to draw graph of $-\log K_{sp}$ vs. $\sqrt{\mu}$.*
Try yourself.

Experiment 7

To determine the pH *of a given solution through matching its colour developed by an indicator.*

Theory: The pH of an aqueous solution of a substance is expressed as

$$pH = -\log_{10} [H^+], \tag{1}$$

where the molar concentration of H^+ ion has been conveniently used for activity of the ion.

Acid-base indicators are weak acids or bases which exhibit two distinct colours in dissociated and undissociated forms of the molecule.

An acid indictor, HIn, may ionise in solution as

$$HIn \rightleftharpoons H^+ + In^- \tag{2}$$

In alkaline solution, the colour is due to In^-, as the equilibrium shifts to the right, while in acid solution, it exhibits the colour of the undissociated form, HIn, as the equilibrium point shifts towards left.

The ionisation constant, K_{In} of the indicator is given by

$$K_{In} = \frac{[H^+][In^-]}{[HIn]} \tag{3}$$

Taking logarithm,

$$\log K_{In} = \log[H^+] + \log\frac{[In^-]}{[HIn]}$$

$$\text{or,} \quad -\log[H^+] = -\log K_{In} + \log\frac{[In^-]}{[HIn]}$$

$$\text{or,} \quad pH = pK_{In} + \log\frac{[In^-]}{[HIn]} \tag{4}$$

$$\text{or,} \quad pH = pK_{In} + \log\frac{[\text{Basic form}]}{[\text{Acid form}]}. \tag{5}$$

In equations (4) or (5), both the forms have different colours. It is generally accepted that human eyes receive that colour which has 10 times concentration over the other.

Hence, for acid colour, i.e. for the colour of HIn,

$$pH = pK_{In} + \log\frac{1}{10}$$

$$\text{or,} \quad pH = pK_{In} - 1$$

Similarly, for basic colour, i.e., for the colour of In^-,

$$pH = pK_{In} + \log 10$$

$$\text{or,} \quad pH = pK_{In} + 1.$$

So for perception of human eyes, the interval is

$$pH = pK_{In} \pm 1 \tag{6}$$

[The table appended in this chapter will show that the pH ranges within two units]

Hence if the given solution has a pH within the pH-range of a suitable acid-base indicator, then matching the colour of this solution, after addition of a definite volume of indicator, with that of a prepared buffer of known pH value and with same volume of the indicator may give the correct pH value of the solution.

A buffer solution is a solution of reserved capacity to resist the change of its pH, due to addition of a small amount of acid or alkali to this solution. A mixture of weak acid and its salt derived from a strong base, or a weak base and its salt resulted from a strong acid is regarded as a buffer. The former is called acid buffer; whose pH, according to Henderson equation, is

$$pH = pK_a + \log_{10} \frac{[\text{salt}]}{[\text{acid}]}, \qquad (7)$$

K_a being the dissociation constant of the weak acid. If the pH of the given solution falls within the pH range of bromocresol green (3.8-5.4), then a buffer solution of acetic acid-sodium acetate buffer will help to determine the pH of the given solution.

Materials : Oxalic acid, NaOH, acetic acid, phenolphthalein, bromocresol green

Equipments: Two 100 ml volumetric flasks, one 50 ml vol. flask, two 250 ml reagent bottles or beakers, ten 15 ml hard-glass test tubes of almost same diameter, 10 ml graduated pipette, burette.

Procedure

1. Prepare 50 ml 0.4 N standard oxalic acid solution; weight required = 1.26 g.
2. Prepare about 150 ml ~0.5 N acetic acid solution. Dissolve 4.5 ml acetic acid from original bottle (~17N) in water (~ 145 ml).
3. Prepare about 200 ml ~ 0.5 N NaOH solution.

   Dissolve about 4 g NaOH in 200 ml water.
4. Standardise NaOH solution by oxalic acid solution and then acetic acid by standardised NaOH solution, taking NaOH solution in burette. Use phenolphthalein indicator.
5. Prepare 100 ml exact 0.4 N solutions each of NaOH and acetic acid by proper dilution from the above standardised solutions.
6. Mark the ten hard glass test tubes with 1-9 and 'unknown' and prepare the buffer solutions as in Table 4.
7. Add 3 drops of bromocresol green indicator in each test tube including the unknown and mix thoroughly.

   Preparation of bromocresol green solution: Dissolve 0.1 g of the indicator in 20 ml dehydrated alcohol and after complete dissolution dilute to 100 ml by water.
8. Match the colour of the unknown solution with other test tubes. The pH of the given solution will be that of calculated pH of the matching buffer.

If the colour does not perfectly match to either of the test tubes, it will be intermediate of two successive test tubes. Take the mean of the pH of the two.

Results and Calculations

Table 1: Preparation of 50 ml 0.4 N oxalic acid solution. Use relevant table.

Table 2: Standardisation of NaOH solution by standard oxalic acid solution. Indicator : Phenolphthalein. Usual table.

Table 3: Standardisation of acetic acid solution by NaOH solution. Indicator : Phenolphthalein. Usual table.

Table 4: Preparation of different buffer solutions

pK_a (CH_3COOH) at 25 °C = 4.74

Test tube No.	Vol. of 0.4 N NaOH soln. (ml)	Vol. of 0.4 N CH_3COOH soln. (ml)	Vol. of water (ml)	Total vol. (ml)	$pH = pK_a + \log\frac{[salt]}{[acid]}$ $= pK_a + \log\frac{\text{vol. of NaOH}}{5 - \text{vol. of NaOH}}$
1	0.5	5.0	4.5	10.0	3.78
2	1.0	5.0	4.0	10.0	4.14
3	1.5	5.0	3.5	10.0	4.37
4	2.0	5.0	3.0	10.0	4.56
5	2.5	5.0	2.5	10.0	4.74
6	3.0	5.0	2.0	10.0	4.91
7	3.5	5.0	1.5	10.0	5.11
8	4.0	5.0	1.0	10.0	5.34
9	4.5	5.0	0.5	10.0	5.69
unknown	—	—	—	10.0	anyone of above values.

Result: The colour of the unknown solution matches with test tube no ···

Hence pH of the given solution is ···

How to select an indicator for colour matching?

i. Take 5 ml of the given solution in a test tube. Add 5 drops of BDH indicator, mix well. Compare the colour with the pH-colour chart. The approximate pH so found may help to select the proper indicator.

ii. Another method is to add 3 drops of the indicator with lowest pH range given in the table. Test one by one indicator in that way. The indicator for which a mixed colour is found, will be the appropriate indicator.

APPENDIX

Table 1

Solubility products (K_{sp}) of some compounds in water at room temperature

Compound	K_{sp}	Compound	K_{sp}
$AgCl$	1.5×10^{-10}	$CaSO_4$	2.3×10^{-4}
$AgBr$	7.7×10^{-13}	$CaCO_3$	4.8×10^{-9}
AgI	9.0×10^{-17}	CaC_2O_4	2.6×10^{-9}
Ag_2CrO_4	2.5×10^{-12}	$MgCO_3$	1.0×10^{-5}
$PbCl_2$	2.4×10^{-4}	$SrSO_4$	2.8×10^{-7}
$PbBr_2$	7.5×10^{-5}	$SrCO_3$	1.6×10^{-9}
PbI_2	8.7×10^{-9}	$BaSO_4$	9.2×10^{-11}
$PbSO_4$	2.2×10^{-8}	$BaCrO_4$	1.6×10^{-10}
$PbCrO_4$	1.8×10^{-14}	CuS	1.0×10^{-35}
CaF_2	3.2×10^{-11}	PbS	5.0×10^{-27}
Hg_2Cl_2	3.5×10^{-18}	CdS	1.4×10^{-28}
$Fe(OH)_3$	3.8×10^{-38}	HgS	4.0×10^{-54}
$Al(OH)_3$	8.5×10^{-33}	NiS	1.4×10^{-19}
$Cr(OH)_3$	2.9×10^{-29}	ZnS	1.0×10^{-23}
Hg_2Br_2	5.2×10^{-23}	CoS	3.0×10^{-26}
Hg_2I_2	1.2×10^{-28}	MnS	1.4×10^{-15}

Table 2

Solubilities of some sparingly soluble salts at 18 °C
(Unit : g/100 ml water)

Compound	$K_{solubility}$	Compound	$K_{solubility}$
$AgCl$	1.4×10^{-4}	$PbCrO_4$	2.0×10^{-5}
$AgBr$	1.0×10^{-3}	$BaCrO_4$	3.8×10^{-4}
AgI	3.0×10^{-7}	Ag_2CrO_4	2.5×10^{-3}
CaF_2	1.6×10^{-3}	$MgCO_3$	0.10
SrF_2	1.2×10^{-2}	$CaCO_3$	1.3×10^{-3}
$CaSO_4$	0.20	$SrCO_3$	1.1×10^{-3}
$BaSO_4$	2.3×10^{-4}	$BaCO_3$	2.3×10^{-3}
$SrSO_4$	1.1×10^{-2}	$PbCO_3$	1.0×10^{-4}
$PbSO_4$	4.1×10^{-3}	CaC_2O_4	5.6×10^{-4}

Table 3
pH-range of some Indicators

Name of the Indicator	pK_{In}	pH-range	Acid-colour	Alkali-colour
Bromophenol blue	4.0	2.8-4.6	Yellow	Blue
Methyl Orange	3.7	3.1-4.4	Red	Yellow
Bromocresol green	4.7	3.8-5.4	Yellow	Blue
Methyl red	5.1	4.2-6.3	Red	Yellow
Thymol blue (1st range)	—	1.2-2.8	Red	Yellow
Thymol blue (2nd range)	8,9	8.2-9.6	Yellow	Blue
Bromocresol purple	6.3	5.2-6.8	Yellow	Purple
Phenolphthalein	9.6	8.3-10.0	Colourless	Pink (Red)

CHAPTER 4

Measurement of Conductances in Solutions

iv. **Expt. 7D: Conductometric standardisation of a given Na_2SO_4 solution by a standard $BaCl_2$ solution (precipitation reaction).**

XI. **Experiment 8**

Conductometric titration of $K_4[Fe(CN)_6]$ solution by $ZnSO_4$ solution and determination of composition of *Zn* (II)-Ferrocyanide complex (for P.G. courses)

XII. **Experiment 9**

Conductometric standardisation of a mixture of acids ($HCl + CH_3COOH$) by a strong alkali

XIII. **Experiment 10**

Conductometric titration of a mixture of halides (HCl, KCl and NH_4Cl) (for P.G. courses)

XIV. **Experiment 11**

Determination of Critical Miceller Concentration (C.M.C.) of a surface-active agent (e.g. Sodium Lauryl Sulphate) Conductometrically (for P.G. courses)

XV. **Experiment 12**

Conductometric determination of rate constant of alkaline hydrolysis of methyl acetate (for P.G. courses)

XVI. **Appendix**

Table 1. **Ionic molar and equivalent conductances of some ions**

Table 2. **Specific conductances of aqueous KCl solutions at different temperatures.**

CHAPTER 4

Measurement of Conductances in Solutions

I. Introductory Words

A conductor is a substance that allows an electric current to flow through it. While in an electronic conductor, electrons are the current-carriers, the ions perform such work in an electrolytic solution. In both cases, the electric current experiences a resistance while flowing. Conductance is the reciprocal of resistance. Larger is the conductance, easier is the passage of current. Mathematically, resistance

$$R = \rho \cdot \frac{l}{a}, \tag{1}$$

l and a are length and area of the conductor respectively and ρ is the 'resistivity' or 'specific resistance' for the conductor of unit length (l) and unit area (a).

$$\text{Conductance,} \quad C = \frac{1}{R} = \frac{1}{\rho} \cdot \frac{a}{l} = \kappa \cdot \frac{a}{l}$$

$$\text{or,} \quad \kappa = C \cdot \frac{l}{a} \tag{2}$$

$$\text{or,} \quad \kappa = C \cdot K_C \tag{3}$$

κ(kappa) is the specific conductance, which is the conductance of a solution kept between two electrodes of unit area of cross section each and at unit distance apart, or it may be regarded as the conductance of a solution of unit volume. K_c is the 'cell constant' of the particular conductivity cell in use.

Unit of specific conductance, κ, generally used is **ohm^{-1} cm^{-1} or S cm^{-1}**.

The S.I. unit is ohm^{-1} m^{-1} or S m^{-1} [S = siemens]

Unit of cell constant, K_c, is cm^{-1} or m^{-1}.

Equivalent conductance (Λ):

This is the most important term used in conductance measurement and is defined as the conductance of a solution in which one gram-equivalent of the experimental electrolyte is present and is placed between two large electrodes at one cm apart so as to acommodate the solution.

As construction of large electrodes and to do work with these are far from reality, the equivalent conductance, Λ, of a solution is expressed with the equation

$$\Lambda = \frac{1000\kappa}{c}, \tag{4}$$

where c is the number of gram-equivalent of the electrolyte dissolved in 1000 cm^3 of the solution, i.e. the **'normality' of the solution. Unit of Λ is S cm^2 g-eqv^{-1}. The S.I. Unit is S m^2 g-eqv^{-1}.**

Molar Conductance (Λ_m):

The molar conductance of a solution is the conductance of that volume of the solution which contains one mole of the electrolyte and is placed between two electrodes at unit distance apart and having enough area of cross-section each to acommodate the solution.

Molar conductance is measured from specific conductance by the relation

$$\Lambda_m = \frac{1000\kappa}{M} \tag{5}$$

where M is the **molarity of the solution.**

Uunit of molar conductance is S cm^2 mol^{-1}. The S.I. unit is **S m^2 mol^{-1}**.

Relation between Equivalent Conductance (Λ) and Molar Conductance (Λ_m)

$\Lambda_m = \Lambda \times$ total charge of cations or anions produced from one molecule of the electrolyte

For example, $\Lambda_m(CaCl_2) = \Lambda \times 2$

$\Lambda_m(Al_2(SO_4)_3) = \Lambda \times 6$, etc.

Variation of κ, Λ and Λ_m with the concentration of the solution

The conductance of an electrolytic solution mainly depends upon the number of ions present in a definite volume of the solution enclosed by the two electrodes, i.e., it depends upon the concentration of the solution. Another factor also plays a role behind the conductance value— the ease with which the ions move overcoming the interionic attraction of two oppositely charged ions. However, the overall effect of concentration on different types of conductances should be considered separately for strong and weak electrolytes.

Strong electrolytes

Specific conductance (κ) decreases with lowering the concentration of the solution (c).

Both equivalent conductance (Λ) and molar conductance (Λ_m) increase smoothly with decreasing concentration, the slope having a constant value.

Weak electrolytes

κ decreases with lowering of c, but not as sharp as for strong electrolytes due to increasing degree of dissociation.

Both (Λ) and (Λ_m) increase with changing slope for lowering c. The curve of Λ_m may be to some extent different for that of Λ, depending upon the valency-type of the electrolyte. However, the nature (i.e., increasing or decreasing tendency) of curves remains same.

The Debye-Onsager equation correlating Λ or Λ_m and concentration is given by

$$\Lambda = \Lambda^o - b\sqrt{c}. \tag{6}$$

Λ^o = equivalent conductance at infinite dilution

c = concentration of the solution

b = a constant, whose value depends upon the nature of the solvent and temperature.

The general characters of the curves are shown here.

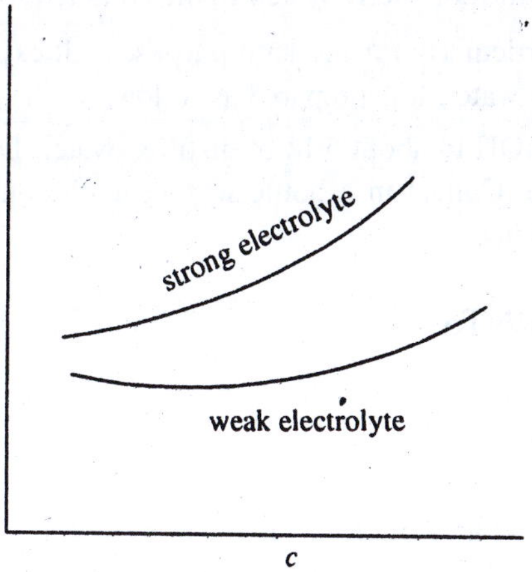

Graph I

Specific conductance vs. concentration

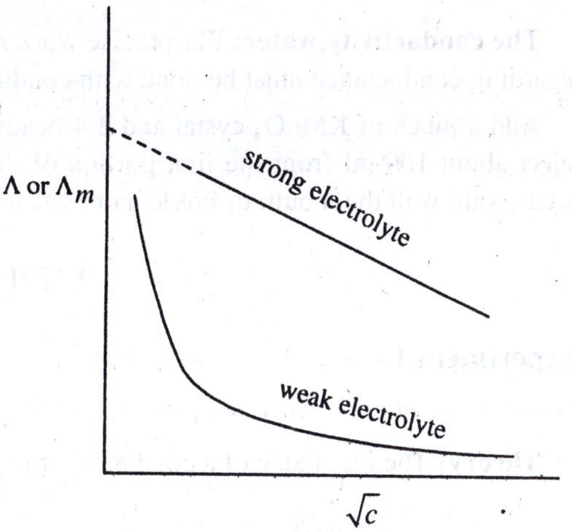

Graph II

Equivalent or molar conductance vs. concentration

The instruments

1. **The conductivity bridge:** It provides a means to know the **resistance** of the solution, an outcome of the famous **Wheatstone Bridge principle**. In old days after its formation, an ear-phone was in use for indication of no current through the detector. For its noisy disturbances, it has completely been replaced by magic-eye electronic detector or cathode ray oscilloscope. At present **Digital conductometers** are largely in use. In either of the arrangements the indicator provides directly the conductance-readings.

An **alternating current** (AC) of high frequency (~ 1000 cycles/s) must be used to avoid electrolysis of the electrolytic solution under study and polarisation.

2. **The conductivity cell:** A different varities of conductivity cells are used in the laboratory. These are constructed generally of Pyrex glass or other very good quality resistance glass in which two platinised platinum electrodes are welded with thick platinum wires fused in the Pyrex glass. The platinum electrodes are coated with Pt-black to minimize the polarisation. These wires are connected through mercury with the wires of the conductivity bridge. The Pt-electrodes are of at least one square cm area each and very stoutly fixed at about one cm apart.

When not in use, the cell is to be kept immersed in conductivity water.

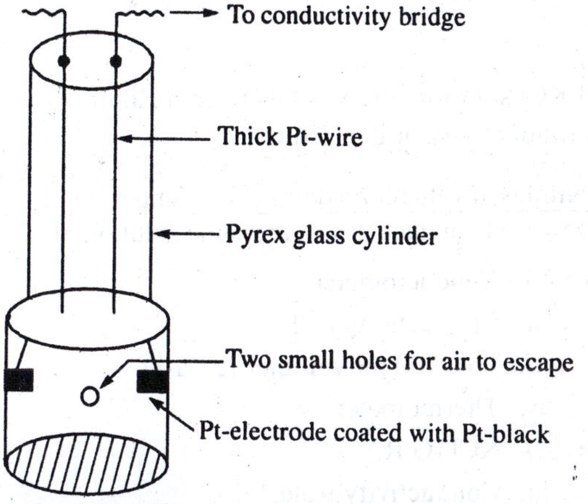

Conductivity Cell

Important: Users are advised to go through the circuit diagram and users' manual supplied by the manufacturers of the instruments before connection of the conductivity cell to the conductivity bridge.

The conductivity water: For precise work and particularly for research purposes, all experiments regarding conductance must be done with conductivity water. It is prepared as follows.

Add a pinch of $KMnO_4$ cystal and 2-4 beads of NaOH to about 5 lit of distilled water. Distill and reject about 100 ml from the first portion of distillate. Collect in a bottle washed with conductivity water, tight well the mouth of bottle and use for experiment.

EXPERIMENTS

Experiment 1

To determine the cell constant of a conductivity cell.

Theory: The resistance of a conductor of length l and area of cross-section a is given by

$$R = \rho \cdot \frac{l}{a}, \tag{1}$$

where ρ, the specific resistance or resistivity, is the resistance of the conductor of unit length and unit area. The conductance (C) is the reciprocal of resistance and is expressed as

$$C = \frac{1}{R} = \frac{1}{\rho} \cdot \frac{a}{l}$$

$$\text{or,} \quad C = \kappa \cdot \frac{a}{l} \tag{2}$$

where κ, the specific conductance, is the conductance of a conductor of unit length and unit area. For electrolytic conductor, the specific conductance is the conductance of the solution of unit volume of the electrolyte, generally of 1 cm^3 of the electrolytic solution.

$$\text{From (2),} \quad \kappa = C \cdot \frac{l}{a}$$

$$\text{or,} \quad \kappa = C \cdot K_c \tag{3}$$

where K_c is the cell constant of the conductivity cell. The conductivity bridge gives conductance, C, of the solution. Hence knowing K_c, specific conductance (κ) and equivalent conductance (Λ) is determined. The unit of K_c is cm^{-1}.

$$\text{From (3),} \quad K_c = \frac{\kappa}{C} \tag{4}$$

Using a 0.1 N or 0.01 N KCl solution, the κ is taken from Kohlrausch's table at a definite temperature and C is observed from conductivity bridge. Thus,

$$K_c = \frac{\text{Kohlrausch value for a definite concentration of KCl}}{\text{Observed conductance of the same solution of KCl}}$$

Apparatus required:
- i. Conductometer
- ii. Conductivity cell
- iii. Two 100 ml volumetric flasks
- iv. Thermometer.

Materials required:
- i. KCl (G.R.)
- ii. Conductivity water.

Procedure

1. Prepare 100 ml exactly 0.1 N KCl solution. Required weight is 0.7456 g. Dissolve a slightly higher amount of required KCl in conductivity water and make up the volume. From this solution prepare exactly 100 ml 0.1 N KCl solution by proper dilution with conductivity water.
2. Pipette out 10 ml of 0.1 N KCl and dilute it to 100 ml to prepare exactly 0.01 N KCl solution.
3. Switch on the conductometer and keep it on for about 15 minutes before taking any reading.
4. Wash the conductivity cell with conductivity water. To ensure the proper cleaning of the electrodes, insert the electrodes in 20-25 ml conductivity water and take reading. Repeat the process for 2-3 times with same volume of water and be sure of two same consecutive values of conductance.
5. Rinse the electrodes with 0.01 N KCl solution. Insert the cell in 25 ml of the solution. It is important that the electrodes are completely immersed in the experimental solution. If required, a larger, definite volume is to be taken. Note the temperature of the solution. Observe the conductance of the solution.
6. Repeat (5) with 0.1 KCl solution with exactly equal volume as in (5).
7. Collect the specific conductance values at your experimental temperature from Kohlrausch's table in the literature for 0.01 N and 0.1 N KCl solution. Calculate the cell constant of your conductivity cell each for the two solutions using equation (4). Take mean of the two values.

Recording of data, Results and Calculations

Table 1: Temperature during the experiment

Temperature at the beginning of the expt. (°C)	Temperature at the end of the expt. (°C)	Average temperature during the expt. (°C)
...	...	...

Table 2: Preparation of 100 ml 0.1 N KCl solution; weight required = 0.7456 g

1st weight of KCl (g) w_1	2nd weight of KCl (g) w_2	Weight taken (g) $w_1 - w_2$	Strength of the KCl solution
...	...	...	$\frac{w_1 - w_2}{0.7456} \times 0.1$ N

[Use $v_1 s_1 = v_2 s_2$ to prepare exact 100 ml 0.1 N KCl solution, if the weight taken is not exactly 0.7456 g. Show the calculation here.]

Table 3: Determination of cell constant (K_c)

No. of observation	Strength of KCl solution	Observed conductance (C) (ohm^{-1})	Specific conductance (κ) from literature at the experimental temp ... °C (ohm^{-1} cm^{-1})	Cell const. $K_c = \frac{\kappa}{C}$ (cm^{-1})	Mean K_c (cm^{-1})
1	0.01 N	...	...	...	...
2	0.1 N	...	...	...	

Conclusion: Hence the cell constant of the given conductivity cell is ... cm^{-1} at ... °C temperature.

Experiment 2

To determine the equivalent conductance at infinite dilution (Λ^o) of a strong electrolyte (say, KCl*).*

Theory: The equivalent conductance (Λ) of an electrolyte is defined as the conductance of a solution containing one g-equivalent of the electrolyte enclosed between two parallel electrodes at 1 cm apart. Its units is ohm^{-1} cm^{-2} $g\text{-}eqv^{-1}$.

Since by definition, the electrodes required to be of enormous size, the equivalent conductance is determined by the relation,

$$\Lambda = \frac{1000\kappa}{c}, \tag{1}$$

where κ is the specific conductance in ohm^{-1} cm^{-1} unit and c is normality of the solution, i.e., number of g-eqv dissolved in 1000 cm^3 of the solution.

κ is determined by the relation,

$$\kappa = C \cdot K_c \tag{2}$$

where C is the observed conductance of the solution available from the conductometer and K_c is the cell constant of the conductivity cell used for the purpose.

The equivalent conductance of a solution decreases with increasing concentration. The quantitative relation is given by the Debye-Hückel-Onsager equation.

$$\Lambda = \Lambda^o - b\sqrt{c}\,. \tag{3}$$

Λ^o is the absolute equivalent conductance, i.e., equivalent conductance at infinite dilution and b is a constant depending on the character of the solvent and the temperature.

Following equation (3), a plot of Λ vs. $\sqrt{c}$, would give a straight line of negative slope. On extrapolation to 'zero' concentration, the intercept will give the value for Λ^o. Hence the equivalent conductances of several dilute solutions of the electrolyte are determined and a plot is drawn for Λ vs $\sqrt{c}$.

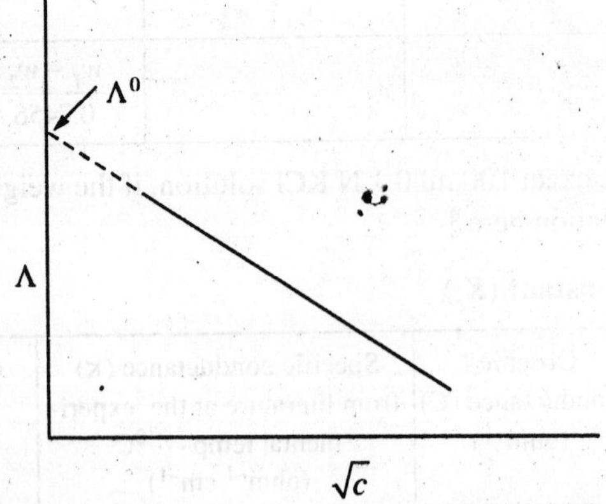

Apparatus required: i. Conductivity bridge
ii. Conductivity cell
iii. One 250 ml and five 50 ml volumetric flasks.

Materials required: i. KCl (G.R.)
ii. Conductivity water.

Procedure

1. Prepare exactly 250 ml 0.2 N KCl solution. Required weight is 3.728 g.
2. From this solution, prepare exactly 50 ml each of 0.1 N, 0.05N, 0.03N, 0.02 N, and 0.01 N KCl solutions by proper dilution. Take the required volume of KCl by burette or graduated pipette.
3. Measure the conductances of all six solutions of KCl (starting from 0.01 N and ending at 0.2 N).
4. Determine the cell constant of the conductivity cell using the 0.1 N KCl solution [Follow expt. 1]
5. Calculate specific conductance (κ) and equivalent conductance (Λ) of the six solutions following equations (2) and (1) respectively.
6. Plot Λ-values against $\sqrt{c}$. Extrapolate the straight line to Λ^{o}-value of KCl at the laboratory temperature.

Results and Calculations

Table 1: Recording of laboratory temperature during experiment: (see table 1 of expt. 1)

Table 1: Preparation of 250 ml. 0.2 N KCl solution

1st weight of KCl (w_1) g	2nd weight of KCl (w_2) g	Weight taken ($w_1 - w_2$)g	concn. of KCl $\frac{w_1 - w_2}{3.728}$(0.2N)
...	...	...	... (N)

[Weight taken should not be less than the required weight. If it exceeds the required weight, which is a common case, properly dilute to prepare 0.2 N solution.]

Table 3: Preparation of different KCl solutions from 0.2 N solution

Vol. flask no.	Conc. of KCl soln. (N)	Volume of 0.2 N KCl soln. (ml)	Final volume by adding water (ml)
1	0.1	25.0	50
2	0.05	12.5	50
3	0.03	7.5	50
4	0.02	5.0	50
5	0.01	2.5	50

Table 4: C, κ, and Λ for different KCl solutions

Cell constant, K_c = ··· cm^{-1}

Conc. of the KCl Soln. (c)	$\sqrt{c}$	C (ohm^{-1})	$\kappa = C \cdot K_c$ ($ohm^{-1}\ cm^{-1}$)	$\Lambda = \frac{1000\kappa}{c}$ $ohm^{-1}\ cm^2\ g\text{-}eqv^{-1}$
0.01	0.1	···	···	···
0.02	···	···	···	···
0.03	···	···	···	···
0.05	···	···	···	···
0.1	···	···	···	···
0.2	···	···	···	···

Conclusion

Plotting Λ vs. $\sqrt{c}$ and then extrapolating, the observed equivalent conductance at infinite dilution (Λ^o) is found to be ··· at ··· °C temperature.

Experiment 3

(a) To determine the equivalent conductance of weak electrolyte (say CH_3COOH or C_6H_5COOH) at infinite dilution (Λ^o) by Kohlrausch's law of independent migration of ions and (b) verification of Ostwald's dilution law.

[Expt 3(a) for B.Sc.(H); complete experiment is for P.G. (2 days' expt.)]

Theory

(a) The equivalent conductance of a strong electrolyte increases linearly with dilution, i.e., decreasing concentration. The plot of Λ vs. $\sqrt{c}$, being a straight line, is extrapolated to Λ-axis, where the intercept gives the value for Λ^o.

In case of weak electrolyte, the equivalent conductance first increases slowly with dilution, and near very low concentration of the electrolytic solution, increases rapidly. This is due to (a) the larger degree of dissociation with dilution as per Ostwald's dilution law and (b) increasing distances between oppositely charged ions for increasing volume of the solution. Thus with changing nature of slope, extrapolation is not possible to have Λ^o-value of a weak electrolyte. (see graph II of page 73)

Kohlrausch's law of independent migration of ions at infinite dilution is successfully used to determine the Λ^o-values of weak electrolytes. Three strong electrolytes are chosen of which the algebraic combination gives none but the sum of the constituent ions of the electrolyte in question giving the required Λ^o-value.

For example, Λ^o of CH_3COOH is computed as

$$\Lambda^o_{CH_3COOH} = \Lambda^o_{CH_3COOK} - \Lambda^o_{KCl} + \Lambda^o_{HCl} \tag{1}$$

All three electrolytes, viz., CH_3COOK, KCl and HCl being strong electrolytes (the first two being salts), their Λ^o -values are estimated by extrapolation method.

Thus Λ^o of CH_3COOH is determined.

(b) The dissociation of a weak electrolyte HA is written as

$$\underset{(1-\alpha)c}{HA} \rightleftharpoons \underset{\alpha c}{H^+} + \underset{\alpha c}{A^-} \tag{2}$$

α and *c* are the degree of dissociation and molar concentration of the weak electrolyte respectively. The dissociation constant,

$$K_a = \frac{a_{H^+} \cdot a_{A^-}}{a_{HA\text{ (undissociated)}}} \simeq \frac{[H^+][A^-]}{[HA]} \tag{3}$$

[*a*'s, the activities are replaced by concentration due to low dissociation concentration]

$$\therefore\ K_a = \frac{\alpha c \cdot \alpha c}{(1-\alpha)c} = \frac{\alpha^2 c}{1-\alpha} \approx \alpha^2 c \tag{4}$$

[for low degree of dissociation, $\alpha << 1$]

$$\therefore\ \alpha = \sqrt{\frac{K_a}{c}} \tag{5}$$

This is the mathematical form of Ostwald's dilution law, which states that the degree of dissociation of a weak electrolyte increases with dilution.

The conductance of a solution depends on the number of ions, which, in turn, is a function of degree of dissociation. The conductance ratio at a particular concentration and that at infinite dilution gives the degree of dissociation (α) at that concentration. Thus,

$$\alpha = \frac{\Lambda}{\Lambda^o}. \tag{6}$$

A number of dilute solutions of the weak electrolyte are prepared. Their Λ-values are measured. Each Λ-value is divided by Λ^o of the electrolyte, giving α for each solution. Plot of α vs. $\frac{1}{\sqrt{c}}$ would give a straight line with positive slope.

Apparatus required: Conductometer, conductivity cell, 250 ml volumetric flasks, required number of 50 ml volumetric flasks, three 100 ml volumetric flasks, graduated pipette (10 ml).

Materials required: KCl, KOH, CH_3COOH, HCl, NaOH, oxalic acid, phenolphthalein.

Procedure

(a) 1. Prepare 250 ml exactly 0.1 N KCl solution. Weight required = 1.865 g.

2. Determine the cell constant of the conductivity cell (K_c) following the procedure of experiment 1, taking only 0.1 N KCl solution.
3. Prepare 50 ml of standard oxalic acid solution of ~ 0.2 N strength. Weight required = 0.63 g.
4. Prepare 250 ml ~ 0.2 N HCl solution. Prepare by proper dilution of required volume of strong HCl.
5. Prepare 250 ml ~ 0.2 N KOH solution. Dissolve about 3 g of KOH in 250 ml conductivity water.
6. Prepare 250 ml ~0.2 N acetic acid (AcOH) solution. Prepare by proper dilution taking required volume from the original bottle of the AcOH.

   [For (1), (4) and (5) & (6) always take a little more weight/volume, so that the solution may not be less than the stipulated concentration.]
7. Standardise KOH solution by standard oxalic acid solution taking 10 ml of oxalic acid each time and using phenolphthalein. From the known strength of KOH solution prepare **100 ml exact 0.1 N** KOH solution.

8. Standardise HCl solution by the standard original KOH solution using phenolphthalein indicator. From the known strength of the HCl solution, prepare **exactly 250 ml 0.1 N** HCl solution.
9. Standardise AcOH solution with the original standard KOH solution. From the known strength of AcOH solution, prepare **exactly 250 ml 0.1 N** AcOH solution.
10. Mix 100 ml each of 0.1 N KOH and 0.1 N AcOH solution to prepare exactly 200 ml 0.05 *N* AcOK solution.
11. Find Λ^o-values of HCl, KCl and AcOK solution by preparing other four 50 ml solutions from each, measuring Λs, and extrapolating the straight lines of the plot of Λ vs. $\sqrt{c}$.
12. Find $\Lambda^\circ_{CH_3COOH}$ using equation (1).

(b) **Verification of Ostwald's dilution law:**

13. Prepare 0.05 N, 0.02 N, 0.01 N and 0.002 N each of 50 ml of AcOH solution from remaining 0.1 N AcOH solution.
14. Measure conductances and calculate Λ-values of each solution.
15. Calculate α of each solution from $\alpha = \frac{\Lambda}{\Lambda^o}$ relation.
16. Plot Λ vs. $\sqrt{c}$.

Recording of data, Results and Calculations

Table 1: Recording of temperature: Same as table 1 of experiment 1.

Table 2: Preparation of 250 ml 0.1 N KCl solution (exact)

Weight required = 1.864 g

1st weight (w_1) g	2nd weight KCl (w_2) g	Weight taken ($w_1 - w_2$)g	Conc. of KCl soln. $\frac{w_1 - w_2}{1.864}\left(\frac{N}{10}\right)$
...	...	...	...

Table 3: Preparation of 50 ml 0.2 N oxalic acid solution

Weight required = 0.63 g

1st weight (w_1) g	2nd weight KCl (w_2) g	Weight taken ($w_1 - w_2$)g	Conc. of oxalic acid solution $\frac{w_1 - w_2}{0.63} \times 0.2$ (N)
...	...	...	...

Table 4: Standardisation of KOH solution by standard oxalic acid solutions

No. of titrations	Vol. of oxalic acid soln. (ml)	Initial burette reading (ml)	Final burette reading (ml)	Vol. of KOH required (ml)	Mean vol. (ml) of KOH	Strength of KOH
1	10.0	0.0		...	...	... (N)
2	10.0	...	...	...		

From this standardised KOH solution, prepare exactly 100 ml 0.1 N KOH solution.

Table 5: Standardisation of HCl solution by standard original KOH solution

No. of titrations	Vol. of HCl soln. (ml)	Initial burette reading (ml)	Final burette reading (ml)	Vol. of KOH required (ml)	Mean vol. KOH (ml)	Strength of HCl
1	10.0	0.0	...	...	...	...(N)
2	10.0	...	...	...		

From this standardised HCl solution, prepare exactly 100 ml 0.1 N HCl solution.

Table 6: Standardisation of AcOH solution by standard original KOH solution

No. of titrations	Vol. of AcOH soln. (ml)	Initial burette reading (ml)	Final burette reading (ml)	Vol. of KOH required (ml)	Mean Vol. of KOH (ml)	Strength of AcOH
1	10.0	0.0	...	...	...	...
2	10.0	...	...	...		

From this standardised AcOH solution, Prepare exactly 250 ml 0.1 *N* AcOH solution.

Table 7: Determination of Λ°-values of KCl, HCl and CH_3COOK

Vol. flask no.	Electrolyte and its conc.	Vol. (ml) taken for dilution to 50 ml	Conc. of the soln, *c* (N)	$\sqrt{c}$	Observed *C* (ohm^{-1})	$\kappa = C \cdot K_c$ ($ohm^{-1}cm^{-1}$)	$\Lambda = \frac{1000\kappa}{c}$ $ohm^{-1}cm^2$ g-eqv^{-1}	Λ° (from graph)
1		*50	0.1	...	...	...	...	
2	0.1N KCl	25	0.05					
3		10	0.02					...
4		5	0.01					
5		1	0.002					
1		*50	0.1	...	...	...	...	
2	0.1N HCl	25	0.05					
3		10	0.02					...
4		5	0.01					
5		1	0.002					
1		*50	0.05	...	...	...	...	
2	0.05N AcOK	25	0.025					
3		10	0.01					...
4		5	0.005					
5		2	0.002					

* Add no water.

$$\Lambda^{o}_{AcOH} = \Lambda^{o}_{HCl} + \Lambda^{o}_{AcOK} - \Lambda^{o}_{KCl}$$

Table 8: Degree of dissociation of AcOH at different concentrations

Vol. flask no.	Vol. of 0.1 N AcOH soln. (ml)	Conc. of AcOH after dilution to 50 ml, c (N)	$\sqrt{c}$	C ohm^{-1}	$\kappa = C \cdot K_c$ ohm^{-1} cm^{-1}	$\Lambda = \frac{1000\kappa}{c}$ ohm^{-1} cm^{-2} g-eqv^{-1}	$\alpha = \frac{\Lambda}{\Lambda^\circ}$
1	*50	0.1	...	...	...	...	...
2	25	0.05					
3	10	0.02					
4	5	0.01					
5	1	0.002					

* Add no water.

Plot α vs. $\sqrt{c}$

Conclusion: From the graph, it is seen that the degree of dissociation of a weak electrolyte varies inversely with the square root of its concentrations.

Experiment 4

To determine the ionization constant of a weak electrolyte (say, CH_3COOH*) [for P.G. courses]*

Theory: The ionization of a weak electrolyte may be represented by the following equilibrium

$$HA \rightleftharpoons H^+ + A^- \tag{1}$$

According to Ostwald's dilution law, the degree of dissociation (α) of a weak electrolyte increases with decreasing concentration (c). Thus for c molar concentration,

$$\begin{array}{ccccc} HA & \rightleftharpoons & H^+ & + & A^- \\ (1-\alpha)c & & \alpha c & & \alpha c \end{array}$$

so that the ionization constant, K_a, is given as

$$K_a = \frac{a_{H^+} \cdot a_{A^-}}{a_{HA}} \simeq \frac{[H^+][A^-]}{[HA]}$$ [for low ionic concentration, activities being replaced by molar concentration]

$$\text{or,} \quad K_a = \frac{\alpha^2 c}{1-\alpha}. \tag{2}$$

From conductometric measurements, the degree of dissociation may be calculated from the relation

$$\alpha = \frac{\Lambda_c}{\Lambda^\circ}, \tag{3}$$

where Λ° is the absolute equivalent conductance of HA and Λ_c is that at concentration c.

Substituting the estimated values of α at different concentrations of HA, K_a may be obtained. The average value of K_a is the required ionization (or dissociation constant) of the weak electrolyte at the experimental temperature.

However, this value of K_a is not the true ionization constant, as α expressed in equation (3) is not 100% correct, though good enough to study the ionic equilibria. The thermodynamic dissociation constant is available from the modified form of Debye-Hückel-Onsager equation (see the next experiment).

Materials and Equipments: Same as mentioned in the previous experiments for determinations of cell constant (K_c), Λ^o of CH_3COOH (or any other weak electrolytes, selecting three appropriate strong electrolytes) and degree of dissociation (α).

Procedure

1. Calculate the cell constant of the conductivity cell.
2. Determine the Λ^o -value of CH_3COOH [Follow expt. no. 3(a)]. (For Λ^o of benzoic acid, find Λ^os of KCl, HCl and C_6H_5COOK).
3. For determination of α at various concentrations of CH_3COOH, follow procedures mentioned in procedures (b) (13)—(15) of expt. no. 3. Here again it is summerized as follow:
   i. Prepare exactly 250 ml 0.1 N CH_3COOH solution. This is done by preparing 250 ml ~ 0.2 N CH_3COOH solution, then standardising against a standard KOH/NaOH solution (which was previously titrated against standard oxalic acid solution) and finally diluting appropriately to 0.1 N.
   ii. Prepare 0.05 N, 0.02 N, 0.01 N and 0.002 N CH_3COOH solutions, each of 50 ml volume from the above 0.1 N AcOH solution. [These solutions may also be prepared directly from the original standard AcOH solution by proper calculation].
   iii. Measure C and then calculate κ and Λ_c for all five solutions of AcOH from 0.1 N - 0.002N.
   iv. Calculate α for the five solutions, using $\alpha = \frac{\Lambda_c}{\Lambda^\circ}$.
4. Calculate K_a for each solution, using $K_a = \frac{\alpha^2 c}{1-\alpha}$.
5. Calculate the mean value of K_a.

Results and Calculations

Put all tables from 1-7 of experiment no. 3.

Table 8: α-and K_a -values of CH_3COOH at differnt concentrations

No. of vol. flask	Vol. of 0.1N AcOH solution (ml)	Conc. of AcOH after dilution to 50 ml. c(N)	Observed conductance C (ohm^{-1})	Sp. cond., $\kappa = C \cdot K_c$ (ohm^{-1} cm^{-1})	Eqv. cond., $\Lambda_c = \frac{1000\kappa}{c}$ (ohm^{-1} cm^2 g-eqv^{-1})	$\alpha = \frac{\Lambda_c}{\Lambda^\circ}$	K_a	Mean K_a
1	*50	0.1	...	...	...	...	...	...
2	25	0.05	...	...	...	...	...	...
3	10	0.02	...	...	...	...	...	...
4	5	0.01	...	...	...	...	...	...
5	1	0.002	...	...	...	...	...	...

* Add no water.

Conclusion: Hence the ionization constant of acetic acid is ··· at ··· °C temperature.

Experiment 5

To determine the Λ^o and K_a of a weak electrolyte at a definite temperature by Debye-Hückel-Onsager equation. [The experiment is for P.G. courses]

Theory: Weak electrolytes are those which do not ionize completely in solution. An equilibrium is established between the ions and the undissociated electroyle in solution which is represented by the corresponding equilibrium constant, K_a. Thus for acetic acid, a weak electrolyte, the ionization in water is expressed as

$$AcOH \rightleftharpoons OAc^- + H^+ \tag{1}$$

(To be more correct, $AcOH + H_2O \rightleftharpoons OAc^- + H_3O^+$). If α be the degree of dissociation at c molar concentration and at a particular temperature, then the equilibrium concentrations of AcOH (undissociated), OAc^- and H^+ are : $[AcOH] = (1-\alpha)c$

$$[OAc^-] = [H^+] = \alpha c,$$

so that the equilibrium constant,

$$K_a = \frac{[OAc^-][H^+]}{[AcOH]} \quad \text{[writing concentration for activities]}$$

$$\text{or, } K_a = \frac{\alpha c \cdot \alpha c}{(1-\alpha)},$$

$$\text{or, } K_a = \frac{\alpha^2 c}{1-\alpha} \tag{2}$$

As the conductance, and hence also the equivalent conductance, Λ, depend on the number of ions, which is an outcome of the degree of dissociation, therefore,

$$\alpha = \frac{\Lambda_c}{\Lambda^o} \tag{3}$$

where Λ_c is equivalent conductance at a molar concentration c and Λ^o is that at infinite dilution.

Putting (3) in (2),

$$K_a = \frac{\left(\frac{\Lambda_c}{\Lambda^o}\right)^2 c}{1-\frac{\Lambda_c}{\Lambda^o}} \qquad \text{or, } \quad 1-\frac{\Lambda_c}{\Lambda^o} = \frac{1}{K_a} \cdot \frac{\Lambda_c^2 c}{\Lambda^{o2}}$$

$$\text{or, } \frac{1}{\Lambda_c} - \frac{1}{\Lambda^o} = \frac{1}{K_a} \cdot \frac{c\Lambda_c}{\Lambda^{o2}}$$

$$\text{or, } \frac{1}{\Lambda_c} = \frac{1}{\Lambda^o} + \frac{1}{K_a} \cdot \frac{c\Lambda_c}{\Lambda^{o2}} \tag{4}$$

This is the modified form of Debye-Hükel and Onsager for the dissociation of a weak electrolyte.

A plot of $\frac{1}{\Lambda_c}$ vs. $c\Lambda_c$ will give a straight line. (Vide figure in graph in the following page)

The slope of the straight line is $\frac{1}{K_a \Lambda^{o2}}$ and the intercept is $\frac{1}{\Lambda^o}$.

$$\therefore \; \Lambda^o = \frac{1}{\text{intercept}}.$$

Putting the value of Λ^o in $\frac{1}{K_a \Lambda^{o2}}$,

the slope will give the thermodynamic equilibrium constant, K_a.

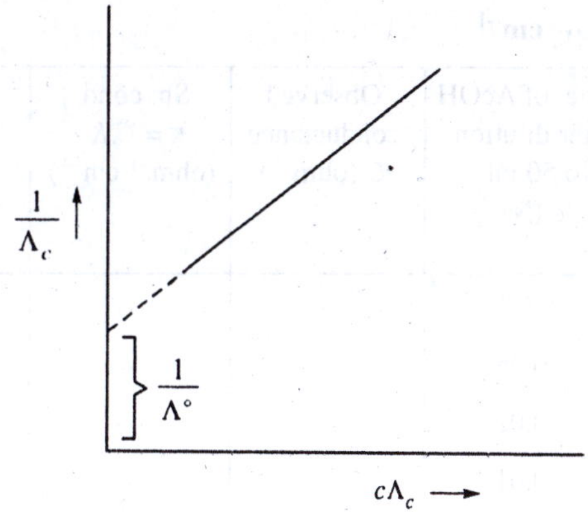

Equipments: Required equipments for determination of cell constant (K_c) and Λ of an electrolyte.

Materials: oxalic acid, NaOH, acetic acid, Phenolphthalein and KCl.

Procedure

1. Determine the cell constant of the conductivity cell.
2. Prepare 50 ml 0.1 N oxalic acid solution.
   Required weight = 0.315 g
3. Prepare 250 ml ~0.1N NaOH solution.
   Required weight = 0.4 g
4. Prepare 250 ml ~ 0.1 N CH_3COOH solution by proper dilution from the original acid bottle (the concentration should be a little bit higher).
5. Titrate 10 ml oxalic acid by NaOH solution taken in the burette. Calculate the strength of the alkali.
6. Titrate 10 ml acetic acid by the above standardised NaOH solution. Calculate the strength of acetic acid. From this standardised solutions, prepare exactly 250 ml 0.1 N AcOH solution. (The dilution may be avoided, if necessary volume be calculated and taken for the preparation of next four dilute solutions).
7. Prepare 50 ml each of 0.05 N, 0.02 N, 0.01 N and 0.002 N AcOH solution either from 0.1 N AcOH (or standard AcOH) solution by proper dilution.
8. Measure and calculate Λ-values of all the above four solutions along with that of 0.1 N solution.
9. Plot $\frac{1}{\Lambda_c}$ vs. $c\Lambda_c$. Calculate Λ° and K_a from intercept and slope.

Results and Calculations

Table 1: Recording of room temperature.

Table 2: Preparation of 50 ml 0.1 N oxalic acid solution. (See table 3 of experiment 3).
Required weight = 0.315 g.

Table 3: Standardisation of NaOH by standard oxalic acid solution. (See table 4 of experiment 3).

Table 4: Standardisation of AcOH by standard NaOH solution. (See table 6 of experiment 3).

Table 5: Cell constant, K_c = ··· cm^{-1}

Vol. flask no.	Vol. of 0.1 N AcOH Soln. (ml)	Conc. of AcOH after dilution to 50 ml. c (N)	Observed conductance C (ohm^{-1})	Sp. cond. $\kappa = C.K_c$ (ohm^{-1} cm^{-1})	Eqv. Cond. $\Lambda_c = \frac{1000\kappa}{c}$ (ohm^{-1} cm^2) g-eqv^{-1}	$\frac{1}{\Lambda_c}$	$c\Lambda_c$
1	50	0.1	···	···	···	···	···
2	25	0.05	···	···	···	···	···
3	10	0.02	···	···	···	···	···
4	5	0.01	···	···	···	···	···
5	1	0.002	···	···	···	···	···

Calculations: From the graph,

$$\text{Intercept}, \ \frac{1}{\Lambda^o} = \ldots; \quad \therefore \ \Lambda^o = \frac{1}{\text{intercept}} = \cdots$$

$$\text{Slope} = \frac{1}{\Lambda^{o2} \cdot K_a}; \quad \therefore \ K_a = \frac{1}{\text{Slope}} \times \frac{1}{\Lambda^{o2}}$$

Conclusion: Hence Λ^o-value of CH_3COOH is ··· and dissociation constant of CH_3COOH is ··· at °C temperature.

[Compare the K_a-value with that availed of from expt. 4].

Experiment 6

To determine the solubility and solubility product of a sparingly soluble salt by conductance measurement.
[both for Hons. and P.G. courses]

Theory: The **solubility** of a substance (the solute) in a particular solvent at a particular temperature is defined as the number of grams of the solute required to prepare a saturated solution of 100 grams of the solvent at that temperature.

As per definition, solubility is a unitless physical quantity or may be expressed as g/100 g. However, an useful unit is 'molarity', i.e. number of moles of the solute in one litre saturated solution.

The **solubility product** of a sparingly soluble salt is defined as the product of activities of the contituent ions after raising them to appropriate index (power) in the saturated solution of the salt at a particular temperature.

If a sparingly soluble salt B_xA_y dissociates as

$$B_xA_y \rightleftharpoons xB^{y+} + yA^{x-},$$

then $$K_{sp} = a^x_{B^{y+}} \times a^y_{A^{x-}}$$

or, $$K_{sp} = C^x_{B^{y+}} \times C^y_{A^{x-}} \qquad (1)$$

[$\because$ the saturated solution is of very low concentration, activity is replaced by concentrations of the ions].

If the solubility be x mol/litre for a sparingly soluble salt, then for (i) 1–1 (e.g., AgCl) or 2–2 (e.g., $BaSO_4$) electrolyles

$$K_{sp} = x^2 \tag{1a}$$

for (ii) 1-2 (e.g., Ag_2SO_4), 2–1 (e.g., $PbCl_2$) electrolytes

$$K_{sp} = 4x^3, \text{ etc.} \tag{1b}$$

The equivalent conductance (Λ) of an electrolyte is related to its specific conductance (κ) as

$$\Lambda = \frac{1000\,\kappa}{c} \tag{2}$$

c being the normality (i.e., g-eqv/litre) of the solution. Since the concentration of the saturated solution of a sparingly soluble salt is very low, A may be replaced by Λ^o, the equivalent conductance at infinite dilution. Again, due to low concentration, κ is also very low, so that the specific conductance of water will have to be taken into account.

Thus for a sparingly soluble salt, equation (2) takes the form

$$\Lambda^\circ = \frac{1000(\kappa_{salt} - \kappa_{H_2O})}{\Lambda^\circ} \tag{3}$$

$$\text{or,} \quad c = \frac{1000(\kappa_{salt} - \kappa_{H_2O})}{\Lambda^\circ}$$

$$\text{or,} \quad c = \frac{1000(\kappa_{salt} - \kappa_{H_2O})}{\Lambda_+^\circ + \Lambda_-^\circ} \tag{4}$$

Λ_+° and Λ_-° are the ion-conductances of the constituent ions of the salt.

Hence, measuring the specific conductance of the saturated solution of a sparingly soluble salt (κ_{salt}), and inserting the literature values of κ_{H_2O}, Λ_+° and Λ_-° in equation (4), the normality (c) of the salt-solution can be determined.

If x **mol/litre** be the solubility of the salt, then

for AgCl, $x = c$,

for $BaSO_4$, $x = \frac{c}{2}$

and for $Ca_3(PO_4)_2$ $x = \frac{c}{6}$ etc.

Finding out x, K_{sp} of the salt is calculated following equation (1).

Equipments: Usual equipments for determination of cell constant (K_c) of the conductivity cell in use and conductance of solution.

Materials: KCl, sparingly soluble salt (AgCl or $BaSO_4$ or $CaSO_4$), conductivity water.

Procedure

1. Note the laboratory temperature.
2. Prepare 100 ml 0.1 N KCl solution and determine the cell constant of the cell to be used following usual procedure (experiment 1).

3. Prepare a saturated solution of the given salt by shaking the salt in conductivity water in a reagent bottle for half an hour. Let settle for 15 minutes. Be satisfied that some salt is sedimented. Filter.
4. Take conductance of the solution (C) and calculate the specific conductance (κ), as

$$\kappa = C \times \text{cell constant.}$$

5. Collect values for $\kappa_{H_2O}, \Lambda^\circ_{Ag^+}, \Lambda^\circ_{Cl^-} \left(\text{or } \Lambda^\circ_{\frac{1}{2}Ba^{2+}}, \Lambda^\circ_{\frac{1}{2}SO_4^{=}}\right)$ as required from literature and put in equation (4) to calculate the normality (c) of the saturated solution.
6. Convert normality into molarity, which will be the solubility of the sparingly soluble salt in mol/litre.
7. Calculate K_{sp} following equations (1a) or (1b).

Results and Calculations

Table 1: Noting of laboratory temperature

Table 2: Preparation of 100 ml 0.1N KCl solution

Table 3: Determination of cell constant (For tables 2 and 3 see experiment 1)

Table 4: Determination of solubility and solubility product

Conductance of the soln. C_{salt} (ohm^{-1})	Cell const. K_c (cm^{-1})	Sp. cond. κ_{salt} (ohm^{-1} cm^{-1}) $\kappa = C \times K_c$	Sp. Cond. of water κ_{H_2O} (ohm^{-1} cm^{-1})	$\Lambda^\circ_{salt} = \Lambda^\circ_+ + \Lambda^\circ_-$	Conc., c (g-eqv/l	Solubility x mol/litre	Solubility product K_{sp}
...	...	...	...	...	...	...	...

Conclusion: The solubility of ... at ... °C temperature is ... mol/litre. The solubility product of ... at ... °C temperature is ...

Experiments relating to Conductometric titrations

General aspects: Acid-base or complexometric titrations involving pipette-burette and indicators are based on the simple formula.

$$v_1 s_1 = v_2 s_2,$$

which comes from the basic idea of the law of equivalent proportion. Without diluting the large scope of such mechanical titrations, some disadvantages may be taken into account.

i. Availability of proper indicator in some titration, e.g., that for a weak acid and a base.
ii. Titration for coloured solutions, though it is overcome in a few cases.
iii. Cautiously and anxiously wait for end point.
iv. Relative volume exactness of the pipette and burette.

Conductometric titrations overcome all these drawbacks. Here also the $v_1 s_1 = v_2 s_2$ relation is followed. The conductance of an electrolytic solution depends on the (i) concentration of ions (ii) charge of the ions and (iii) ionic mobilities. One gram-equivalent of all electrolytes produces same number of same valence-type of ions and equal charge by cations and anions. Hence concentration becomes a function of the mobility (or the ion-conductances) of the ions.

The concept of conductometric titrations originates from the change of ionic mobilities during progress of titrations—acid-base, complexometric or precipitation—the replacing ion may have nearly same or different ion conductance from that of the replaceable ion. Thus when H^+ ($\Lambda^o_{H^+}$ ~ 350 ohm^{-1}cm^2 mol^{-1}) is replaced by Na^+ $\left(\Lambda^o_{Na^+} \sim 50 \text{ ohm}^{-1}\text{cm}^2\text{mol}^{-1}\right)$, there is sharp decrease of conductance. But when Cl^-($\Lambda^o_{Cl^-}$ ~ 76.3 ohm^{-1}cm^2 mol^{-1}) is substituted by NO_3^-($\Lambda^o_{NO_3^-}$ ~ 71.4 ohm^{-1}cm^2 mol^{-1}), there is not much appreciable change in the conductance.

With the addition of the titrant, the plot of fall or rise of conductance vs. the drop (i.e., volume) of the titrant gives two straight lines with different slopes before and after the end point. The point of intersection of the two straight lines give the end point of the titration. The corresponding drops of the titrant is converted into the required volume for completion of the titration.

The following points are important

i. To avoid much increase in volume, which affects the conductance value, the titrant should be 5-10 times stronger than the solution to be titrated.

ii. It is preferable to use microburette.

iii. For precise curve and its explanation, the hydrolysis of the titration-product, if any, is to be considered.

iv. Much care is to be taken, not necessarily near the end point, but some readings at the beginning and a little bit away from the end point.

v. Write the electrolyte to be titrated, in ionic form, titrant either in molecular or ionic form and the products, if strong electrolyte in ionic form; if weak, then in molecular form. Compare the ion-conductances of those which are being changed.

Experiment 7

To determine the strength of a hydrochloric acid solution by titrating against a standard caustic soda solution conductometrically (or vice versa).

Theory: When a definite volume of HCl solution is titrated with a standard caustic soda solution from the burette, H^+ ions with large ionic mobilities are gradually replaced by Na^+ ions with much lower ionic mobilities. The result is the sharp fall of conductance of the whole solution. When all the H^+ ions are exactly replaced by the equivalent number of Na^+ ions, the end point is achieved, showing the lowest conductance value of the resulting solution mixture. On further addition of alkali, a rise in conductance will be exhibited due to mainly OH^- ions from alkali. The changes in conductance can be explained by the following equations:

$$\left(H^+ + Cl^-\right) \xrightarrow{\left(Na^+ + OH^-\right)} \left(Na^+ + Cl^-\right) + H_2O$$

Thus two straight lines are obtained, one before and the other after the end point, when conductances of the solution are plotted against drops of alkali added. However, the slopes of the two straight lines are different, the latter being flatter. The intersection of the two lines indicates the neutral point and the corresponding drops of alkali, converted to volume is the required volume of the titrant.

Instruments and equipments: Conductometer, conductivity cell, one 50 ml volumetric flask, two 250 ml beakers, one 100 ml beaker for conductometric titration, one 100 ml conical flask, funnel, 10 ml pipette, microburette.

Materials: Oxalic acid, NaOH, HCl (conc.), phenolphthalein solution as indicator, conductivity water.

Procedure

1. Prepare 50 ml ~ 0.1N standard oxalic acid solution.
   Required weight = 0.315 g.
2. Prepare about 100 ml ~0.5 N NaOH solution in beaker
   Required weight = 2.0 g (take slightly more).
3. Prepare about 100 ml ~ 0.1 N HCl solution in a beaker or volumetric flask by properly diluting the concentrated HCl in the laboratory.
4. Standardise the NaOH solution taking it in burette against 10 ml standard oxalic acid solution using phenolphthalein.
5. Wash the conductivity cell with conductivity water.
6. Pipette out 10 ml HCl solution in a 100 ml beaker and insert the cell in it. Conductivity water is to be added, if required, to immerse the electrodes completely in the solution [The dilution will not affect the end point for a conductometric titration].
7. Connect the cell with the conductometer and take reading of the condustance of the HCl solution.
8. Fill up the previous burrete with the same standardised NaOH solution.
9. Add 2 drops of NaOH solution on HCl solution in the cell, shake slowly and measure the conductance. Carry on the process of addition of definite number of drops of NaOH solution, shaking and conductance reading until a minimum value of conductance reaches and thereafter 5-6 readings of increasing conductances are taken. Stop titration. [If addition of 2 drops of NaOH solution does not change the conductance value appreciably, 4 drops or even more may be added, but the counting of drops must be accurate.]
10. Count the number of drops of 1 ml NaOH solution falling from the same burette. Repeat once more. Take average number of drops for 1 ml solution. Calculate the volume of one drop of the solution.
11. Plot conductance against the number of drops of alkali (the titrant), find the required number of drops for end point, convert it into volume.
12. Use $v_1 s_1 = v_2 s_2$ to calculate the strength of the HCl solution.
13. Perform a mechanical titration using phenolphthalein indicator, calculate the strength of HCl and compare it to that obtained conductometrically.

Results and Calculations

Table 1: Recording of temperature.

Table 2: Preparation of 50 ml 0.1 N oxalic acid solution. See table 3 of experiment 3.
Required weight = 0.315 g.

Table 3: Standardisation of NaOH solution by standard oxalic acid solution
See table 4 of experiment 3. Put NaOH in place of KOH.

Table 4: Standardisation of 10 ml of HCl solution by standard NaOH solution conductometerically
Strength of NaOH solution = ... (N)

No. of readings	No. of drops of NaOH solution	Conductance (ohm^{-1})
1	0	...
2	2	...
3	...	...
...	...	...
...	...	...

Table 5: Estimation of the volume of one drop of alkali solution

Volume of NaOH solution (ml)	No. of drops	Volume of one drop of NaOH solution (ml)	Mean vol. of one drop of NaOH soln. (ml)
1.0	...	...	...
1.0	...	...	

Calculation

From the graph, it is seen the number of drops of NaOH solution required for complete neutralisation of HCl solution = ··· drops.

Hence the volume of NaOH = No. of drops × vol. of one drop = ··· ml

$$\therefore \text{ strength of HCl solution} = \frac{V_{NaOH} \times S'_{NaOH}}{10}(N) = \cdots (N)$$

Conclusion: Hence the strength of the given HCl solution estimated conductometrically is ··· (N).

The strength estimated through mechanical tiration is ··· (N)

Nature of the curve

N.B. Note the slope of the two straight lines. The first line is steeper due to the replacement of H^+ ion by much less mobile Na^+ ion. The second line is flatter due to OH^- ion which is not too less mobile than H^+ ion, comparable to Na^+ ion.

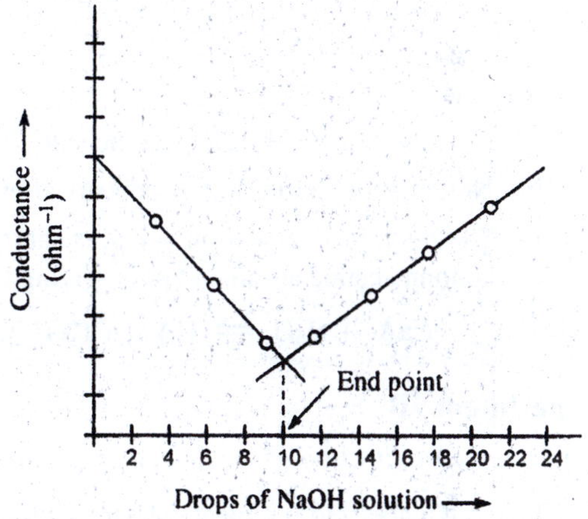

Some similar experiments

Experiment 7A

Standardisation of CH_3COOH *solution by standard* NaOH *solution.*

Theory: Acetic acid, being a very weak acid shows, much low conductance compare to strong acids like HCl. On addition of 1 or 2 drops of NaOH solution in a definite volume of acetic acid solution, the conductance falls due to further suppression of dissociation of CH_3COOH in presence of common CH_3COO^- ion resulting from CH_3COONa formed thereon [This fall of conductance may not be observed, if initially 6-8 drops of alkali is added instead of 1-2 drops]. On further addition of drops of alkali, the conductance increases slowly due to the dissociation of CH_3COONa, a strong electrolyte, overcoming the decrease in dissociation of CH_3COOH. After the equivalence point, the conductance rises more sharp due to OH^- ions coming from excess alkali. From the end point, the strength of the acetic acid solution is calculated. The relevant equation is

$$CH_3COOH \xrightarrow{Na^+ + OH^-} (CH_3COO^- + Na^+) + H_2O$$

Instruments, Materials, Process, Results and Calculations

Same as experiment no. 7, only CH_3COOH is taken for HCl.

Nature of the graph:

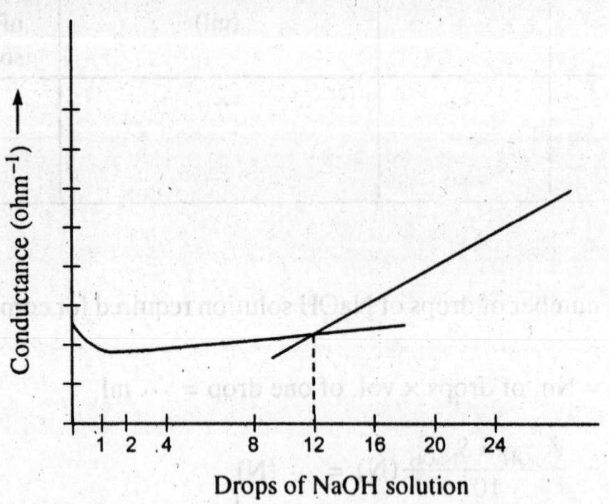

N.B.

i. Note the low conductance at the start. This is true for all weak electrolytes.

ii. Note that the initial fall disappears after a few drops of alkali is added.

iii. The point of intersection is not as sharp as for a strong acid. This is due to the hydrolysis of sodium acetate at the end point giving Na^+ and OH^- ions as the current-carriers.

$$NaAc + H_2O \rightleftharpoons HAc + (Na^+ \ OH^-)$$

Experiment 7B

Conductometric titration of oxalic acid by a standard NaOH *solution.*

Theory: Oxalic acid is a weak dibasic acid. It dissociates successively in two stages, such that K_1 is much greater than K_2.

$$H_2C_2O_4 \rightarrow H^+ + HC_2O_4^- ; K_1 = \text{first dissociation constant}$$

$$HC_2O_4^- \rightarrow H^+ + C_2O_4^= ; K_2 = \text{second dissociation constant.}$$

With the successive addition of drops of NaOH solution, two breaks will be obtained for complete titration. The corresponding drops (and hence volume) of alkali for the second break give the end point of the titration.

The process is like that of experiment no. 7, with HCl being replaced by oxalic acid.

The nature of the graph is shown in the next page.

N.B.

1. The first dissociation constant being much greater than the second, the first part of the graph (*AB*) appears close to that for a strong acid, like HCl.
2. After the first break, the cruve (*BC*) rises slowly due to the formation of the monoacidic salt, which behaves like a weak acid.
3. After the second break, the curve rises steeply due to excess OH^- ions.

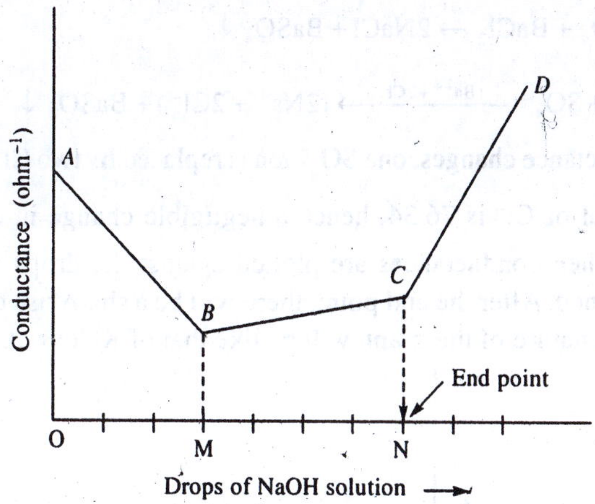

4. The total amount of alkali required for the complete titration is exactly double that required for the first break. (i.e. ON = 2 OM).

For calculation, convert the corresponding number of drops to volume of second break point.

Experiment 7C

Conductometric titration of KCl solution by $AgNO_3$ solution (Precipitation reaction).

Theory: When silver nitrate solution is added dropwise to a definite volume of potassium chloride solution, sparingly soluble salt silver chloride is formed, leaving potassium nitrate in the solution.

$$(K^+ + Cl^-) \xrightarrow{(Ag^+ + NO_3^-)} AgCl \downarrow + (K^+ + NO_3^-),$$

thus Cl^- ions being gradually replaced by NO_3^- ions. As the two ions possess nearly same conductance value, a straight line results till the end point, when conductance of the reaction mixture is plotted against the drops or volume of titrant ($AgNO_3$) added. After the end point, there will be a sharp rise of conductance due to the conductances of Ag^+ and NO_3^- coming from excess $AgNO_3$ solution added.

The nature of the graph

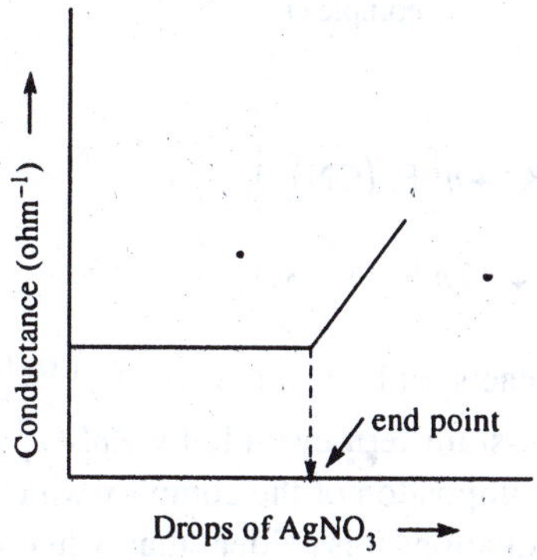

Experiment 7D

Conductometric standardisation of a given Na_2SO_4 *solution by a standard* $BaCl_2$ *solution (Precipitation reaction).*

Theory: Same as 7C.

With successive addition of $BaCl_2$ solution from the microburette, the following reaction takes place:

$$Na_2SO_4 + BaCl_2 \rightarrow 2NaCl + BaSO_4 \downarrow$$

$$\text{i.e., } 2Na^+ + SO_4^{=} \xrightarrow{(Ba^{++}+2Cl^-)} (2Na^+ + 2Cl^-) + BaSO_4 \downarrow$$

So, in regard to conductance changes, one $SO_4^{=}$ ion is replaced by two Cl^- ions. Since ion conductance of $\frac{1}{2}SO_4^{=}$ is 80 and that of Cl^- is 76.34, hence a negligible change in conductance of the reaction mixture results. Thus, when conductances are plotted against the drops of $BaCl_2$ solution, almost a straight line will be obtained. After the end point, there will be a sharp rise of conductance due to highly ionisable $BaCl_2$ salt. The nature of the graph will be like that of KCl vs. $AgNO_3$.

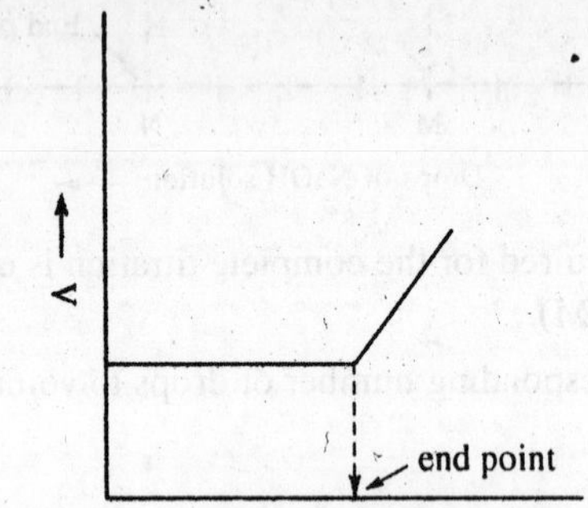

Experiment 8

Conductometric titration of potassium ferrocyanide solution with a solution of Zinc sulphate and also determination of the composition of Zn(II)—ferrocyanide complex which is formed.

Theory: When zinc sulphate solution is added to a solution of potassium ferrocyanide, a white precipitate of Zn (II)-ferrocyanide complex is formed. The reaction may be expressed by the following equation,

$$mZnSO_4 + nK_4[Fe(CN)_6] \rightarrow K_nZn_m[Fe(CN)_6]_n \downarrow + mK_2SO_4$$

(white ppt. of Zn (II)-ferrocyanide complex)

or, in ionisation form,

$$(mZ_n^{2+} + mSO_4^{2-}) + \left(4nK^+ + n\left[Fe(CN)_6\right]^{4-}\right)$$

$$\rightarrow K_nZn_m\left[Fe(CN)_6\right]_n \downarrow + 2mK^+ + mSO_4^{2-}$$

wehre '*m*' moles of $ZnSO_4$ solution reacts with '*n*' moles of $K_4[Fe(CN)_6]$ solution. The aim of the present experiment is to estimate potassium ferrocyanide by $ZnSO_4$ solution and also to find out the moles ratio *m* : *n*, i.e., to find out the compositon of the complex which is formed as a precipitate, both conductometrically. From the above equations it is evident that, when Zn(II)-sulphate solution is added to a solution of potassium ferrocyanide, ferrocyanide ions are removed as a precipitate and these ions are replaced by sulphate ions. The values of ionic conductivities at infinite dilution of $\frac{1}{4}[Fe(CN)_6]^{4-}$ and $\frac{1}{2}SO_4^{2-}$ are 111 ohm^{-1} cm^2 $eqv.^{-1}$ and 80 ohm^{-1} cm^2 $eqv.^{-1}$ at 25 °C respectively. Thus, with addition of $ZnSO_4$ solution to $K_4[Fe(CN)_6]$ solution the conductance of the solution will decrease up to the equivalence point. However, after the equivalence point, the excess of added salt ($ZnSO_4$) causes a sharp rise in conductance as shown in the following figure.

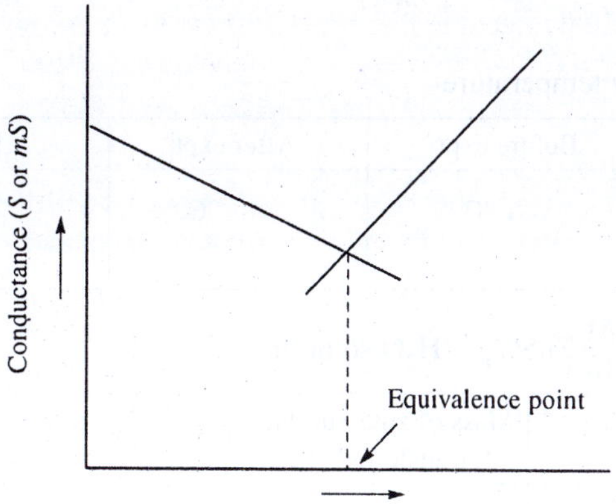

Thus, strength of potassium ferrocyanide solution can be calculated as the strength of $ZnSO_4$ solution is known (it is taken as a primary standard). Again number of moles (m) of $ZnSO_4$ that has reacted with n moles of potassium ferrocyanide can be calculated from the equivalance point. Hence, the moles ratio $m : n$, i.e., the composition of the complex may be determined.

Materials required

1. $ZnSO_4 \cdot 7H_2O$ crystals.
2. $K_4[Fe(CN)_6] \cdot 3H_2O$ crystals
3. Conductivity meter with conductivity cell.
4. Pipette (10 ml), microburette, beaker (100 ml), volumetric flasks (100 ml), funnel, etc.

Procedure

1. Prepare 100 ml each of $\frac{M}{10}$ $ZnSO_4$ solution and $\frac{M}{100}$ $K_4[Fe(CN)_6]$ solution by accurate weighing (Relative molar masses of $ZnSO_4 \cdot 7H_2O$ and $K_4[Fe(CN)_6] \cdot 3H_2O$ are 287.35 and 422.26 respectively).
2. Wash and clean the conductivity cell. Take 10 ml of $K_4[Fe(CN)_6]$ solution $\left(\sim \frac{N}{100}\right)$ in a clean beaker of 100 ml capacity with a pipette and place the conductivity cell into the solution. Add some distilled water so that the cell (platinum black) is immersed into the solution. Measure the conductance of the solution before addition of $ZnSO_4$ solution to it. Add 1-2 drops of $ZnSO_4$ solution preferably from a microburette. A white precipitate will be formed. Shake well and note the conductance of the solution. Continue this titration up to the equivalence point and note the conductance after each addition of $ZnSO_4$ solution. Note 5-6 more readings after passing the equivalence point. Wash for the cell immediately with distilled water for several times and keep immersed in distilled water.
3. Calibrate the number of drops of $ZnSO_4$ solution into volume (ml)
4. Plot conductance vs. no. of drops of $ZnSO_4$ solution added and calculate the strength of $K_4Fe(CN)_6$ solution. Also calculate the ratio $m : n$, and conclude about the composition of the complex.

Results and Calculations

i. Recording of laboratory temperature

	Before expt.	After expt.	Mean
Temperature ⟶	... °C	... °C	... °C

ii. Preparation of 100 ml $\frac{M}{10}$ $ZnSO_4 \cdot 7H_2O$ solution

Initial mass (g)	Final mass (g)	Mass of zine Sulphate transferred (g)	Mass to be taken (g)	Strength of $ZnSO_4$ solution
...	...	...	2.8735	...(M)

iii. Preparation of 100 ml $\frac{M}{100}$ $K_4[Fe(CN)_6] \cdot 3H_2O$ solution

Initial mass (g)	Final mass (g)	Mass of ferrocyanide transferred (g)	Mass of ferrocycyanide to be taken (g)	Strength of ferrocyanide solution
...	...	...	0.4222	...(M)

iv. Titration result and data table for graph plotting:

No. of observations	Volume of pot. ferrocyanide taken (ml)	No. of drops of Zn-sulphate solution added (n)	Strength of $ZnSO_4$ solution	Observed conductances (mS or S)
1		0		⋮
2		2		⋮
3		4		⋮
4	10	6	S (M) (say)	⋮
⋮		⋮		⋮

v. Calibration of no. of drops of $ZnSO_4$ solution into volume:

No. of observations	Volume of 20 drops of $ZnSO_4$ solution (ml)	Mean vol. of 20 drops of $ZnSO_4$ solution (ml)	No. of drops for 1 ml solution
1	v_1	$\frac{v_1 + v_2 + v_3}{3}$	$\frac{20}{v'}$ drops
2	v_2		
3	v_3	$= v'$ (say)	

vi. **Calculation of Strength of $K_4Fe(CN)_6$ solution**: From the graph, the plot of observed conductance vs. n (no. of drops of $ZnSO_4$ solution added), it is found that the number of drops of $ZnSO_4$ solution corresponding to the equivalence point = ...

= ... ml = v ml (say)

Hence the strength of $K_4[Fe(CN)_6]$ solution = $\frac{v \times S}{10}$[M].

vii. **Comparison of strengths of $K_4[Fe(CN)_6]$ solution prepared by weighing and obtained conductometrically**

Strength of pot. ferrocyanide solution by weighing S_1	Strength of pot. ferrocyanide solution conductometrically (S_2)	The ratio S_1/S_2
S_1 (M) ...	S_2 (M) ...	...

viii. **Calculation for the determination of the composition of the complex**

The volume of $ZnSO_4$ solution of strength S(M) required at the equivalence point = v ml (say)

$\therefore$ number of moles of $ZnSO_4 = \frac{v \times S}{1000} = m$ (say).

Again, if 10 ml solution of $K_4[Fe(CN)_6]$ be taken for titration, the concentration of $K_4[Fe(CN)_6]$ solution = $\frac{v \times S}{10}$(M)

$\therefore$ number of moles of $K_4[Fe(CN)_6]$ that combine with m moles of $ZnSO_4$ is

$$= \frac{v \times S \times 10}{10 \times 1000} = \frac{v \times S}{1000} = n \text{ (say)}$$

Hence the ratio $\frac{m}{n} = \frac{\text{No. of moles of } ZnSO_4}{\text{No. of moles of } K_4[Fe(CN)_6]}$

ix. **Conclusion**: The expected value of the above ratio m/n is = $\frac{3}{2}$

If it is, then we may conclude that the reaction should be as follows:

$$3ZnSO_4 + 2K_4[Fe(CN)_6] \rightarrow \underset{\text{Zn(II)-ferrocyanide complex}}{K_2Zn_3[Fe(CN)_6]_2} \downarrow + 3K_2SO_4$$

Experiment 9

Conductometric standardisation of a mixture of acids (HCl + CH_3COOH) *by a strong alkali* (NaOH).

Theory: When drops of solution of a strong alkali, say caustic soda, are added to a mixture of a strong acid, like HCl and a weak acid like CH_3COOH, replacement of H^+ ions coming from HCl by Na^+ takes first. The dissociation of CH_3COOH almost inhibits due to the presence of a large number of H^+ ions resulting from complete dissociation of HCl. Hence the conductance of the mixture continues to fall till complete neutralization of HCl.

After removal of all H^+ ions of HCl, dissociation of CH_3COOH starts and addition of NaOH solution slowly increases the conductance due to formation of highly ionizing salt CH_3COONa. After complete neutralization of CH_3COOH, the conductance of the mixture increases sharply for further addition of alkali.

The nature of the graph is shown below, where the first break indicates the neutralization of HCl and the second, that for CH_3COOH.

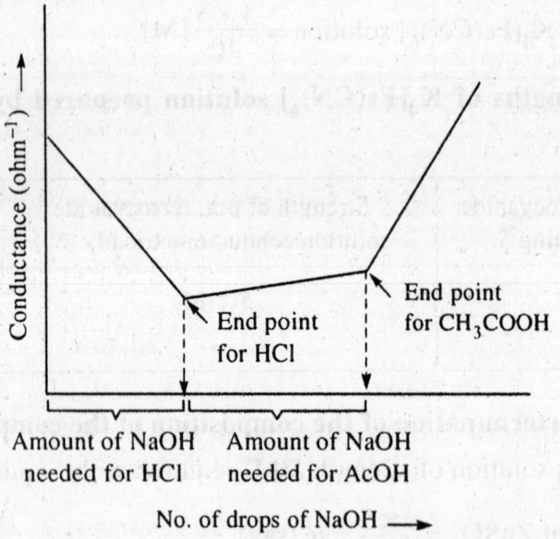

Total number of drops of NaOH solution required for complete neutralisation of the acid mixture minus that required for HCl, gives the amount of alkali needed for neutralization of acetic acid. Converting the drops into volume and knowing the strength of NaOH solution by standardising against standard oxalic acid solution, the individual strength of HCl and CH_3COOH in the mixture can be calculated.

Equipments: All those needed for a conductometric titration, as mentioned in the previous experiments.

Materials

i. ~ 0.1 N NaOH solution (100 ml)

ii. ~ 0.1 N standard oxalic acid solution (50 ml).

iii. The given acid mixture. For practice, mix about 25 ml each of ~ $\frac{M}{50}$ HCl and CH_3COOH solution to prepare a 50 ml acid mixture in which individual normality of the acids will be approximately 0.01 N.

Procedure

1. Prepare 50 ml ~ 0.1 N standard oxalic acid solution. Required weight = 0.315 g.
2. Prepare 100 ml 0.1 N NaOH solution. Required weight = 0.4 g.
3. Preparation of the acid mixture—as noted above (iii).
4. Standardise NaOH solution against 10 ml oxalic acid solution either mechanically using phenolphthalein indicator or conductometrically. If intended to do conductometrically, dilute the oxalic acid solution ten times quantitatively (i.e., 5 ml of the 0.1 N acid is to be diluted to 50 ml with conductivity water).
5. Take 10 ml of the acid mixture by pipette and standardise against the standardised NaOH solution and after each addition of NaOH drops, note the conductance.
6. Plot conductance against drops of NaOH added.

7. Note the number of drops of NaOH solution required for neutralisation of HCl and CH_3COOH in the mixture.
8. Convert the no. of drops into volume.
9. Use $v_1 s_1 = v_2 s_2$ to estimate HCl and CH_3COOH.

Results and Calculations

Table 1: Recording of temperature.

Table 2: Preparation of 50 ml 0.1 N oxalic acid solution. Required weight = 0.315 g. See table 3 of experiment 3.

Table 3: Standardisation of NaOH solution by standard oxalic acid solution.

(a) See table 4 of experiment 3 for mechanical titration. Use NaOH in place of KOH.

(b) For conductometric standardisation, follow experiment 7B, and table 4 of experiment 7.

Table 4: Estimation of the volume of one drop of the alkali solution. Use table 5 of experiment 7.

Calculations: From the graph, it is seen that the number of drops of NaOH solution required for neutralization of HCl = ···

∴ required volume of NaOH = no. of drops × vol. of each drop = ··· cm^3.

Number of drops of NaOH required for neutralization of CH_3COOH

= total no. of drops – no. of drops for HCl

= ··· drops

= ··· cm^3.

Hence strength of HCl in the given mixture = ···(N) and

strength of CH_3COOH in the given mixture = ···(N).

Experiment 10

Conductometric titration of a mixture of halides (HCl, KCl *and* NH_4Cl): [for P.G. courses]

Theory: When a solution of a mixture of three halides, viz., HCl, KCl and NH_4Cl is conductometrically titrated by a strong base, like NaOH, HCl and NH_4Cl can be estimated, leaving KCl unaffected. If the equal volume of the same mixture is titrated conductometrically by a standard $AgNO_3$ solution, the total halide contents can be estimated. Total halide concentration minus the sum of concentrations of HCl and NH_4Cl gives the amount of KCl present in the mixture.

When a definite volume of the solution mixture is gradually neutralised by drops of NaOH solution, HCl will response first, being the strong acid. Hence conductance falls steeply due to replacement of highly mobile H^+ ions by Na^+ ions of much lower ion-conductance as per the equation

$$(H^+ + Cl^-) \xrightarrow{(Na^+ + OH^-)} (Na^+ + Cl^-) + H_2O$$

After all the H^+ ions of HCl are used up, NH_4Cl takes part in the reaction as per the equation

$$(NH_4^+ + Cl^-) \xrightarrow{(Na^+ + OH^-)} NH_4OH + (Na^+ + Cl^-)$$

Thus NH_4^+ ions are gradually replaced by Na^+ ions. As NH_4^+ ions possess ionic conductance value of about 73.0 and that for Na^+ ion is about 50.0, hence a slow decrease of conductance results. After nuetralisation of all the NH_4Cl in the mixture, there will be a sharp rise of conductance due to addition of excess NaOH.

A plot of conductance vs. drops of NaOH gives two break points—the first corresponds to the amount of HCl and the second for NH_4Cl present in themixture.

Now the volume of the mixture when titrated with $AgNO_3$ solution, all the Cl^- ions are replaced gradually by NO_3^- ions, leaving practically no change in conductance (as $\lambda^o_{Cl^-} \sim \lambda^o_{NO_3^-}$). After complete precipitation of all the chlorides, further addition of $AgNO_3$ solution will enhance the conductance rapidly. The break in the graph of conductance versus drops of $AgNO_3$ solution corresponds to the total halide contents in the mixture.

$$(K^+ + Cl^-) + (H^+ + Cl^-) + (NH_4^+ + Cl^-) \xrightarrow{(Ag^+ + NO_3^-)}$$
$$AgCl \downarrow + (K^+ + NO_3^-) + (H^+ + NO_3^-) + (NH_4^+ + NO_3^-)$$

Subtracting the total amount of HCl and NH_4Cl from the total halide, amount of KCl can be estimated.

Equipments: As needed for a conductometric titration.

Materials: Oxalic acid, NaOH, HCl, NH_4Cl, KCl, $AgNO_3$ and conductivity water.

Procedure

1. Prepare 50 ml 0.1 N standard oxalic acid solution. From this solution prepare 50 ml 0.01 N solution by diluting ten times (dilute 5 ml to 50 ml). Weight required = 0.315 g.
2. Prepare 100 ml ~ 0.1 N NaOH solution. Weight required = 0.4 g.
3. Prepare 50 ml ~ 0.1 N $AgNO_3$ solution. Weight required = 0.8490 g. (25 ml solution may also be prepared dissolving 0.4245 g $AgNO_3$ in 25 ml water with the help of measuring cylinder, as this solution is to be standardised by KCl solution).
4. Prepare 50 ml $\frac{N}{30}$ standard KCl solution. Weight required = 0.124 g.
5. The experimental mixture of halides will be supplied.

   However, for own work, the following solution may be prepared and the strengths may be verified after conductometric estimation.

   (a) Prepare 50 ml ~ $\frac{N}{30}$ HCl solution. Dilute from a 0.1 N HCl solution.

   (b) Prepare 50 ml ~ $\frac{N}{30}$ NH_4Cl solution. Weight required = 0.089 g.
6. Standardise NaOH solution, taking it in microburette, by oxalic acid solution conductometrically, taking 10 ml acid solution by pipette. Do not throw away NaOH solution from burette after titration.
7. Take 15 ml of the supplied solution. For practice, mix 5 ml each of the prepared HCl, KCl and NH_4Cl solutions. Also prepare a second set with same composition.

   Standardise the mixture by NaOH solution.
8. Standardise $AgNO_3$ solution by standard KCl solution. Take 5 ml KCl and add $AgNO_3$ solution from microburette. Do not throw $AgNO_3$ solution after titration.
9. Take 15 ml of the supplied solution or the second set you have just kept ready.

   Standardise against $AgNO_3$ solution from burette.

Results and Calculations

Table 1: Recording of laboratory temperature.

Table 2: Preparation of 50 ml 0.1 N standard oxalic acid solution: Required weight = 0.315 g. See **table 3** of **experiment 3**.

Table 3: Preparation of 50 ml N/30 standard KCl solution. Weight required = 0.124 g.

1st weight w_1 (g)	2nd weight w_2 (g)	Weight of KCl transferred $(w_1 - w_2)$ (g)	Strength of KCl (N)
...	...	...	$\frac{w_1 - w_2}{0.124}\left(\frac{N}{30}\right)$

Table 4: Standradisation of NaOH solution by oxalic acid solution. See **Table 4** of **experiment 3.**

Table 5: Estimation of volume of one drop of NaOH solution. See **Table 5** of **experiment 7.**

Table 6: Standardisation of 15 ml of supplied solution of halides by standard NaOH solution. Usual table of conductometric titration noting conductance after definite number of drops of alkali added. See **table 4** of **experiment 7.**

Table 7: Estimation of volume of one drop of $AgNO_3$ solution.

Table 8: Standadisation of $AgNO_3$ solution by standard KCl solution. Usual table of conductance after addition of drops of $AgNO_3$ (see table 9).

Table 9: Standardisation of 15 ml of supplied solution of halide mixture by $AgNO_3$ solution.

No. of drops of $AgNO_3$ soln.	Conductance (ohm^{-1})
0	...
2	...
4	...
...	...
...	...
...	...

Nature of graphs and related calculations

Graph 1. Standardisation of NaOH by oxalic acid: See graph of experiment 7B.

Convert the number of drops of the second break point into volume and find out the strength of NaOH solution.

Graph 2. Titration of halide mixture by NaOH solution:

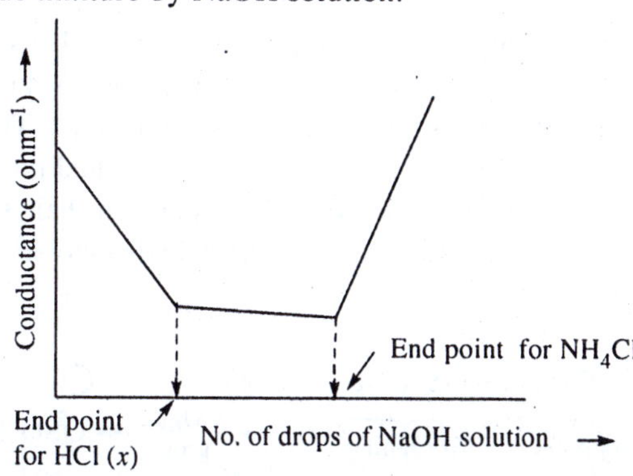

Graph 3: Standardisation of $AgNO_3$ solution by KCl solution: See graph of experiment 7C. Convert drops related to end point into volume and calculate strength of $AgNO_3$.

Graph 4: Titration of halide mixture by $AgNO_3$ solution: Same as graph 3.

Calculation of the strengths of the halides in the mixture.

From graph 2,
the number of drops of NaOH solution required for neutralisation of HCl

$= x = x \times$ volume of 1 drop of NaOH = $\cdots$ cm^3

$\therefore$ strength of HCl in the mixture $= \dfrac{\text{vol. of NaOH} \times \text{strength of NaOH}}{15}$ (N).

Again, the number of drops of NaOH required for neutralisation of $NH_4Cl = (y - x)$ drops = $\cdots$ cm^3.

$\therefore$ strength of NH_4Cl in the mixture $= \dfrac{\text{vol. of NaOH} \times \text{strength of NaOH}}{15}$ (N).

From graph 4,
number of drops of $AgNO_3$ solution for equivalence of total chlorides in the mixture = z drops (say)

$= z \times$ volume of 1 drop of $AgNO_3$ solution = $\cdots$ cm^3.

$\therefore$ strength of total chlorides in the mixture $= \dfrac{\text{vol. of } AgNO_3 \times \text{strength of } AgNO_3}{15}$ (N)

$\therefore$ strength of KCl in the mixture

= (total concentration of Cl^-) – (strength of HCl + strength of NH_4Cl) = $\cdots$ (N).

Conclusion

In the supplied solution,

strength of HCl = $\cdots$ (N)

strength of NH_4Cl = $\cdots$ (N)

strength of KCl = $\cdots$ (N)

[Since NH_4Cl and KCl have been taken by weighing the salts, you may compare their analytical strengths to that obtained conductometrically].

Experiment 11

Determination of Critical Miceller Concentration (C.M.C.) of a surface-active agent (e.g., Sodium Lauryl Sulphate) conductometrically. [for P.G. courses]

Theory: Soaps, synthetic detergents, and a number of dyes are called surface-active agents. When added a very small amount to water, they decrease the surface tension of water (or a solvent) appreciably, retaining themselves mostly on the surface of the solvent than in the bulk. They form either emulsion (a colloidal system) at low dilution or in most cases remain as individual molecules forming ions. Such type of compounds have a hydrophobic and a hydrophilic parts in their structures. For example, in sodium lauryl sulphate, a detergent, the hydrocarbon part is hydrophobic and the sodium sulphonate part is hydrophilic.

$\underbrace{CH_3-(CH_2)_{11}}_{\text{hydrophobic}}-\underbrace{O\bar{S}O_3Na^+}_{\text{hydrophilic}} \equiv$ 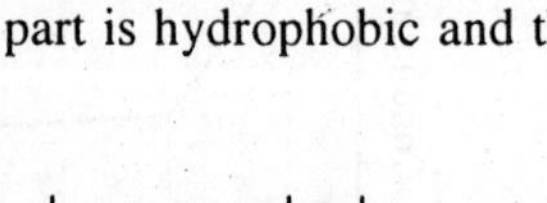

Cetyltrimethylammonium bromide (CTAB), Cetylpyrridiumchloride (CPC) are some examples of detergents, while sodium stearate, sodium oleate, etc are examples of soaps.

With increasing concentration, the hydrophilic portion are readily attracted by water, while the hydrophobic parts gradually stay away from water and approach each other to form an almost spherical ionic cluster inside the bulk, which is called **micelle**—an **association colloid.** A minimum concentration of the surface-active agent is required to form micelle. In the micelle, the hydrocarbon ends remain in the interior and the ionic functional groups project outward in the solution.

At a definite temperature, the minimum concentration at which a surface-active agent begins to form micelle, is called the 'critical micellar concentration' or **'critical micellization concentration' (CMC) of the particular agent at that temperature.**

For example, a soap is unlikely to form micelle below 10^{-3} M concentration of it at normal temperature.

Micelle may be cationic or anionic in nature. At the point of formation of micelle, many properties of the solution like surface tension, conductances, osmotic coefficient, etc suffer a sudden change, which so far changed regularly with concentration. Measuring the conductance, the CMC value can be estimated.

Increase in temperature lowers the CMC value. A number of solutions of SLS is prepared and their equivalent conductances (Λ) are measured following the usual procedure. Now a plot of Λ vs $\sqrt{c}$ is drawn, which shows a minima. The corresponding concentration is the value of CMC for SLS.

Formation of micelle

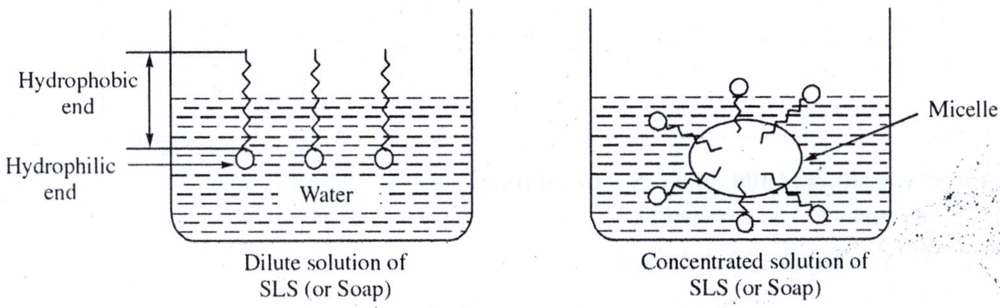

Appraratus: Conductivity bridge, 7-8 numbers of 50 ml volumetric flasks, one 500 ml volumetric flask, a 50 ml burette.

Materials: Sodium Lauryl Sulphate (SLS) AR, conductivity water, KCl.

Procedure

1. Prepare a standard 500 ml 0.05 M SLS solution. Weight required = 7.2095 g (MW 288.38).
2. Prepare a number of SLS solutions (50 ml each) of different concentrations from the stock solution.

   [You may use the stock solution as it is, even its concentration may not be 0.05 M, but the concn. should be exact one].
3. Prepare 100 ml 0.1 N KCl solution and determine the cell constant (K_c) of the experimental conductivity cell following experiment 1.
4. Determine the conductance of each solution by conductometer.
5. Calculate the specific condustance ($\kappa = C \times K_c$) and equivalent conductance $\left(\Lambda = \frac{1000\,\kappa}{c}\right)$ of each solution.

6. Plot Λ vs. $\sqrt{c}$.

7. CMC value will be $\left(\sqrt{c}\right)^2$ at the corresponding point.

Results and Calculations:

Table 1: Preparation of 500 ml standard 0.05 M SLS solution

Weight required = 7.2095 g

1st weight of SLS w_1 (g)	2nd weight of SLS w_2 (g)	Weight taken $w_1 - w_2$ (g)	Strength of SLS solution, (M)
...	...	...	$\frac{w_1 - w_2}{7.2095}(0.05)$ M

Table 2: Preparation of various SLS solutions from the stock solution

Volumetric flask no.	Vol. of stock solution (ml)	Final volume after making up with water (ml)	Conc. of the soln. (c) (M)
1	1.0	50	...
2	2.0	50	...
3	4.0	50	...
4	8.0	50	...
5	10.0	50	...
6	15.0	50	...
7	20.0	50	...

Table 3: Preparation of 100 ml 0.1 N KCl solution

See Table 2 of experiment 1.

Table 4: Determination of Cell Constant.

See Table 3 of experiment 1 (use only 0.1 N KCl solution).

Table 5: Determination of *C*, κ and Λ of different solutions. Cell constant, K_c = ... cm⁻¹.

No. of Vol. flask	Conc. of SLS solution (c) (M)	$\sqrt{c}$	Conductance (C) ohm^{-1}	Sp. Cond. (κ) (ohm^{-1} cm^{-1}) $\kappa = C \times K_c$	Eqv. conductance (Λ) (ohm^{-1}cm^2 g-eqv^{-1}) $\Lambda = \frac{1000\,\kappa}{c}$
1	...	...	...	...	...
2	...	...	...	...	...
3	...	...	...	...	...
4	...	...	...	...	...
5	...	...	...	...	...
6	...	...	...	...	...
7	...	...	...	...	...

Table 6: Recording of room temperature.

Conclusion: From the graph, it is seen that the CMC for Sodium Lauryl Sulphate is ⋯ M at ⋯ °C temperature.

The nature of the graph:

N.B.: 1. First, there is a gradual fall of equivalent conductance with increase in concentration (*AB*).

2. With the formation of micelle, Λ falls falls suddenly (*BC*).

3. The *CD* portion indicates the completion of micellization. Due to absence of any free molecule in ionic form, no change in conductance is observed.

[CMC may also be determined by surface tension measurement, See Expt. 3 in the surface chemistry section].

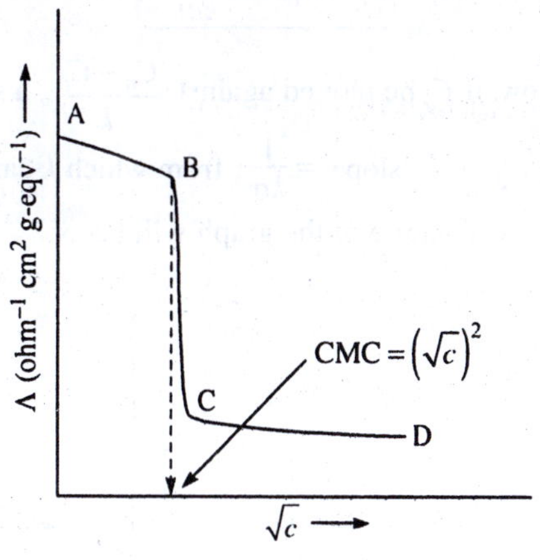

Experiment 12

Conductometric determination of the rate constant of alkaline hydrolysis of methylacetate. [for P.G. courses]

Theory: In presence of a strong alkali, like NaOH, the hydrolysis of methyl acetate takes place as follows:

$$\text{MeCOOMe} + \text{OH}^- \rightarrow \text{MeCOO}^- + \text{MeOH}$$

Such alkaline hydrolysis of ester is called *saponification reaction.*

Thus OH^- ions are successively replaced by acetate ions, having much lower ion conductance (41.2) compare to that of OH^- ion (198). Hence there will be a decrease of conductance of the reaction mixture with the progress of the reaction.

The reaction is second order — first order each with respect to ester and alkali. If the reaction is carried out with same initial concentrations of both the reactants, then the rate constant may be written as

$$k = \frac{x}{ta(a-x)}, \quad (1)$$

which is the result of integration of the second order rate equation, $\frac{dx}{dt} = k(a-x)^2$,

where a = initial molar concentration of the ester and the alkali,

x = concentration of the product at time t

Let the conductance of the reaction mixture at $t = 0$ is C_0 (which is in fact the conductance of only NaOH solution); that at any intermediate time t be C_t and that after completion of the reaction be C_α.

Hence equation (1) may be written as

$$k = \frac{C_0 - C_t}{ta(C_t - C_\alpha)}$$

$$\text{or,} \quad (C_t - C_\alpha) = \frac{C_0 - C_t}{kat}$$

or, $$C_t = C_\alpha + \frac{C_0 - C_t}{kat} \tag{2}$$

Now, if C_t be plotted against $\frac{C_0 - C_t}{t}$, a straight line will be obtained with positive slope, where

slope $= \frac{1}{ka}$, from which k can be calculated.

The nature of the graph will be as:

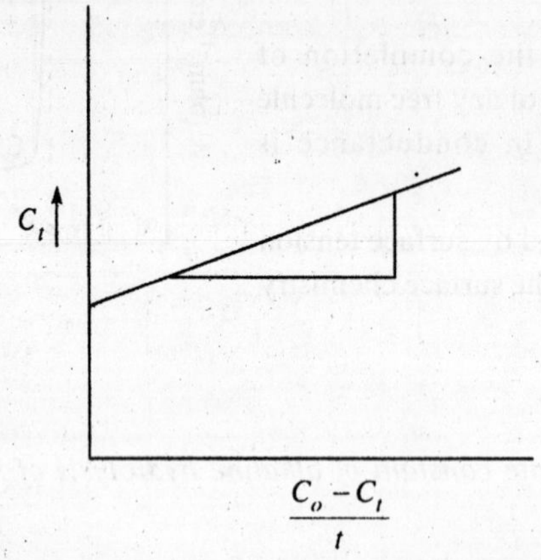

[Compare the value of k with that obtained from analytical *kinetic* experiment]

Instrument & equipments: Conductometer, two 100 ml and two 50 ml volumetric flasks, 500 ml beaker, conical flask, pipette, burette, stopwatch.

Materials: Methyl acetate (GR), NaOH, oxalic acid, phenolphthalein and conductivity water.

Procedure

1. Prepare standard 50 ml 0.1 N oxalic acid solution. Weight required = 0.315 g.
2. Prepare 250 ml ~0.1N NaOH solution. Weight required = 1.0 g.
3. Standardise the alkali solution against standard oxalic acid solution using phenolphthalein indicator.
4. From the known strength of NaOH solution, prepare exactly 50 ml of each of $\frac{N}{25}$ and $\frac{N}{50}$ NaOH solution by quantitative dilution.
5. Prepare 100 ml exact $\frac{N}{25}$ methyl acetate solution. Calculate the required volume from density/specific gravity and molecular weight of the ester from the label of the bottle.
6. Measure the conductance of $\frac{N}{50}$ NaOH solution. The value corresponds to C_o.
7. Wash the conductivity cell well.
8. Take a clean and dry 100 ml beaker. Pipette out 25 ml $\frac{N}{25}$ NaOH solution and keep in it. With another pipette, take 25 ml $\frac{N}{25}$ methyl acetate and add in the alkali. During addition, start the stopwatch when about half of the pipette is emptied.

   [Note that the strength of both the alkali and the ester is now $\frac{N}{50}$]
9. Take the conductance readings (C_t) of the mixture at an interval of about one minute and note the correct time (t). Stop taking readings when 2-3 constant value of conductance are achieved.
10. Plot C_t vs. $(C_o - C_t)/t$. Take slope and calculate k.

Results and Calculations

Table 1: Recording of room temperature.

Table 2: Preparation of 50 ml 0.1 N oxalic acid solution. Weight required = 0.315 g.

Table 3: Standardisation of NaOH solution by oxalic acid solution.
Strength of oxalic acid = ··· (N)

No. of titrations	Vol. of oxalic acid taken (ml)	Initial burette reading	Final burette reading	Vol. of NaOH required (ml)	Mean vol. of NaOH (ml)	Strength of NaOH (*N*)
1	10.0	0.0	···	···		
2	10.0	···	···	···	···	···
3	10.0	···	···	···		

Table 4: Measurement of conductances: C_0 = ··· ohm^{-1}

Time (t) s	Conductance (C_t) (ohm^{-1})	(C_0-C_t) (ohm^{-1})	$\frac{C_0-C_t}{t}$ ($\text{ohm}^{-1}s^{-1}$)
···	···	···	···
···	···	···	···
···	···	···	···
···	···	···	···
···	···	···	···
···	···	···	···

Calculation: From the plot of C_t vs. $\frac{C_0-C_t}{t}$, slope = ···

$\therefore$ the rate constant, k = ··· litre mol^{-1} s^{-1}.
[Compare the value of k with that obtained from analytical kinetic experiment]

N.B.

1. The saponification reaction can be studied with ethylacetate also. The hydrolysis can be represented as $CH_3COOC_2H_5 + NaOH \rightarrow CH_3COONa + C_2H_5OH$
   or, ionically, $CH_3COOC_2H_5 + OH^- \rightarrow CH_3COO^- + C_2H_5OH$
2. We have

$$k = \frac{C_0 - C_t}{ta(C_t - C_\alpha)}$$

or, $$kat = \frac{C_0 - C_t}{C_t - C_\alpha} \quad (3)$$

Hence, a plot of $\frac{C_0 - C_t}{C_t - C_\alpha}$ vs. t will give a straight line, passing through the origin.

The slope is positive and slope = ka, $\therefore$ $k = \frac{\text{slope}}{a}$.

This way of plotting requires the C_α-value, *which is not required in the experiment discussed above. Here is the advantage of the previous method.*

However, one can find out C_α-value as follows:

Mix exactly equal volumes of $\frac{N}{25}$ acetic acid solution and $\frac{N}{25}$ NaOH solution. The conductance of this mixture is C_α.

APPENDIX

Table 1

Ionic molar conductivity (Λ^o_m) and equivalent conductivity (Λ^o) at infinite dilution at 298 K in water of some ions

Cations	Λ^o_m (S cm^2 mol^{-1})	$\Lambda^o = \frac{\Lambda_m}{\text{Valency}}$ (ohm^{-1} cm^2 mol^{-1})	Anions	Λ^o_m (ohm^{-1} cm^2 mol^{-1})	$\Lambda^o = \frac{\Lambda^o_m}{\text{Valency}}$ (ohm^{-1} cm^2 mol^{-1})
H^+	349.81	349.81	OH^-	198.3	198.3
Li^+	38.68	38.68	F^-	55.4	55.4
Na^+	50.11	50.11	Cl^-	76.34	76.34
K^+	73.52	73.52	Br^-	78.14	78.14
NH_4^+	73.4	73.4	I^-	76.84	76.84
Rb^+	77.8	77.8	NO_3^-	71.44	71.44
Ag^+	61.92	61.92	CN^-	82.0	82.0
Mg^{2+}	106.12	53.06	SO_4^{2-}	160.04	80.02
Ca^{2+}	119.0	59.5	$C_2O_4^{2-}$	148.3	74.15
Sr^{2+}	118.9	59.45	$HCOO^-$	54.6	54.6
Ba^{2+}	127.26	63.63	CH_3COO^-	40.9	40.9
Cu^{2+}	107.2	53.6	$[Fe(CN)_6]^{4-}$	442.0	110.5
Zn^{2+}	105.6	52.8	$[Fe(CN)_6]^{3-}$	302.7	100.9
Cd^{2+}	108.0	54.0	$C_6H_5COO^-$	32.3	32.3
Pb^{2+}	139.0	69.5	ClO_4^-	67.4	67.4
Ni^{2+}	108.0	54.0	CO_3^{2-}	138.6	69.3
Co^{2+}	110.0	55.0			
Fe^{3+}	204.0	68.0			
Al^{3+}	189.0	63.0			

Table 2

Specific conductances (κ) of aqueous KCl solutions (in S m^{-1})

Strength of KCl soln.	15 °C	18 °C	20 °C	22 °C	25 °C	30 °C
0.1 N	0.01048	0.01119	0.01167	0.01215	0.01288	0.01412
0.01 N	0.001147	0.001225	0.001278	0.001332	0.001413	—

CHAPTER 5

Spectrophotometric / Colorimetric Experiments

I. **Introductory Words**

II. **The Instruments**

III. **Experiment 1**

(a) Verification of Beer-Lambert's law with a given solution and
(b) the determination of the concentration of a solution

IV. **Experiment 2**

To determine the composition and stability constant of Fe (III)-Salicylic acid complex colorimetrically by Job's method of continuous variations (for P.G.)

V. **Experiment 3**

Determination of
(a) Isobestic point of bromocresol green indicator and
(b) Indicator constant of the indicator (for P.G.)

CHAPTER 5

Spectrophotometric / Colorimetric Experiments

I. Introductory words

The interaction of light with matters and thereby collection of information regarding structure, electronic arrangements, bond lengths, bond energy, composition, stability constant of a complex, etc. of a chemical species constitute the subject matter of a spectroscopy.

A chemical entity—an ion, atom or a molecule has its characteristic energy states, which are quantised. Impact with light energy causes a definite change in these energy states, which provides ample information regarding the species as mentioned above.

When light falls on a system, it may be reflected or refracted, or may be absored or may be emitted as a part of absorbed radiation. The extent of absorption depends on the energy levels of the system, its polarity, population of atoms or molecules at different molecular energy levels, i.e., it is characteristic of the chemical species. In spectorscopy, only absorption or emission of light by a chemical system is considered.

In absorption spectroscopy, the fraction of light intensity of a particular wavelength absorbed by an absorbing medium of the substance is measured. In emission spectroscopy, the emitted extra energy by an atom or a molecule in coming down to the ground state from an excited state is measured. Both give the same information. Here we will discuss the absorption spectroscopy.

We know that light possesses **dual nature**—the **wave** and **particle**. While wave, light is considered as an electromagnetic wave—having an electric as well as a magnetic field perpendicular to each other. While particle, the smallest form is known as **photon** which is the tiny packet of energy, known as **quantum** of energy, given by the equation,

$$E = h\nu = hc/\lambda = hc\bar{\nu}$$

where ν = frequency, λ = wavelength, $\bar{\nu}$ = wave number and c = velocity of light.

Unit of ν = Hz (s^{-1}), λ = m or cm,

$\bar{\nu}$ = m^{-1} or cm^{-1} and c = ms^{-1} or cms^{-1}.

The Electromagnetic spectrum

The broad range of electromagnetic radiation with respect to frequency, wavelength or wave number, is known as electromagnetic spectrum. Starting from radiowave with longest wave length ~ 30 m the spectrum is extended up to gamma rays with the shortest wavelength ~ 0.003 nm. As the velocity remains constant with that of light over the whole of the spectrum, the frequency changes accordingly.

Spectroscopy concerns the results of interaction of the light wave with matters, where frequency is of prime importance. However, we will confine our interests in the **uv-visible zone of the electromagnetic spectrum**. Still preceding regions may also be informed for interest of students.

Spectral Region	Wavelength (nm)	Frequency (Hz)	Wave number (cm^{-1})	Energy J/mole.	Types of Interaction with atom or molecule	Information from spectral data
Microwave	3×10^8–3×10^6	10^9–10^{11}	3.3×10^{-2}–3.3	0.4–40	Rotation of polyatomic molecules	Inter-nuclear distances
Far infrared	3×10^6–3×10^4	10^{11}–10^{13}	3.3–330	40–4×10^3	Rotation of simple molecules	Inter-nuclear distances of small molecules
Infrared	3×10^4–700	10^{13}–10^{14}	330–14300	4×10^3–4×10^4	Vibration of chemical bond	Force constant, Amplitude of vibration
Visible	700–350	10^{14}–10^{15}	14300–28500	4×10^4–4×10^5	Transition of outer shell electrons of atoms or molecules	Electronic arrangements of atoms or molecules
Ultraviolet	350–190	10^{15}–1.6×10^{15}	28500–52000	4×10^5–6.4×10^5	Transition of electrons reading to ionisation of atoms or molecules	Bond energies-electronic structure of atoms or molecules

The Instruments:

In the instrumental device a monochromatic light is allowed to fall on a solution in a cell of definite length, it passes through the solution, a portion of the intensity is absorbed and then transmitted. The instrument provides an arrangement to measure the fraction of the intensity of the light of a particular wave length absorbed by the solution—colour or colourless, or the amount transmitted and also to measure the concentration of the compound. Other than % of absorbance or transmittance a very useful parameter, 'optical density' (O.D), can be measured directly.

O.D. of the medium is given as

$$\text{O.D.} = \log\frac{I_0}{I_t},$$

where I_o = intensity of the light of particular wave length before incidence and I_t = intensity of the transmitted light.

For measurement of O.D., absorbance or transmittance, the instruments used are **colorimeter** and **spectrophotometer**. A colorimeter uses only visible light and thereby only coloured solutions can be studied satisfactorily. It costs much less than the spectrophotometer, easy to handle and enough for experiments in General or Hons. courses.

A spectrophotometer is a very sophisticated instrument and has the provisions to use UV, visible and IR light sources.

A schematic diagram is shown below in Fig. 5.1.

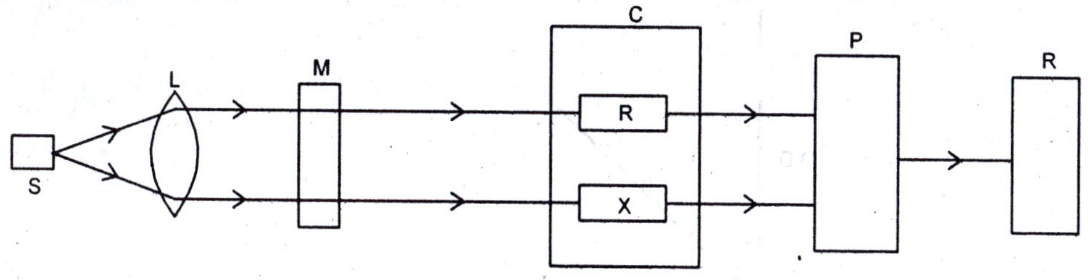

Fig. 5.1

S = Light source; generally tungsten filament under low voltage and high current density

L = Collimating lens

M = Wavelength selector or Monochromator

C = Cell chamber in which R = reference solvent and X = experimental solution

P = Photo detector

R = Measuring device.

Experiment 1

(a) *Verification of Beer-Lambert's law with a given solution and* (b) *determination of the concentration of a solution.*

Theory: According to Lambert's law, when a monochromatic light passes through a transparent absorbing species of thickness l, the optical density of the medium is proportional to l, i.e.,

$$\text{O.D.} = \log\frac{I_o}{I_t} \propto l \tag{1}$$

Beer's law applies when the absorbing medium is a solution. According to Beer's law the optical density is proportional to the concentration (c) of the solution, i.e.,

$$\log\frac{I_o}{I_t} \propto c \tag{2}$$

I_o is the intensity of incident light before being absorbed and I_t is that after transmission.

Combining (1) and (2)

$$\text{O.D.} \propto cl$$

$$\text{or,}\quad \text{O.D.} = \varepsilon cl \tag{3}$$

which is Beer-Lambert's law, where ε is a constant, specific for a given substance present in the solution at a given wavelength of the incident light.

When c is expressed in molarity, ε is called 'molar absorption coefficient' or 'molar extinction coefficient'.

Unit of l is generally cm, so that unit of ε is litre mol^{-1} cm^{-1} or dm^3 mol^{-1} cm^{-1}.

(a) According to equation (3), when O.D. is plotted against c, a stright line is obtained passing through the origin. Preparing a number of solutions with different concentrations from the given solution and measuring their O.D.s, a graph may be drawn of O.D. vs. c. If a straight line is obtained, then Beer-Lambert's law will be verified.

The nature of the graph will be as shown in Fig. 5.2.

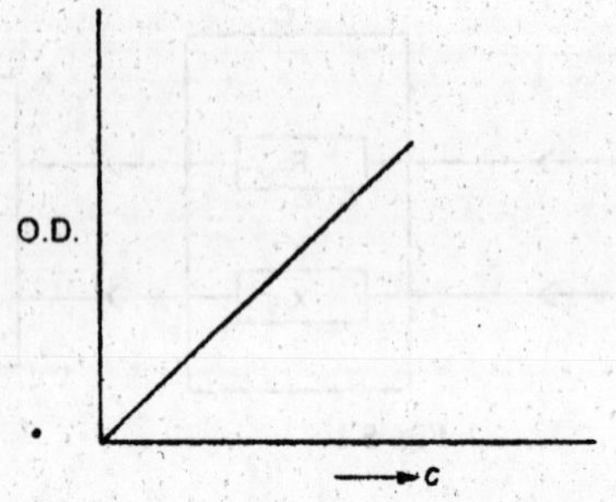

Fig. 5.2

(b) To determine the conc. of a given $KMnO_4$ or $K_2Cr_2O_7$ solution, the O.D. of the solution is measured. The corresponding conc. will be the conc. of the given solution.

Equipments: Colorimeter or spectrophotometer, ten 15 ml or 20 ml perfectly cleaned and dried test tubes, 10 ml graduated pipette, three 100 ml volumetric flasks, 25 ml pipette, conical flask, beaker, funnel, Bunsen burner.

Materials: Oxalic acid, $KMnO_4$ (AR) (or $K_2Cr_2O_7$), 2N H_2SO_4, glass or cotton wool.

Procedure

1. Prepare 100 ml 0.1 N standard oxalic acid solution.
2. Prepare 100 ml 10^{-2} M $KMnO_4$ solution (or take the supplied $KMnO_4$ solution).
   Dissolve about 0.16 g $KMnO_4$ crystal in 100 ml water in a 250 ml beaker by heating and stirring with a glass rod. Cool and filter through glass or cotton wool in a 100 ml volumetric flask. Make up the volume.
3. Prepare 250 ml 2N H_2SO_4 solution.
   Add 14 ml conc. H_2SO_4 dropwise and slowly to ~236 ml water.
4. Titrate $KMnO_4$ solution against 25 ml standard oxalic acid solution.
   Pipette out 5 ml oxalic acid solution in a 100 ml conical flask. Add ~ 25 ml 2N H_2SO_4 solution. Heat up to ~ 70 °C (do not boil at all) over a Bunsen burner. Add $KMnO_4$ solution from burette dropwise and stir. Do not allow the temperature to fall below 60 °C. Warm again, if necessary. A permanent light pink colour will indicate the end point.
   Calculate the strength of $KMnO_4$ solution.
   [For $K_2Cr_2O_7$, a standard solution is prepared from direct weight of $K_2Cr_2O_7$. Weight required for 100 ml 10^{-2}M $K_2Cr_2O_7$ solution = 0.2942 g]
5. Prepare exactly 100 ml 10^{-3}M $KMnO_4$ solution from the standardised solution (the normality will be ~ 5×10^{-2}, i.e., molarity ~ 10^{-2}) by proper dilution in a 100 ml volumetric flask.
   [For $K_2Cr_2O_7$, the molarity should be exactly 10^{-3}M also.]
6. Prepare $KMnO_4$ solution of different concentrations (10^{-4}M – 10^{-3}M) as per table 3.

7. Take the most dilute or any intermediate dilute $KMnO_4$ solution in a cell. Fill up another cell with water. Place properly in the colorimeter and take O.D. of the solution at 450 nm. Continue study of O.D. readings of the same solution at increasing interval of 10 or 15 nm. Follow the results till the O.D. value becomes maximum and then falls.
8. Draw graph of O.D. vs. wavelength and select the filter (wavelength) for maximum O.D. (λ_{max}).
9. Take O.D. readings of other $KMnO_4$ solutions at the selected wavelength (λ_{max}).
10. Plot O.D. vs. c—a straight line through origin verifies the Beer-Lambert's law.

Results and Calculations

Table 1: Preparation of 100 ml 0.1 N standard oxalic acid solution

1st weight of oxalic acid w_1 (g)	2nd weight of oxalic acid w_2 (g)	Weight transferred $w_1 - w_2$ (g)	Weight required (g)	Strength of oxalic acid soln.
...	...	...	0.63	$\frac{w_1 - w_2}{0.63}(0.1)$ N

Table 2: Standardisation of $KMnO_4$ solution

No. of Titrations	Vol. of oxalic acid (ml)	Burette reading		Vol. of $KMnO_4$ soln. required (ml)	Mean vol. (ml)	Strength of $KMnO_4$ solution
		Initial (ml)	Final (ml)			
1	5	...	...	...	...	x (N)
2	5	...	...	...		

Strength of $KMnO_4$ solution $= x$ (N) $= \frac{x}{5}$(M)

Table 3: Preparation of $KMnO_4$ solutions of different concentrations from 10^{-3} M solution

Test tube No.	Vol. of 10^{-3} M $KMnO_4$ soln. (ml)	Vol. of water (ml)	Total Vol. of solution (ml)	Strength of $KMnO_4$ solution (M)
1	1.0	9.0	10.0	1×10^{-4}
2	2.0	8.0	10.0	2×10^{-4}
3	3.0	7.0	10.0	3×10^{-4}
4	4.0	6.0	10.0	4×10^{-4}
5	5.0	5.0	10.0	5×10^{-4}
6	6.0	4.0	10.0	6×10^{-4}
7	7.0	3.0	10.0	7×10^{-4}
8	8.0	2.0	10.0	8×10^{-4}
9	9.0	1.0	10.0	9×10^{-4}

Table 4: Measurement of O.D. of 5×10^{-4} M $KMnO_4$ solution for selection of λ_{max} (any other lower concentration may be chosen also).

Wave length (nm)	O.D.
450	...
460	...
470	...
...	...
...	...
...	...

The nature of the graph will be as shown in Fig. 5.3.

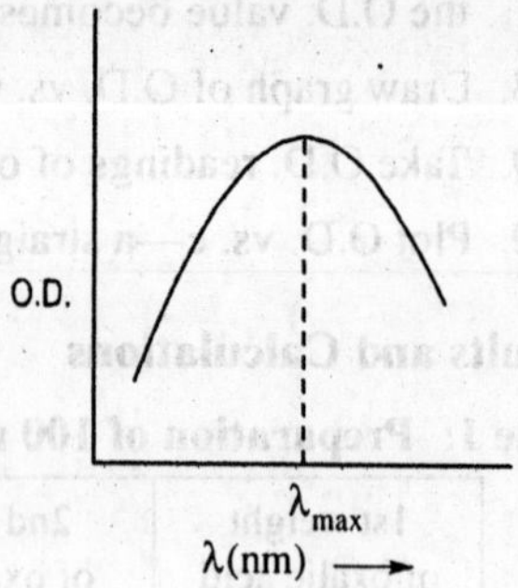

Fig. 5.3

Table 5: Measurement of O.D. of different $KMnO_4$ solutions at ... nm

Test tube	Conc. of $KMnO_4$ solution ($c \times 10^4$) (M)	O.D.
1	1.0	
2	2.0	...
3	3.0	...
4	4.0	...
5	5.0	...
6	6.0	...
7	7.0	...
8	8.0	...
9	9.0	

Conclusion: As the O.D. vs. *c* graph is a straight line and it passes through the origin, hence Beer-Lambert's law is verified.

If conc. of a given $K_2Cr_2O_7$ or $KMnO_4$ solution (supplied) is to be found out, perform the above experiment and take the O.D. of the given solution. Draw O.D. vs. *c* graph. Locate the O.D. value of the unknown solution on the straight line. The corresponding conc. will be the concentration of the given solution as shown in Fig. 5.4.

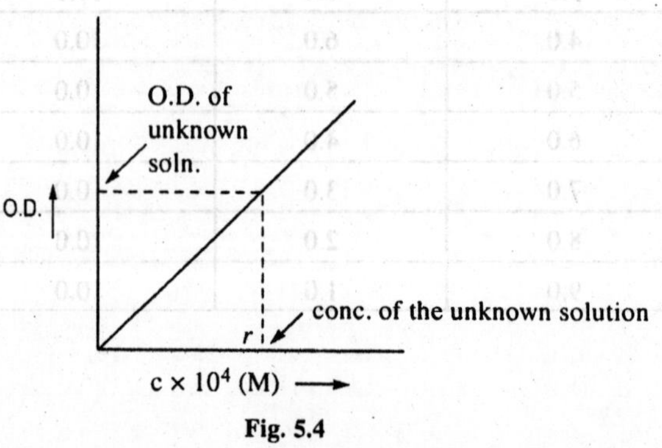

Fig. 5.4

Conclusion: The conc. of the given solution is ... (M)

Additional parameter:

From equation (3),

$O.D. = \varepsilon cl$

$$\therefore \ \varepsilon = \frac{O.D.}{cl} = \frac{O.D.}{c} \text{ (when path length, } l\text{, of the absorbing medium = 1 cm)}$$

∴ the slope of either of the graphs gives ε, the molar absorption coefficient.

Experiment 2

To determine the composition and stability constant of Fe (III)—*salicylic acid complex colorimetrically by Job's method of continuous variation.*

Theory: Iron (III) forms a number of coloured complexes with some phenolic compounds, like salicylic acid. Job's method provides an easy way to determine the stoichiometry of such complexes.

The principle of this method is that, if the metal ions and the ligand are taken in equal molarity and mixed in varied proportions, the maximum amount of the complex will be formed at equilibrium, when the proportion of the reactants correspond to the empirical formula of the complex. At this stage the absorption will be maximum also.

Let the empirical formula of the complex be A_mB_n, which is formed as

$$mA + nB \rightleftharpoons A_mB_n \tag{1}$$

Let the molarity of A and B be x. According to the principle mentioned above, if m parts of A and n parts of B are mixed, then amount of A_mB_n will be maximum at equilibrium. And in this condition, the absorption will be maximum at a selected wavelength.

pH plays an important role for the stability of the complex. In this case, the optimum pH-range is 2.6-2.8. This is achieved by using a 0.002M HCl solution. Otherwise the complex may hydrolyse.

The stability constant (K) of the complex is given by

$$K = \frac{[\text{Fe (III)-salicylic acid}]_{\text{complex}}}{[\text{Fe (III)}]_{\text{unreacted}}\ [\text{salicylic acid}]_{\text{unreacted}}} \tag{2}$$

For this purpose, a fixed amount of S.A. is complexed with different amount of Fe (III). From the graph of O.D. vs. [Fe (III)], the slope = ε, i.e., the molar absorptivity is found out.

Now from the equation,

$$\text{O.D.} = \varepsilon cl = \varepsilon c \quad (l = 1 \text{ cm}),$$

$$c = \frac{\text{O.D.}}{\varepsilon} \tag{3}$$

i.e., concentration of Fe (III) in each solution is determined. This concn. corresponds to the concn. of the complex (numerator). Let the original concn. of Fe (III) in the composition determining concn. (i.e. for maximum absorbance) be a.

Then $[\text{Fe (III)}]_{\text{unreacted}} = (a - c)$.

Since iron : S.A. is 1 : 1, ∴ $[\text{S.A.}]_{\text{unreacted}} = (a - c)$.

$$\therefore \ K = \frac{c}{(a-c)(a-c)}. \tag{4}$$

Equipments: Colorimeter, 100 ml volumetric flasks, 10 ml graduated pipette, 500 ml beaker, ten 15/20 ml test tubes (preferably stoppered).

Materials: Ferric-ammonium sulphate $[Fe_2(SO_4)_3, (NH_4)_2SO_4, 24H_2O]$, salicylic acid, HCl.

Procedure

1. Prepare 400 ml 0.002M HCl solution. (Calculate required volume of conc. HCl, taking its molarity 11.3).
2. Prepare 100 ml 0.002 M ferric-ammonium sulphate solution in 0.002M HCl solution.
   Dissolve exactly 0.1927g ferric-alum in 20-25 ml 0.002 M HCl solution in a 100 ml volumetric flask. After complete dissolution add 10 ml water, shake well and then make up the volume with 0.002 M HCl. (If larger amount of salt is taken, dilute properly with 0.002 M HCl).
3. Prepare 100 ml 0.002 M salicyclic acid in 0.002M HCl solution.
   Dissolve exactly 0.0276 g salicylic acid in 10 ml water in a 100 ml volumetric flask. After complete dissolution, make up the volume with 0.002 M HCl solution. (If some larger amount of SA is taken, dilute properly with 0.002 M HCl).
   [**N.B.** In (2) the salt is dissolved in HCl. Water is added in equal amount in both (2) and (3), to keep the same molarity. Generally S.A. does not dissolve in HCl. So it is first dissolved in minimum amount of water, so that the required pH-range is maintained.]
4. Label nine test tubes 1-9 and prepare the solution as in table 3.
5. Take test tube no. 4 or 6 and determine its O.D. using different filters (wave lengths) from 420-660 nm at about 40 nm intervals.
6. Plot O.D. vs λ and find λ_{max}.
7. Set the spectrophotometer at the λ_{max} and take O.D. readings of all the solutions.
8. Plot O.D. vs. volume of salicylic acid and find the maximum absorbance. Find the ratio of vol. of Fe (III) and S.A. for this solution and determine the composition of the complex.
   Let the original concn. of Fe (III) in this solution be *a*.
9. Prepare a saturated solution of S.A. in 0.002M HCl.
10. **For determination of stability constant** (equilibrium constant) mark another five stoppered test tubes I-V and prepare solutions as per table 6 and take O.D. readings of the solutions.
11. Plot O.D. vs. concn. of Fe (III) solutions. The slope gives the molar absorptivity, ε.
12. Find concn. of $[Fe (III)]_{complex}$, $c = \dfrac{O.D.}{\varepsilon}$
13. Find concn. of unreacted Fe (III) and S.A.
14. Calculate stability constant as per equation (4).

Results and Calculations

Table 1: Preparation of 100 ml exact 0.002 M Ferric-ammonium sulphate solution in 0.002 M HCl weight required = 0.1927 g.

Table 2: Preparation of 100 ml exact 0.002 M salicylic acid solution in 0.002 M HCl.
Weight required = 0.0276 g
(You may first prepare 0.02 M and then dilute to 0.002 by adding 0.002 M HCl quantitatively).

Table 3: Preparation of solutions for determination of composition of complex

Test tube No.	Vol. of Fe^{3+} (0.002M) (ml)	Vol. of S.A. (0.002 M) (ml)	Total vol. (ml)
1	9	1	10
2	8	2	10
3	7	3	10
4	6	4	10
5	5	5	10
6	4	6	10
7	3	7	10
8	2	8	10
9	1	9	10

Table 4: Determination of λ_{max} (taking test tube no. 4 or 6)

λ (nm)	O.D.
420	...
460	...
500	...
...	...
...	...

The nature of the graph will be as in experiment 1.

Table 5: Determination of the composition of the complex

Test tube no.	O.D. at λ_{max} = ... nm
1	...
2	...
⋮	
8	
9	

The plot of O.D. vs vol. of salicylic acid is shown in Fig. 5.5.

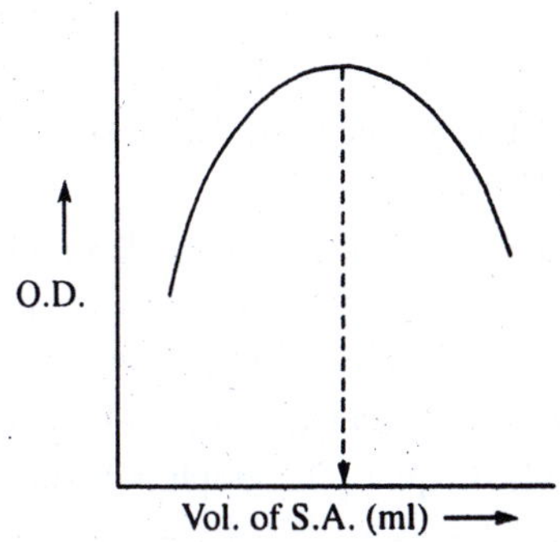

Fig. 5.5

From the plot, the vol. ratio of Fe (III) and S.A. at maximum absorbance = ...:...

∴ empirical formula of the complex is ...

Original concn. of Fe (III) in this solution ... (M)

Table 6: Preparation of solutions for determination of stability constant and O.D. readings at λ_{max} = ... nm

Test tube no.	Vol. of 0.002 M Fe^{3+} soln. (ml)	Vol. of 0.002M S.A. soln. (ml)	Vol. of 0.002 M HCl soln. (ml)	Total vol. (ml)	Conc. of Fe^{3+} (M)	O.D.
I	1	5	4	10	...	...
II	2	5	3	10	...	...
III	3	5	2	10	...	...
IV	4	5	1	10	...	...
V	5	5	0	10	...	...

The plot of O.D. vs. conc. of Fe (III) is shown in Fig. 5.6.

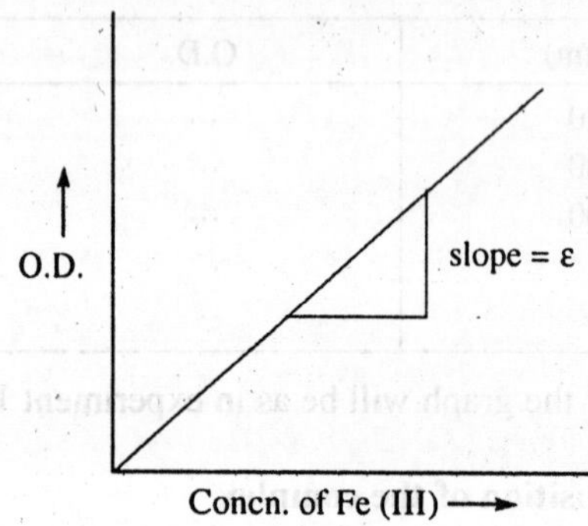

Fig. 5.6

Experiment 3

Determination of—(a) Isobestic point of bromocresol green indicator, and (b) Indicator constant of the indicator.

Theory: An acid-base indicator is a weak acid or base which exhibits two distinct colours in dissociated and undissociated forms of the molecule.

If the indicator be represented as HIn, then its ionisation in aqueous solution may be shown as

$$HIn(aq) \rightleftharpoons H^+(aq) + In^-(aq) \quad (1)$$

In strong alkaline solution, the colour is due to $In^-(aq)$ form and in strong acid medium, the colour is due to the undissociated form of $HIn(aq)$.

When two different spectra of the indicator in strong alkali and in strong acid medium is studied using long range of filters (in visible region) and the O.D.s are plotted against wavelengths, the two curves will cut at a point. This point is called as the **isobestic point** of the indicator.

Hence isobestic point of an indicator is defined as a wavelength at which the absorbance of the two forms (dissociated and undissociated) is equal. The graph will appear as shown in Fig. 5.7.

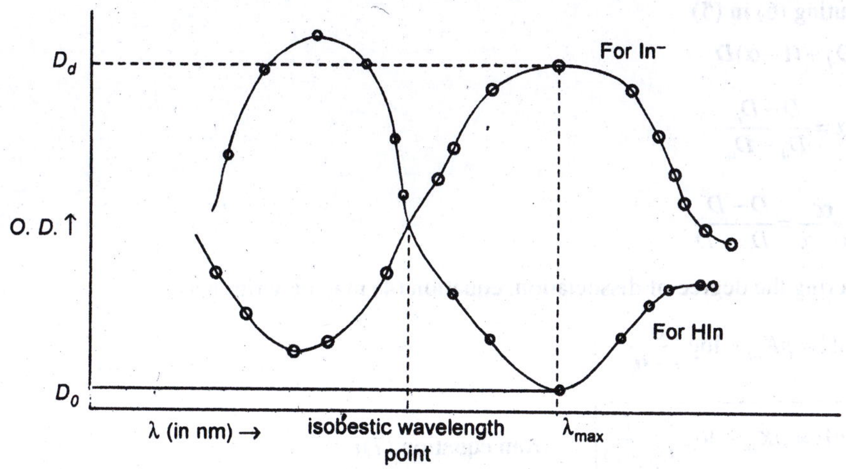

Fig. 5.7

From equation (1), the indicator constant, K_{In} may be written as

$$K_{In} = \frac{[H^+][In^-]}{[HIn]} \tag{2}$$

Taking logarithm on both sides

$$\log K_{In} = \log[H^+] + \log\frac{[In^-]}{[HIn]}$$

or $$-\log[H^+] = -\log K_{In} + \log\frac{[In^-]}{[HIn]}$$

or, $$pH = pK_{In} + \log\frac{[In^-]}{[HIn]} \tag{3}$$

or, $$pH = pK_{In} + \log\frac{[\text{basic form}]}{[\text{acidic form}]} \tag{4}$$

Let us consider ε_d = molar extinction coefficient of the species In^-,
and ε_u = molar extinction coefficient of the species HIn,
α = degree of dissociation of the indicator in the basic In^- form, and
$(1 - \alpha)$ = amount of undissociated HIn.

Then according to Lambert-Beer's law, the optical density (D) of the indicator, when both In^- and HIn forms are present, is given by

$$D = \varepsilon_d cl\alpha + \varepsilon_u(1-\alpha)cl \tag{5}$$

when $\alpha = 0, (1-\alpha) = 1$, then

$$D_u = \varepsilon_u cl$$

And when $\alpha = 1, D_d = \varepsilon_d cl$

Thus, $$\varepsilon_u = \frac{D_u}{cl} \text{ and } \varepsilon_d = \frac{D_d}{cl} \tag{6}$$

Substituting (6) in (5),

$$D = \alpha D_d + (1-\alpha) D_u$$

or, $$\alpha = \frac{D - D_u}{D_d - D_u}$$

or, $$\frac{\alpha}{1-\alpha} = \frac{D - D_u}{D_d - D} \quad (7)$$

Considering the degree of dissociation, equation (4) may be written as

$$pH = pK_{In} + \log \frac{\alpha}{1-\alpha}$$

or, $$\boxed{pH = pK_{In} + \log \frac{D - D_u}{D_d - D}} \quad \text{from equation (7)]} \quad (8)$$

So, if we plot $\log \frac{D - D_u}{D_d - D}$ vs. pH, a straight line with positive slope, having intercept, pK_{In}, will be obtained. Thus K_{In} can be calculated from the intercept.

Equipments and materials: Spectrophotometer, ten clean test tubes (20 ml), 100 ml beakers, conical flasks, burette, graduated pipette (10 ml), bromocresol green, oxalic acid, NaOH, HCl, CH_3COOH.

Procedure

1. Prepare 50 ml 0.4N standard oxalic acid solution. Weight required = 1.26 g.
2. Prepare about 150 ml ~ 0.5N acetic acid solution. Dissolve 4.5 ml acetic acid from original bottle (~17N) in water (~ 145 ml).
3. Prepare about 200 ml ~ 0.5 N NaOH solution. Dissolve ~4.0 g NaOH in 200 ml water.
4. Prepare 100 ml ~0.1 M HCl solution. Dissolve ~ 1 ml conc. HCl in 100 ml water.
5. Standardise NaOH solution by oxalic acid solution and then acetic acid solution by standardised NaOH solution taking NaOH solution in burette. Use phenolphthalein indicator.
6. Prepare 100 ml exact 0.4 N solution each of NaOH and AcOH by proper dilution from the above-standardised solutions.
7. Mark the nine hard glass test tubes with 1-9 and prepare the buffer solutions as in table 4.
8. Prepare 50 ml 0.1N NaOH solution from 0.4 N solution by proper dilution.
9. Prepare ~ 0.1% bromocresol green indicator solution.
   Dissolve ~ 0.1 g solid indicator in 100 ml alcohol.
10. Take two beakers—one containing 10 ml of 0.1 N NaOH solution and the other 0.1 N HCl (need not be standardised as HCl is a strong acid).
11. Add 2 drops of indicator solution to each beaker, mix well and take O.D. reading starting from 450 nm-650 nm of each solution, at an interval of 10 nm.
12. Plot in a same graph paper, O.D. vs. λ and find the isobestic point (Fig. 5.7). Choose λ_{max} from the graph where the O.D. values-difference is maximum after the isobestic point.

    λ_{max} = ... nm

    Isobestic point = ... nm.

13. Measure O.D. values of the indicator in nine buffer solutions at λ_{max}. The corresponding values are 'D' in equation (8).

14. Select D_d and D_u from Fig. 5.7.

15. Plot $\log \frac{D - D_u}{D_d - D}$ vs. pH and calculate pK_{In} and hence K_{In} from the intercept (Fig. 5.8).

The nature of graph will be as shown in Fig. 5.8

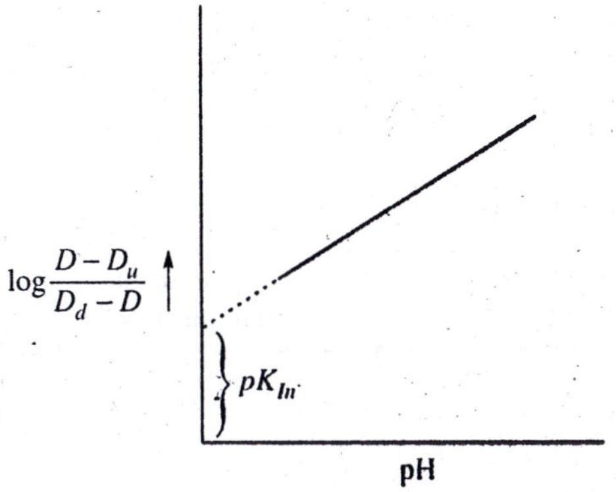

Fig. 5.8

Results and Calculations

Table 1: Recording of laboratory temperature—usual table.

Table 2: Preparation of 50 ml 0.4 N oxalic acid solution—usual table.

Table 3: Standardisation of NaOH solution by standard oxalic acid solution—usual table.

Table 4: Standardisation of AcOH solution by NaOH solution—usual table.

Preparation of 100 ml 0.2 N AcOH solution: Take 50 ml 0.4 N AcOH solution and dilute to 100 ml with distilled water.

Preparation of 100 ml 0.2 N AcONa solution: Mix 50 ml each of 0.4N AcOH and 0.4N NaOH solution together.

Table 5: Preparation of different buffer solutions

$pK_{AcOH} = 4.74$ at 25 °C temperature

Test tube no.	Vol. 0.2 N AcONa soln. (ml)	Vol. of 0.2 N AcOH soln. (ml)	pH of the buffer	No. of drops of 0.1% BCG
1	1	9	3.78	2
2	2	8	4.14	2
3	3	7	4.37	2

Test tube no.	Vol. of 0.2 N AcONa soln. (ml)	Vol of 0.2 N AcOH soln. (ml)	pH of the buffer	No. of drops of 0.1% BCG
4	4	6	4.56	2
5	5	5	4.74	2
6	6	4	4.91	2
7	7	3	5.11	2
8	8	2	5.34	2
9	9	1	5.69	2

Table 6: O.D. readings for Isobestic point

λ (nm)	O.D. for alkaline solution (In^-)	O.D. for acidic solution (HIn)
450	...	...
460	...	...
470		
⋮		
⋮		
650		

Table 7: O.D. readings for K_{In} value

$D_d = \cdots$ $D_u = \cdots$ $\lambda_{max} = \cdots$ nm

Test tube no.	pH	D	$D - D_u$	$D_d - D$	$\log \frac{D - D_u}{D_d - D}$
1	3.78	...	...	...	...
2	4.14	...	...	...	...
3	4.37	...	...	...	...
4	4.56	...	...	...	...
5	4.74	...	...	...	...
6	4.91	...	...	...	...
7	5.11	...	...	...	...
8	5.34	...	...	...	...
9	5.69	...	...	...	...

Results

(a) From graph 1 (Fig. 5.7), the isobestic point of bromocresol green = ... nm

(b) From graph 2 (Fig. 5.8), the intercept = $pK_{In} = \cdots$,

$\therefore\ K_{In} = \cdots$.

CHAPTER 6

Potentiometry and pH-metry

I. Introductory Words

- Electrode potentials and formal potentials
- Reference electrodes: standard hydrogen electrode
- Calomel electrode
- Quinhydrone electrode
- Silver-silver chloride electrode
- Glass electrode
- The convention of representation of cells and calculation of cell e.m.f.
- Measurement of cell e.m.f.
- Standard cell
- Salt bridge
- Calibration or standardisation of a digital potentiometer
- Standardisation or calibration of a pH-meter
- Regeneration of a 'bad' or 'dead' glass electrode.

II. Experiments with Potentiometer

Experiment 1

Potentiometric titration of a solution of a strong acid (say, hydrochloric acid) with a solution of strong alkali (say, caustic soda solution) using quinhydrone electrode.

Experiment 2

Potentiometric titration of a weak acid (say, acetic acid) with caustic soda solution and determination of the dissociation constant of the acid using quinhydrone electrode at room temperature.

Experiment 3

Determination of standard reduction potential (E^0) of quinhydrone electrode at room temperature (for U.G./P.G).

Experiment 4

Potentiometric titration of a solution of silver nitrate with a solution of potassium chloride (i.e., standardisation of silver nitrate solution by standard potassium chloride solution potentiometrically).

Experiment 5

Potentiometric titration of a solution of chloride ion with silver nitrate solution.

Experiment 6

Potentiometric titration of a mixture of halides (KCl and KI) [for P.G.]

Experiment 7

Determination of activity solubility product of silver chloride by e.m.f. measurements (for P.G.)

Experiment 8

Determination of solubility and solubility product of silver chloride by e.m.f. measurement of a concentration cell (for P.G.)

Experiment 9

Potentiometric estimation of Mohr salt solution with standard potassium dichromate solution and also determination of formal potential (reduction) of ferric-ferrous system.

Experiment 10

Potentiometric titration of Zn (II) solution by potassium ferrocyanide solution and also determination of the composition of Zn (II)-ferrocyanide complex. [for P.G.].

Experiment 10A

Determination of composition of Pb (II)—ferrocyanide complex. [for P.G.]

Experiment 10B

Determination of composition of Cd (II)—ferrocyanide complex potentiometrically. [for P.G.]

Experiment 11

Determination of standard reduction potential of ferricyanide—ferrocyanide ion system at room temperature. (for P.G.).

III. Experiments with pH-meter

Experiment 12

Determination of strengths of hydrochloric acid and acetic acid in their mixture pH-metrically.

Experiment 13

Determination of ionic product (K_w) of water at room temperature pH-metrically.

Experiment 14

Determination of hydrolytic constant (K_h) of ammonium chloride solution pH-metrically. [for P.G.]

Experiment 15

Determination of hydrolytic constant (K_h) of aniline hydrochloride using a *pH*-meter.

Experiment 16

Determination of pK_a-values of a dibasic acid (e.g., oxalic acid) pH-metrically at room temperature [for P.G.].

Experiment 16A

Determination of pK_a-values of phosphoric acid pH-metrically at laboratory temperature. [for P.G.]

IV. APPENDIX

Table 1. Standard reduction potentials (E^o) of some common redox systems at 25 °C and 1 atm. in volts.

Table 2. Formal potentials ($E^{o'}$) of some redox systems in different medium at 25 °C.

Table 3. Ionic equations for use in the calculation of relative equivalent mass (equivalent weight) of some oxidising and reducing agents.

Table 4. Dissociation constants of some acids in water at 25 °C.

Table 5. pH of some standard buffer solutions at 25 °C.

Table 6. pH range of some buffer solutions.

CHAPTER 6

Potentiometry and pH-metry

I. Introductory Words

Electrode Potentials

When a metal is immersed in a solution containing its own ions, a potential difference is established between the metal and the solution. The potential difference E for an electrode

$$M^{n+} + ne \rightleftharpoons M$$

is given by the expression:

$$E = E^{\circ} + \frac{RT}{nF} \ln a_{M^{n+}}, \tag{1}$$

where R is the universal gas constant, T is the absolute temperature, F is the Faraday constant, n is the number of electrons transferred, $a_{M^{n+}}$ is the activity of the ions in the solution, and E° is a constant dependent upon the metal. Equation (1) can be simplified by introducing the known values of R and F, and converting natural logarithms to base 10 by multiplying 2.3036; it then becomes

$$E = E^{\circ} + \frac{0.0001984T}{n} \log a_{M^{n+}}.$$

For a temperature of 25 °C (T = 298 K)

$$E = E^{o} + \frac{0.0591}{n} \log a_{M^{n+}}. \tag{2}$$

For many purposes in quantitative analysis, it is sufficiently accurate to replace $a_{M^{n+}}$ by $C_{M^{n+}}$ (the concentration of the ions in moles per litre), so equation (2) becomes

$$E = E^{o} + \frac{0.0591}{n} \log C_{M^{n+}}. \tag{3}$$

It is the form of the Nernst equation.

If in equation (2), $a_{M^{n+}}$ is put equal to unity, E is equal to E^{o}. E^{o} is called the *standard electrode potential of the metal*. Both E and E^{o} are expressed in volts.

In order to determine the potential difference between an electrode and a solution, it is necessary to have another electrode and solution of accurately known potential difference. The two electrodes can then be combined to form a Voltaic cell, the *e.m.f.* of which can be directly measured. The *e.m.f.* of the cell is the difference of the electrode potentials at zero current. The value of the unknown potential can then be calculated. By potentiometric measurements, however, we determine the *e.m.f* of a cell, not the potential of an electrode.

Formal potentials

In the Nernst equation (3), the standard electrode potentials (E^o) are evaluated with respect to the unit activity and with all ions expressed in simple forms. But during the experimental condition, the solution may be quite concentrated making activity and concentration unequal and further may contain the ions not in their simple forms but in the form of complex ions. Under such conditions the activities of the species are smaller than their concentrations. The actual species involved may often differ from those to which the standard electrode potentials refer. Thus, standard electrode potentials which are determined under the actual experimental conditions, called formal potentials. These are denoted as $E^{o\prime}$. The factors which affect the formal potentials are the concentrations of the species, pH of the solution, any complexation in solution, etc. For example, Fe^{3+}/Fe^{2+} system has $E^o = +0.77$ volt. But the formal potentials in different acid media are: + 0.73 V in 1(M) $HClO_4$, + 0.70 V in 1(M) HCl, + 0.68 V in 1(M) H_2SO_4 and + 0.61 V in 0.5(M) H_3PO_4 + 1(M) H_2SO_4.

Reference electrodes

A reference electrode is that which has an accurately known electrode potential at 298 K and is used to find the single electrode potentials of other electrodes. The choice of a particular reference electrode depends upon the nature of solution used. It is often desirable to choose an electrode which will not require a salt bridge. In general, there are two types of reference electrodes.

1. Primary reference electrode.
2. Secondary or subsidiary reference electrode.

1. Primary reference electrode (Standard hydrogen electrode)

Standard hydrogen electrode whose electrode potential is arbitrarily fixed as 0.00 V at 298 K, is employed as primary reference electrode. Different types of hydrogen electrodes are available but the simple and most common type of hydrogen electrode is due to Hildebrand type, shown in Fig. 6.1. It consists of a glass tube, in lower end of which a platinum wire with a platinum foil, coated with platinum black, is used. This platinum foil is surrounded by hydrogen gas at one atmosphere (at unit activity) and it is dipped into a solution containing H^+ ions, say HCl, of unit activity (i.e., $a_{H^+} = 1$). For all practical purposes, unit activity of hydrogen gas is taken in approximation to be one bar pressure and unit activity of H^+ ions is replaced by 1.0 (M) concentration.

Thus standard hydrogen electrode is represented as

$$H^+(aq, a=1)\,|\,H_2(g, a=1)\,|\,Pt$$

$$\text{or, } H^+(1.0M)\,|\,H_2\,1\text{ bar}\,|\,Pt$$

The hydrogen gas is passed through the side inlet as shown in Fig. 6.1 and bubbled into the acid solution. After 10-15 minutes the electrode assumes the constant potential independent of the rate of bubbling. The platinum foil absorbs a part of hydrogen and the rest comes out through the holes. Thus, an equilibrium between the absorbed hydrogen on the electrode surface and the ions in the solution, is established, i.e.,

$$H_2(g) \rightleftharpoons 2H^+(aq) + 2e$$

By convention the potential of this standard hydrogen electrode is arbitrarily fixed at zero. Standard hydrogen electrode is used as primary reference electrode to determine the relative values of electrode potentials (reduction) of other electrodes as there is no method to find out the absolute value of single electrode potential. Also this electrode finds its use in potentiometric titrations.

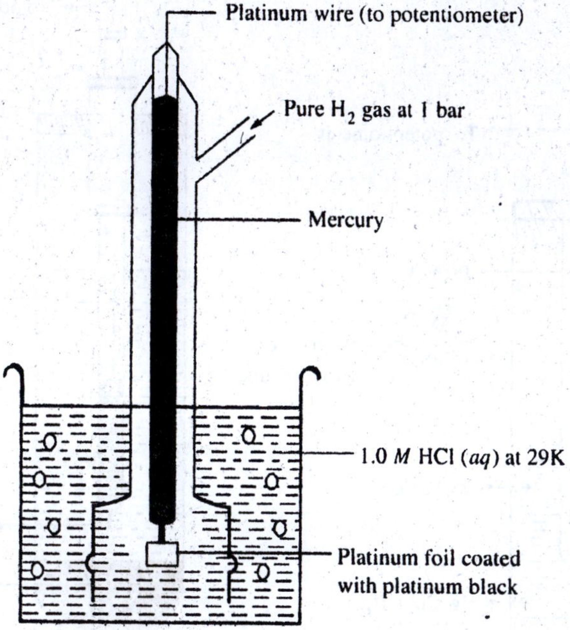

Fig. 6.1: Standard hydrogen electrode

2. **Secondary reference electrodes**

In using standard hydrogen electrode as a reference electrode in the laboratory, many difficulties arise. For example, it is very difficult to set up and maintain unit activity of H^+ions in solution, and pressure of H_2 gas at 1.0 bar. It cannot be used in presence of air, oxygen, oxidising and reducing agents. It is potential fire and explosion harzard. The coating of Pt-black on Pt-electrode is easily poisoned due to absorption of impurities and it disturbs the equilibrium between $H_2(g)$ and ions in the solution.

For the sake of simplicity, many other secondary or subsidiary reference electrodes whose single electrode potentials are accurately known on the standard hydrogen electrode scale are used in place of standard hydrogen electrode. It is always better to select an electrode in which salt bridge is not used. A few secondary reference electrodes are: (i) Calomel electrode, (ii) The quinhydrone electrode, (iii) Ag-AgCl electrode, (iv) Glass electrode.

Here, we are giving brief description of these electrodes.

- ***Calomel electrode***

A common form of calomel electrode is shown in Fig. 6.2. It consists of a glass tube having a side at each end. A small quantity of pure mercury is placed at the bottom of the vessel, which is then covered with a calomel paste of pure mercurous chloride, pure mercury and potassium chloride. The calomel paste is prepared by rubbing mercury and calomel (Hg_2Cl_2) in a glass mortar with a small quantity of saturated KCl solution. The paste is washed many times with saturated KCl solution, the mixture is allowed to stand until the calomel has settled and the clear liquid is decanted off.

The vessel is now filled with the saturated KCl solution, by sucking it through the bent tube fitted with a rubber tube and the rubber tube is closed by folding and fitting a pinch-cock. The concentration of KCl solution may be normal, deci-normal or saturated. A glass tube with a platinum wire sealed into lower end, is fitted into the electrode vessel so that platinum wire dips in mercury layer. The electrical connection is made by the help of this platinum wire.

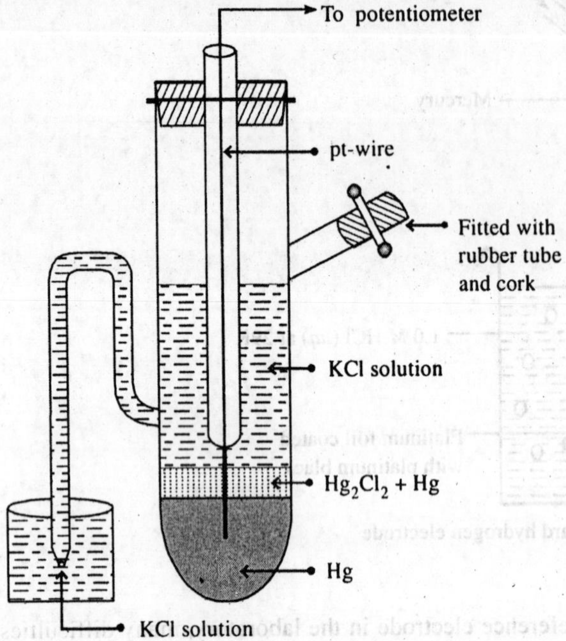

Fig. 6.2a: Calomel electrode

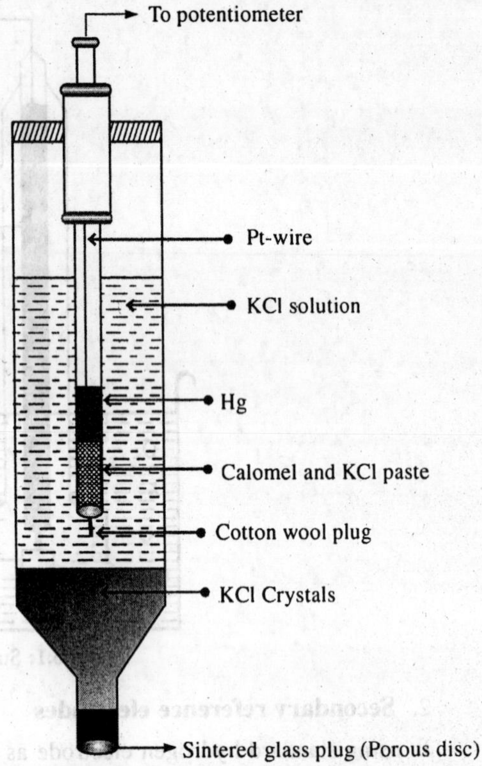

Fig. 6.2b: Dip type calomel electrode

A dip type calomel electrode which is very convenient and now widely used is shown in Fig. 6.2b. The calomel electrode of the type 6.2b is ready-made and commercially available.

Thus, a calomel electrode is represented as

$Hg(l) \mid Hg_2Cl_2(s) \mid KCl\,(aq)$, for which the half-cell reaction is

$$Hg_2Cl_2(s) + 2e \rightleftharpoons 2Hg(l) + 2Cl^-(aq).$$

According Nernst equation the electrode potential (e.m.f.) is given by

$$E_{cal} = E^o - \frac{RT}{F}\ln a_{Cl^-}.$$

Thus the potential of the calomel electrode dependents on the activity of Cl^- ions, i.e., concentration of KCl solution used for preperation of the electrode. The potentials on hydrogen scale of three reference calomel electrodes are given below:

$$Hg|Hg_2Cl_2(s)|0.1(M)\,KCl;\ E_{cal} = 0.3338 + 0.00007\,(t - 25)\text{ volt}$$

$$Hg|Hg_2Cl_2(s)|1.0\,(M)\,KCl;\ E_{cal} = 0.2800 + 0.00024\,(t - 25)\text{ volt}$$

$$Hg|Hg_2Cl_2(s)|\text{Saturated KCl};\ E_{cal} = 0.2415 + 0.00076\,(t - 25)\text{ volt}$$

where E_{cal} is the reduction potential of calomel electrode and t is the temperature at °C.

The deci-molar calomel electrode [0.1 (M) KCl solution is used] has the lowest temperature coefficient and is preferred for accurate work. However, saturated calomel electrode is the most convenient due to the ease of replacing the solution and is thus frequently employed. Commercially available calomel electrodes [Dip type shown in Fig-6.2b] are nomally supplied with saturated KCl solution.

Preservation of calomel electrode in the laboratory: It is very important work to preserve calomel electrodes in the laboratory after use. Generally, saturated calomel electrode, after use, should be washed with distilled water and then the outer end should be immersed into a saturated KCl solution. Otherwise the electrode will be dried and it will not function. Never it is kept immersed into distilled water. If it is done, KCl solution will enter from the electrode to the distilled water. So, the concentration of KCl solution in the electrode will change and its potential will be inaccurate.

- ***Quinhydrone electrode***

The quinhydrone electrode consists of an inert metal like platinum or gold dipped into a solution of unknown pH which is saturated with quinhydrone. The latter is an organic compound which sparingly dissolves in water to give an equilibrium mixture of two non-electrolytic compounds, hydroquinone and quinone i.e.,

$C_{12}H_{10}O_4$	$\rightleftharpoons$	$C_6H_4O_2$	+	$C_6H_6O_2$
$(C_6H_4O_2 \cdot C_6H_6O_2)$		Quinone		Hydroquinone
Quinhydrozne		(Q)		(H_2Q)

Since the aqueous solution of quinhydrone of pH < 7 contains equimolar amounts of quinone and hydroquinone, we have

$$[\text{Quinone}] = [\text{Hydroquinone}]$$

As the solution is dilute, we can say that in the solution of quinhydrone,

$$a_{\text{Quinone}} = a_{\text{Hydroquinone}} \qquad \text{or,} \qquad \frac{a_{\text{Quinone}}}{a_{\text{Hydroquinone}}} = 1.$$

Now hydroquinone, being a weak acid, it dissociates in solution and the system involves the following reversible oxidation-reduction equilibrium,

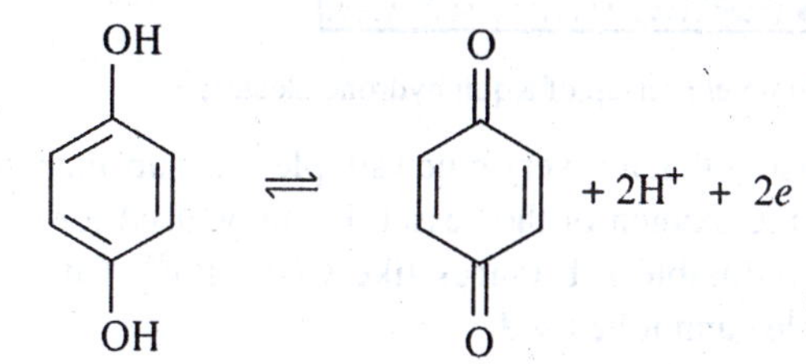

Hydroquinone Quinone

i.e. $H_2Q \rightleftharpoons Q + 2H^+ + 2e$

Reversing the above equation the electrode reaction (reduction) is given by

$$Q + 2H^+ + 2e \rightleftharpoons H_2Q$$

If a clean Pt-electrode is dipped into the solution the electrode is

$$Q/H_2Q\,(\text{Pt})$$

The potential developed on platinum electrode dipped in this system is given by

$$E = E^o + \frac{RT}{2F}\ln\frac{a_Q \cdot a_{H^+}^2}{a_{H_2Q}}$$

$$= E^o + \frac{RT}{F}\ln a_{H^+} \qquad \left[\because \frac{a_Q}{a_{H_2Q}} = 1\right]$$

$$= E^o + 0.0591 \log a_{H^+} \text{ at } 25\ ^\circ C$$

$$= E^o - 0.0591 \text{ pH}$$

The value of standard reduction potential (E^o) of quinhydrone electrode is found to be 0.6994 volt at 25 °C. Hence,

$$E = 0.6994 - 0.0591 \text{ pH at } 25\ ^\circ C.$$

This shows that e.m.f. of the quinhydrone electrode depends upon pH value of the solution, i.e., quinhydrone electrode behaves as a reversible hydrogen electrode.

This electrode may be set up easily by adding a pinch of quinhydrone (e.g., 1 gm per 100 ml) to the solution whose pH is to be determined and insert a clean platinum electrode in the solution as shown in the Fig. 6.3.

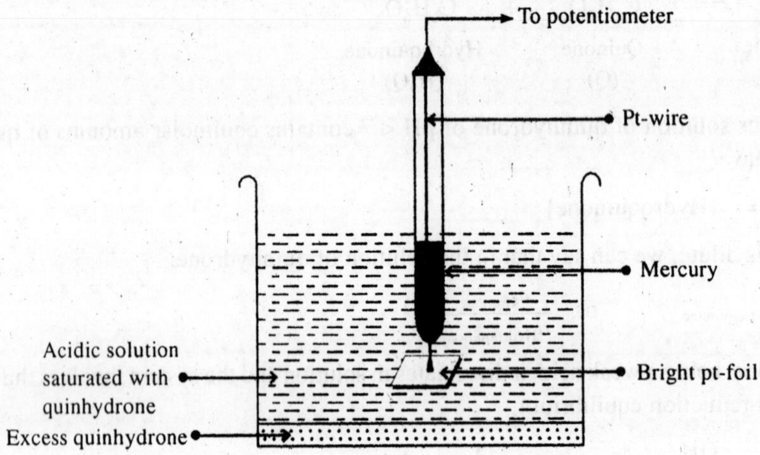

Fig. 6.3: Complete set-up of a quinhydrone electrode

Advantages of quinhydrone electrode is that it is very much simple to set up and it attains equilibrium rapidly. It is not necessary to remove the oxygen of the vessel. It can be used for determining the pH-values of solutions containing even reducible substances like Cu^{2+}, Pb^{2+}, Cd^{2+}, HNO_3, alkaloids, unsturated acids for which H_2 electrode cannot be used.

Quinhydrone electrode gives accurate pH values for solution having pH less than 8. Above this pH, quinone may be attacked by OH^- and hydroquinone may be attacked in O_2. As a result of this, the redox equilibrium is disturbed and the ratio $\frac{a_Q}{a_{H_2Q}}$ remains no longer unity. In fact, for solution of pH more than 8, it acts as an anode with respect to saturated colomel electrode and the potentiometer shows negative e.m.f. values.

- **Silver-silver chloride electrode**

More recently silver-silver chloride electrode has been used as a reference electrode, specially in connection with accurate measurements in chloride solutions and thus the use of liquid junction potential is avoided.

It consists of a silver wire or silver plated platinum wire, coated electrolytically with a thin layer of silver chloride, dipping into a solution of KCl or HCl of known concentration which is saturated with silver chloride. This is achieved by the addition of two or three drops of 0.1(M) silver nitrate solution and shaking.

The electrode may be represented as,

$Ag/AgCl(s)/Cl^-(eq)$ which is shown in the following Fig. 6.4

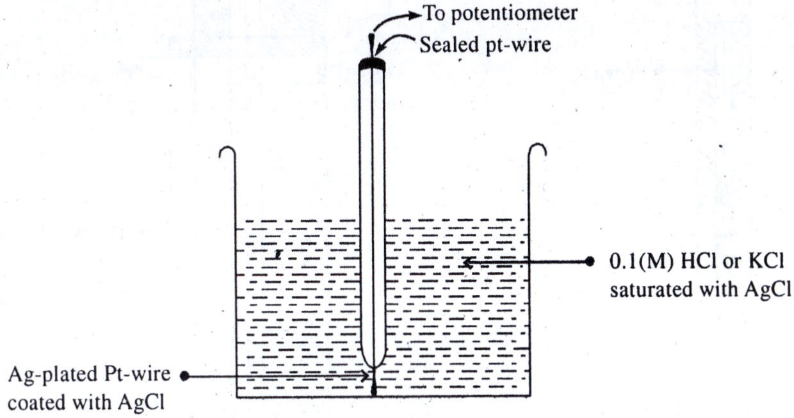

Fig. 6.4: Silver-silver chloride electrode

The net electrode reaction is

$$AgCl(s) + e \rightleftharpoons Ag(s) + Cl^-$$

for which the potential is given by

$$E = E^o - \frac{RT}{F}\ln a_{Cl^-}$$

Thus, this electrode is also reversible (like calomel electrode) with respect to Cl^- ions (anions) and the potential of the electrode is governed by the activity of chloride ions in the potassium chloride solution. The standard reduction potential (E^o) of Ag-AgCl(s) electrode is 0.223 volt at 25 °C.

Commercial forms of the electrode are available and in general are similar to the calomel electrode depicted in Fig. 6.2b with the replacement of the mercury by silver electrode, and calomel by silver chloride. The use of the sintered disc and ion exchange membranes are equally applicable to the silver-silver chloride electrode.

- **Glass electrode**

The glass electrode is most widely used hydrogen ion responsive electrode, and its use is dependent upon the fact that when a glass membrane is immersed in a solution containing hydrogen ions, a potential is developed which is a linear function of the hydrogen ion concentration, i.e., pH of the solution.

The glass electrode consists of a very thin-walled glass bulb, the glass being of relatively low melting point and high electrical conductivity, and it is blown at the end of the glass tube. A solution of 0.1(M) HCl, which furnishes a constant hydrogen ion concentration is kept inside the bulb and a silver-silver chloride electrode (or calomel) is immersed into it to make electrical contact as shown in Fig. 6.5(a).

Glass electrodes are now available as combination electrodes which contain the indicator electrode (a thin glass bulb) and a reference electrode (silver-silver chloride) combined in a single unit as shown in Fig. 6.5(b). The thin glass bulb and the narrow tube to which it is attached are filled which 0.1 (M) hydrochloric acid and carry a silver-silver chloride electrode. The wide tube is fused to the lower end of the narrow tube and contains saturated potassium chloride solution which is also saturated with silver chloride. The wide tube contains saturated KCl solution, also carries a silver-silver chloride electrode. The assembly is sealed with an insulating cap.

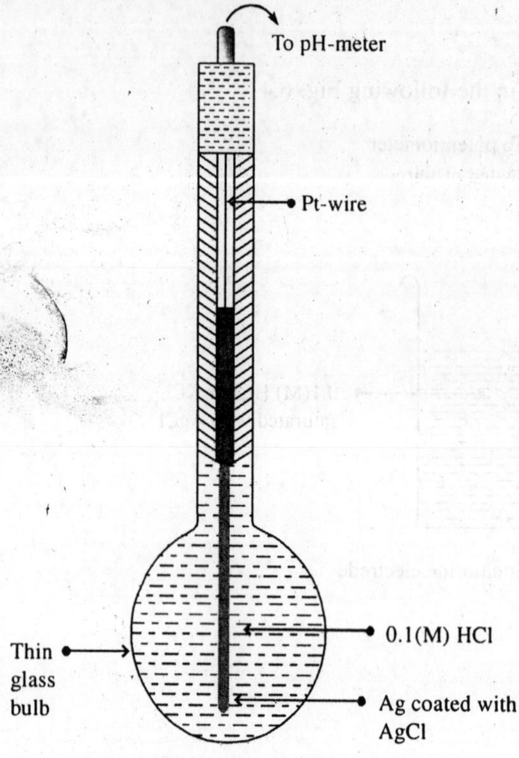

Fig. 6.5(a): Glass electrode

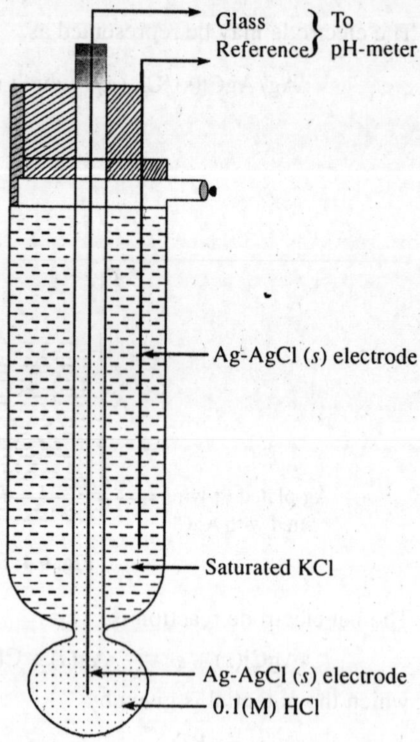

Fig. 6.5(b): Combined electrode (glass + reference)

Combination electrodes may be purchased in which the glass electrode and the reference electrode (silver-silver chloride or saturated calomel) are combined into a single unit, thus giving a more robust piece of equipment, and the convenience of inserting and supporting a single probe in the test solution instead of the two seperate components.

Before use of a freshly made glass electrode, it should be immersed in distilled water for some hours (at least twelve hours) and it is concluded that a hydrated layer is formed on the glass surface, where an ion exchange process can take place. If the glass contains sodium, monovalent $Na^{\oplus}$ ions at and near the surface of the glass are replaced by $H^{\oplus}$ ions. As a result of this, when a glass electrode is immersed in a solution, an equilibrium between the $H^{\oplus}$ ions in the glass surface and the $H^{\oplus}$ ions in the solution is attained. This equilibrium can be represented as

$$H^+_{\text{solution}} + Na^+_{\text{glass}} \rightleftharpoons H^+_{\text{glass}} + Na^+_{\text{solution}}$$

This exchange of charge produces a potential difference between the glass and the solution, and this difference in potential is the potential of the glass electrode which depends on pH of the solution. The e.m.f. of glass electrode is given by

$$E_G = E_G^o + 0.059 \log a_{H^+}$$

$$= E_G^o - 0.059\text{pH at } 25\ ^oC \qquad (1)$$

E_G^o is unknown, and depends partly upon the nature of the glass used in the construction of the membrane, and partly upon the individual character of each electrode. So in each case, the value of E_G^o is to be ascertained. Any way, the glass electrode acts as reversible with respect to $H^{\oplus}$ ions in the same manner as standard hydrogen electrode. Thus it is obviously used for pH measurements in combination with a suitable reference electrode, e.g., saturated calomel electrode. The pH of a solution may be measured using a cell

$$\text{Glass electrode}\,|\,\text{test solution}\,\|\,\underset{\text{(saturated)}}{\text{KCl},\ Hg_2Cl_2(s)}\,|\,\text{Hg}$$

$$\text{i.e.,}\quad \text{Ag}\,|\,\text{AgCl}(s),\ 0.1\text{M HCl}\,|\,\text{Glass}\,|\,\text{test solution}\,\|\,\underset{\text{(saturated)}}{\text{KCl},\ Hg_2Cl_2(s)}\,|\,\text{Hg}$$

The E^{o}_{G} is ascertained by measuring the e.m.f. using buffer solutions of known pH, e.g., pH-7 and pH-4 or 9.2. This is called calibration or standardisation of the instrument.

For the above cell,

$$E = E_{cal} - E_G = 0.2415 - E^{o}_{G} + 0.059\ \text{pH at } 25\ ^{o}\text{C} \qquad (2)$$

Then, the solution of unknown pH is used and pH of that solution is ascertained. Since glass is a material of high resistance, ordinry potentiometer cannot be used. Instead, high resistance electronic digital voltmeters (called pH-meters) are used which directly read pH of the solution.

- **Preservation of glass electrode in the laboratory**

The glass electrode should be thoroughly washed with distilled water after use and then kept immersed into a solution of 4(M) KCl covering the mouth of the vessel (it is easy to use a 100 ml stoppered reagent bottle so that the electrode is dipped into 4 (M) KCl in the bottle and the mouth of the bottle is sealed with cork to prevent evaporation of water from KCl solution).

It may be kept immersed into a buffer solution of pH-4 but buffer solution should be changed and fresh buffer solution must be used after seven days. Otherwise a fungus is grown into the buffer solution and longevity of the electrode will fall. Do not keep it dipping in alkaline solution (pH > 7) for a long time. If it is done, the function of glass electrode will be lost.

The glass electrode should not be allowed to become dry. If it becomes dry, it will not return to its responsive condition when immersed in distilled water even for a long time. For details about preservation and technique of its use, follow the manual supplied by the manufacturing company from where the electrodes have been purchased.

- **The convention of representation of cells and calculation of cell e.m.f.**

All cells are constituted of two half-cells, i.e., two electrodes. Of the two electrodes, the electrode having higher standard reduction potential value should be placed on the right-hand side and it is regarded as positive electrode and the electrode having lower standard reduction potential value should be placed on the left-hand side and it is regarded as negative electrode. The two half cells are connected through a salt bridge to eliminate liquid junction potential and it is indicated by a double vertical lines (||). Thus a complete cell may be represented as

$$\overset{(-)}{\text{Left-Hand Electrode (L.H.E.)}}\ \|\ \overset{(+)}{\text{Right-Hand Electrode (R.H.E.)}}$$

Here, $E^{o}_{R} > E^{o}_{L}$, where E^{o}_{R}, E^{o}_{L} are the standard reduction potential values of the R.H.E. and L.H.E. respectively.

If E_R and E_L be the electrode potentials (in reduction form) of the R.H.E. and L.H.E., then the observed e.m.f. of cell (E_{cell}) will be

$$E_{Cell} = R_R - E_L$$

E_R and E_L may be calculated using Nernst equation. The idea will be clear to us if we take one example.

We know $E^o_{Zn^{2+}/Zn} = -0.76$ volt, and $E^o_{Cu^{2+}/Cu} = +0.337$ volt at 25 °C.

Since $E^o_{Cu^{2+}/Cu} > E^o_{Zn^{2+}/Zn}$, so the cell may be represented as

$$\underset{L}{\overset{(-)}{Zn|Zn^{2+}}} \| \underset{R}{\overset{(+)}{Cu^{2+}|Cu}}$$

At L.H.E. $Zn \rightarrow Zn^{2+} + 2e$

$\therefore$ from Nernst equation $E_L = E^o_{Zn^{2+}/Zn} + \frac{0.059}{2} \log a_{Zn^{2+}}$ at 25 °C

At **R.H.E.** $Cu^{2+} + 2e \rightarrow Cu$

$$\therefore \quad E_R = E^o_{Cu^{2+}/Cu} + \frac{0.059}{2} \log a_{Cu^{2+}}$$

$$\therefore \quad E_{Cell} = E_R - E_L = \left(E^o_{Cu^{2+}/Cu} - E^o_{Zn^{2+}/Zn}\right) + \frac{0.059}{2} \log \frac{a_{Cu^{2+}}}{a_{Zn^{2+}}}.$$

Thus the observed e.m.f. of the cell will depend upon the activities or concentrations of Zn^{+2} and Cu^{+2} ions in the two half-cells.

• Measurement of cell e.m.f.

The electromotive force (e.m.f.) of a cell is the potential difference responsible for the flow of current from one electrode of higher potential to the other of lower potential. It is expressed in volts.

The most satisfactory method for measurement of the e.m.f. of a cell was, for many years, known as Poggendorff's compensation method. The principle of the method is that the e.m.f. of the cell is just balanced by an equal and opposite difference so that no current flows. This consists in balancing the unknown e.m.f. against a known potential drop which can be varied. This involves the use of potentiometer circuit (Fig. 6.6).

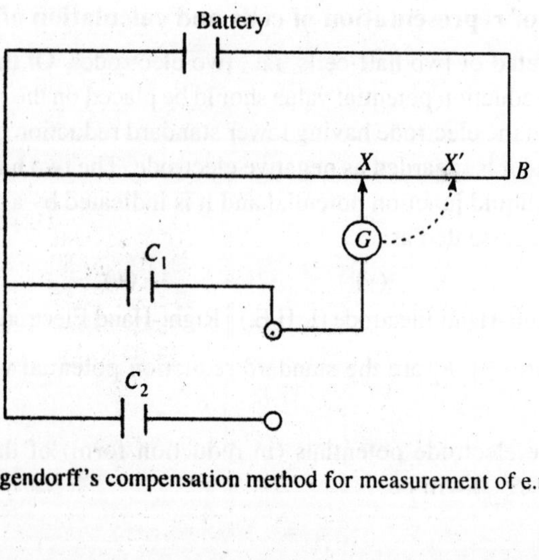

Fig. 6.6: Poggendorff's compensation method for measurement of e.m.f.

Battery, usually a storage cell (2 volts), is the steady source of e.m.f. (large cell) greater than the e.m.f. to be measured, which sends a current in the calibrated resistance AB with a movable contact X. AB is known as potentiometer wire, made up of platinum or platinum-iridium of high resistance. If no current passes through the contact X, then there is uniform potential gradient along AB and the potential difference across any section of it depends upon the resistance of that section.

Now, if the cell C_1 of e.m.f. E_x is joined through a galvanometer G to the sliding contact X in AB and point X is adjusted so that there is no flow of current which will be indicated by no deflection state in galvanometer. In this stage, the fall of potential along AX due to battery is exactly compensated by the e.m.f. of C_1.

It is also essential to use a standard cell C_2 of correctly known e.m.f. (E_s), since the e.m.f. of battery is not accurately known. By means of suitable switch the cell C_1 is replaced by standard cell C_2 and the point of balance X' is noted. Here, the fall of potential along AX' will be equal to E_s. Thus, we have the relation

$$\frac{\text{e.m.f. of } C_1}{\text{e.m.f. of } C_2} = \frac{E_x}{E_s} = \frac{\text{Resistance } AX}{\text{Resistance } AX'}$$

This relation can be used for calculation of e.m.f. (E_x) of unknown cell. However, during measurement of E_x, calculation using the above relation is not necessary. In the ordinary manual potentiometer, a scale (in volts or mV) is fitted in connection with the sliding contact which is actually a movable knob. After balancing the standard cell (E_s), the unknown cell (E_x) is connected to the potentiometer and the galvanometer will show deflection. The movable knob is then adjusted at balance point. At this point e.m.f. (E_x) can be directly read from the scale.

• Standard Cell

The standard cell of accurately known e.m.f. is the essential feature of the Poggendorff's compensation method for measuring e.m.f. The standard cell cannot be used for getting supply a current but used for a comparison of e.m.f.

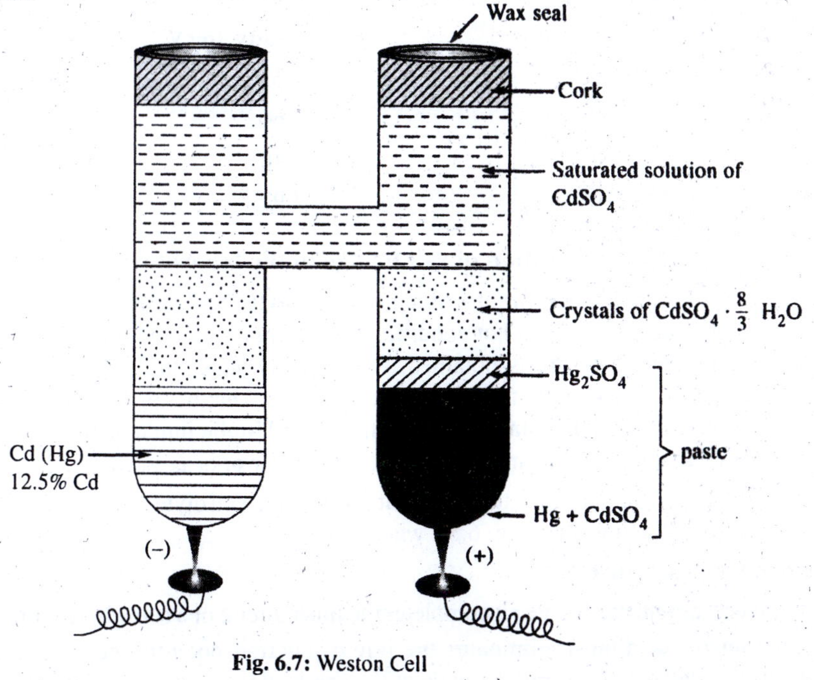

Fig. 6.7: Weston Cell

A standard cell is that which is capable of giving constant and reproducible e.m.f. independent of temperature changes.

The most convenient and one generally employed standard cell is the Weston cell (Fig. 6.7) in which the mercury forms the positive pole and a 12.5% cadmium amalgam is the negative pole. The electrolyte is a saturated solution of hydrated cadmium sulphate ($3CdSO_4 \cdot 8H_2O$). In order to prevent the loss of water by evaporation the limbs are sealed with wax. The cell reaction may be respresented as below:

$$Cd + Hg_2^{++} \rightleftharpoons Cd^{++} + 2Hg$$

The positive pole is usually covered with a paste of sparingly soluble mercurous sulphate and cadmium sulphate in mercury.

The cell may be represented as

$$12.5\%\ Cd\ in\ Hg \mid CdSO_4 \cdot {}^8\!/_3 H_2O\,(s)\ \text{saturated solution: } Hg_2SO_4\,(s) \mid Hg$$

The e.m.f. of cell is 1.0183 volt at 20 °C and at any other temperature (t) it may be calculated by

$$E = 1.0183 - 0.0000406\,(t - 20)\ \text{volt.}$$

The standard cell must never be kept in circuit for a long time, otherwise its e.m.f. will no longer be constant.

Salt bridge

A salt bridge is a U-shaped glass tube containing a set jelly of agar-agar and viscous concentrated solution (saturated) of a suitable strong electrolyte whose ions are of nearly same ionic mobility or of almost equal transport numbers in aqueous solution and not reached with the two electrolytic solutions of the cell.

The strong electrolytes which are generally used in the preparation of salt bridge are potassium chloride or nitrate or ammonium nitrate. These salts are used in the preparation of a salt brigde, as the cations and anions of those salts have almost equal ionic mobilities as evident from the following data:

Ions	Mobility ($m^2V^{-1}\ sec^{-1}$) × 10^8
K^+	7.62
NH_4^-	7.60
NO_3^-	7.40
Cl^-	7.92

- **Preparation and preservation of salt bridge in the laboratory**

Prepare 50 ml saturated solution of a suitable electrolyte (KCl or KNO_3 or NH_4NO_3) by heating. Add 1.5 gm of agar-agar powder in hot condition into it. Again boil the mixture carefully and stir on a steam bath till the clear solution is obtained. Immediately suck the viscous liquid into the U-tube through a rubber tube in one end of the U-tube carefully so that no air bubbles are present inside the U-tube and the tube is filled up completely. Dip the filled U-tube into a vessel contains cold water for half a minute so that the jelly becomes solidified. The tube works as a salt bridge.

However, it should be kept in saturated solution of the electrolyte which is used in the salt bridge preparation in the following way (Fig. 6.8), when not in use.

- **Function of a salt bridge**

i. It connects the electrodes, i.e., it completes the inner circuit of a cell whose e.m.f. is to be measured.
ii. It minimizes or almost eliminates the liquid junction potential occurring at the interface of electrolyte salt solutions and the ends of the salt bridge due to diffusion of ions.

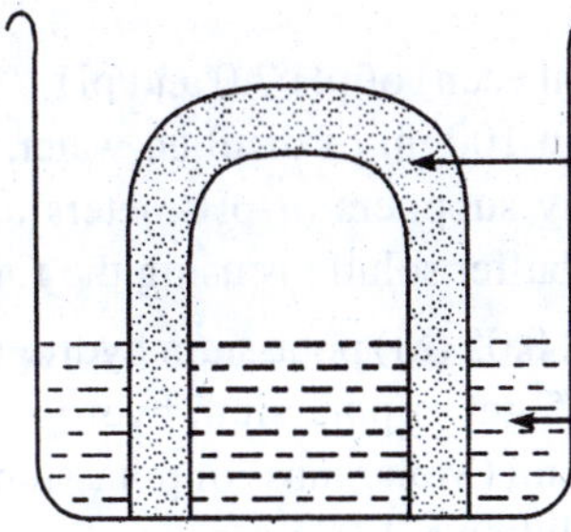

Fig. 6.8: Preservation of a salt bridge

However, in potentiometric titration, the type of salt bridge that should be used depends upon the nature of titration. For example, in acid-base titration KCl-agar salt bridge or KNO_3-agar salt bridge or NH_4NO_3-agar salt bridge may be used. But when chlorides or halides are titrated KCl-agar salt bridge can never be used. Instead, KNO_3-agar or NH_4NO_3-agar salt bridge may be used.

• Calibration or standardisation of a digital potentiometer

Nowadays, digital electronics potentiometer is available in which Poggendorff's compensation method of measurement of e.m.f. of a cell is somewhat modified by electronic circuit. These digital potentiometers should be standardised before measuring e.m.f. of a cell. The methods of standardisation or calibration of these meters are different for different potentiometers supplied by the manufacturers.

However, two general methods, which follow the manual of most of potentiometers are briefly described:

i. Connect the meter's plug to 220-230 V electric line and on the switch of the line. Then if available set the ON/OFF switch of the meter. Connect the electrodes of the cell to the meter through jack and then set the switch-in/switch-out of the meter. The display of the meter will show 1.018 V. If the reading deviates from 1.018V, then adjust it exactly 1.018 V by moving the knob "adjust" or "calibrate". After adjustment of the meter, again set the switch-in/switch-out of the meter. The display will show 0.000V if the electrodes are immersed into distilled water. Then the instrument is ready for measurement of e.m.f. of the unknown cell. Just connect the cell and note the readings from the display.

ii. Some instruments require standard buffer solution of pH 7.0 for calibration. Prepare 100 ml of buffer solution by dissolving a buffer capsule completely. Connect a glass electrode and a calomel (saturated) electrode to the potentiometer and switch on. Immerse the electrode assembly into the buffer solution (pH 7.0). The display will show e.m.f. 0.000 V ($\pm$ 0.001 is acceptable). If the readings differ from 0.000 V, adjust the 'calibrate' knob to a 0.000 V. If it is done, then the meter is ready for measurement of e.m.f. of the test cell. Remove the buffer solution and electrode assembly and connect the test cell (whose em.f. is to be measured). The display will show direct e.m.f. readings of the cell in volts or millivolts.

However, some instruments have scale expansion facility by which the same e.m.f. values can be read up to one digit, two digits, three digits. Obviously three-digit reading will be more accurate. For details about the standardisation or calibration and procedure of e.m.f. measurement, follow manual booklet supplied by the manufacturing company along with the instrument. If necessary take a demonstration about the instrument for handling it.

Standardisation or calibration of a pH-meter and measurement of pH of unknown (test) solution

There are two general procedures for standardisation of a pH-meter and measurement of pH:

i. For manual pH-meters

Prepare two buffer solutions (100 ml each) of pH 7.0 and pH 4.0 or 9.2 by dissolving respective buffer capsules (or buffer tablets) in 100 ml of distilled water. These buffer capsules or buffer tablets may be obtained from many suppliers of pH-meters and laboratory supply houses. If these are not available prepare the buffer solutions using the following procedure:

a. **pH 4.00 buffer solution:** Exact 0.05 (M) potassium hydrogenphthalate solution has pH 4.00 at 25 °C. Dissolve exactly 1.025 gm of potassium hydrogenphthalate in distilled water and the final volume is made to 100 ml (Molar mass of potassium hydrogenphthalate, $C_8H_5O_4K$, is 205. So, 100 ml 0.05 (M) solution = 1.025 gm).

b. **pH 7.00 buffer solution:** A mixture of 6.008 gm of citric acid, 3.893 gm of potassium dihydrogenphosphate, 1.769 gm of boric acid, 5.266 gm of pure diethylbarbituric acid is dissolved in distilled water and the volume is made up to 1000 ml. If we prepare 100 ml of the mixture, then the above components are mixed in 1/10th amount of each. Prepare an exact 0.2 (M) sodium hydroxide solution (100 ml) by usual procedure. Mix 50.6 ml of 0.2 (M). NaOH solution with 49.4 ml of the above acid mixture to prepare 100 ml of Universal buffer mixture whose pH is 7.00 at 25 °C.

c. **pH 9.2 buffer solution:** Exact 0.05 (M) borax aquous solution has pH 9.18 at 25 °C Relative molar mass of borax, $Na_2B_4O_7 . 10H_2O$, is 381.28. So, dissolve exactly 1.9064 gm of borax in distilled water and volume is made up to 100 ml. The pH of the resulting solution is 9.18 at 25 °C.

Switch on and allow the instrument to warm up for about 15 minutes. Adjust the temperature control to the temperature of the solution. If automatic control is available, then the display will show the solution temperature when temperature probe is connected.

Insert the electrode assembly into the buffer solution of pH = 7 in such a way that the glass bulb of the glass electrode is completely dipped into the solution, and adjust the set buffer control so that the meter reading agrees with the known pH of the buffer solution. Remove the electrode assembly and wash well with distilled water. Rinse the electrode assembly with the second buffer solution of known pH (say, 4.00 or 9.2) and then immerse into the second buffer solution. If the meter reading does not agree exactly with the known pH, adjust the 'slope' control until the required reading is obtained. Remove the electrode assembly, rinse in distilled water, place in the first buffer solution and confirm that the correct pH reading is shown on the meter. If not, repeat the calibration procedure.

If the calibration is satisfactory, wash the electrodes with distilled water and introduce into test solution contained in a small beaker. Read out the pH of the solution.

Remove the electrodes, wash well with distilled water, and leave standing in 4(*M*) potassium chloride solution.

ii. For automatic μ pH systems

Insert the electrode assembly into a buffer solution of PH = 7.0 contained in a small beaker. Then push the 'STD' switch. The display of the meter will show 'PUT pH 4.0/9.2'. (That means pH 7.0 is adjusted by the meter. If not adjusted, the display will show 'Error'. This 'Error' is due to the bad electrode or improper buffer solution). Remove the electrode assembly, first wash well with distilled water and then rinse with buffer solution of pH 4.0. Immerse the electrode assembly into the buffer solution of pH 4.0 contained in a small beaker. Again push the 'STD' switch. After sometime, the instrument will show 'STD OVER'. (It means standardisation procedure is over satisfactorily. Here also the instrument may show 'Error' due to improper buffer solution).

Any way, if standardisation is satisfactory, then leave the electrode assembly, wash well with distilled water first and then rinse with the test solution. Immerse the electrode assembly into the test solution contained in a small beaker and push the 'pH' switch of the instrument. The meter will directly show pH of the test solution along with the temperature of the solution.

- **Regeneration of a 'bad' or 'dead' glass electrode**

A glass electrode should be kept immersed in 4(M) KCl solution when not in use. If its function is lost, it may be regenerated in two ways:

a. Allow the glass electrode to immerse in 0.1 (N) HCl solution for 24 hours. After that the electrode is removed from HCl solution, wash with distilled water. It may work again.

b. If the above method is not satisfactory, immerse the glass bulb in 2% hydrofluoric acid (HF) solution for 5 to 10 seconds. Immediately wash it with distilled water. Remember that it is the last stage of regeneration of a glass electrode.

II. Experiments with Potentiometer

Experiment 1

Potentiometric titration of a solution of a strong acid (say hydrochloric acid) with a solution of strong alkali (say, caustic soda solution) using quinhydrone electrode.

Theory: Quinhydrone, which is a compound of equimolar composition of quinone (Q) and hydroquinone (H_2Q) undergoes dissociation in solution according to the following electrochemical equation (reduction).

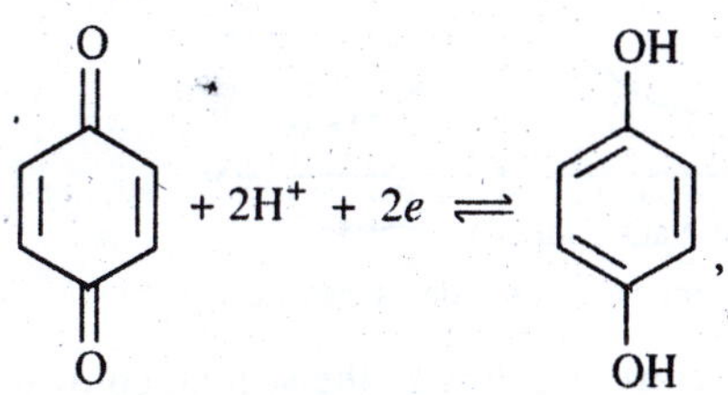

, i.e.,

$$Q + 2H^+ + 2e \rightleftharpoons H_2Q$$

If a clean Pt-electrode is dipped into an acidic solution of quinhydrone, it will serve as an electrode of the type quinhydrone H^+ | (Pt). The reduction potential of quinhydrone electrode is given by Nernst equation as

$$E = E_Q^o + \frac{RT}{2F} \ln \frac{a_Q \cdot a^2_{H^+}}{a_{H_2Q}} \quad [a \text{ represents activity}]$$

$$= E_Q^o + \frac{RT}{F} \ln a_{H^+} \quad [\because \text{ in solid state } a_Q = 1 \text{ and } a_{H_2Q} = 1] \qquad (1)$$

If this electrode is coupled with a standard saturated calomel electrode, the set up of the experimental cell will be

(–) (+)
Saturated Calomel Electrode ‖ Solution of H^+, Quinhydrone | Pt
(L) (R)

Now, the observed e.m.f. at the cell will be

$$E_{cell} = E_R - E_L = E_R^o + \frac{RT}{F} \ln a_{H^+} - E_{cal}$$

$$\text{or,}\quad E_{cell} = E^{o}_{Q} - E_{cal} + 0.059 \log a_{H^+} = \left(E^{o}_{Q} - E_{cal}\right) - 0.0591\,\text{pH}\ \ (\text{at } 25\ ^{\circ}\text{C}) \tag{2}$$

Here E^{o}_{Q} and E_{cal} are the standard reduction potentials of quinhydrone electrode and the saturated calomel electrode respectively.

Thus, from equation (2) it is evident that after complete setting of cell, if alkali is added to an acid solution (saturated with quinhydrone), pH of the solution will increase and thus E_{cell} will decrease. At equivalence point there will be a sharp decrease in E_{cell}, from which equivalence point can be found out. After pH = 7 the e.m.f. of the cell will be negative, i.e., polarity of the cell will change.

The diagrammatic representation of cell set up is almost same as the experiment no. 3"Determenation of standard reduction potential of quinhydrone electrode at room temperature". Only the difference is that take an acid solution instead of buffer solution.

The experiment can be divided into two parts:

a. Standardisation of NaOH solution by standard oxalic acid solution (Potentiometrically). Here the nature of curve, if $\Delta E_{cell}/\Delta n$ vs. n (n is the number of drop of NaOH solution added) is plotted, will be as shown below:

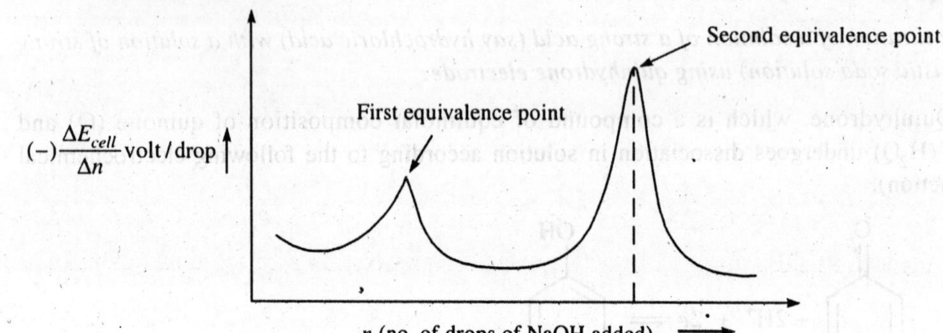

Fig. 6.9: Potentiometric titration of oxalic acid solution vs. caustic soda solution

The number of drops of NaOH required, corresponding to the second equivalence point, gives volume of NaOH solution to calculate the strength of NaOH solution as the strength and volume of oxalic acid are known.

b. Titration of hydrochloric acid solution against standardised NaOH solution.

Here also if we plot $\Delta E_{cell}/\Delta n$ vs n, the nature of curve will be as follows (Fig. 6.10)

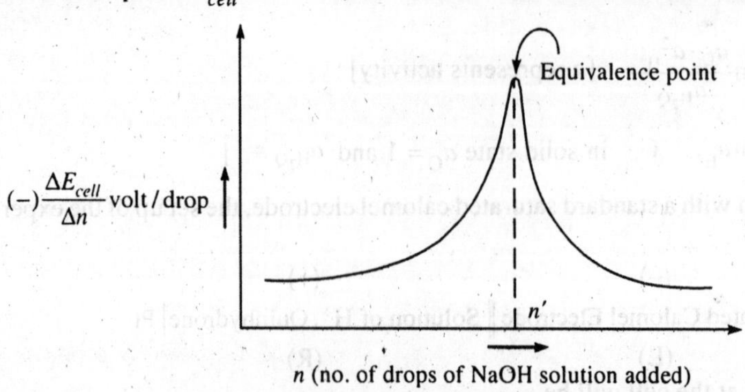

Fig. 6.10: Potentiometric titration of HCl solution vs. NaOH solution

Calculation of strength of hydrochloric acid solution

Let the number of drops of NaOH required = n' (say) = V ml (say)

Strength of NaOH solutions = S(N).

Volume of hydrochloric acid solution = 10 ml (say)

$\therefore$ strength of hydrochloric acid $= \frac{S \times V}{10}$(N).

[**N.B.** : Quinhydrode must be used in very small amount, otherwise the two equivalence points of oxalic acid (i.e., two peaks) will not be sharp].

Materials required: (a) Quinhydrone, (b) 0.01 (N) hydrochloric acid (c) 0.01 (N) oxalic acid solution (d) 0.1 (N) NaOH solution (e) Pt and saturated calomel electrode (f) KCl-Agar salt bridge, (g) Microburette, pipette, beaker, volumetric flasks, etc, (h) Potentiometer.

Short procedure

i. Prepare 0.01(N) oxalic acid solution, 0.01 (N) hydrochloric acid solution and 0.1 (N) NaOH solution (100 ml each).

ii. Take 10 ml of 0.01 (N) oxalic acid solution in a 100 ml beaker. Add pinch of quinhydrone to it and shake gently to make it saturated. Insert a clean Pt-electrode into the solution and connect this quinhydrone electrode with the saturated calomel electrode through a KCl-Agar salt bridge. Note the initial meter reading after shaking the solution for some time. Continue this titration up to the negative potertential (Take 5 to 6 negative readings)

iii. Wash the beaker well. Take 10 ml 0.01 (N) hydrochloric acid (instead of oxalic acid) and repeat the titrantion just like before.

iv. Calculate $\Delta E / \Delta n$ for each titration and then plot it against number of drops of NaOH added. Calculate the strength of NaOH solution and hydrochloric acid solution.

Results and Calculations

1. Recording of laboratory temperature

	Before experiment	After experiment	Mean
Temperature →	°C	°C	°C

2. Preparation of 100 ml 0.01 (N) oxalic acid

Relative equivalent mass of COOH . $2H_2O$ = 63

Initial mass (g)	Final mass (g)	Mass of oxalic acid transferred (g)	Mass of oxalic acid to be taken (g)	Strength of oxalic acid solution
...	...	...	0.063	...(N)

3. Preparation of 100 ml 0.1 (N) hydrochloric acid

Dissolve ≈ 0.9 ml concentrated hydrochloric acid (≈ 11.3 (N)) in 100 ml water. Titrate it analytically and then prepare 100 ml 0.01 (N) HCl solution.

4. Preparation of 100 ml 0.1 (N) NaOH solution: Dissolve about 0.4 gm (≈ 3 beads) NaOH (solid) in 100 ml water.

5. Standardisation of NaOH solution by standard oxalic acid solution (Potentiometrically) :
Volume of oxalic acid taken = 10 ml.

Observation No.	No. of drops of NaOH added n	Observed e.m.f. E_{cell} in volt or mV
1	0	
2	1	
3	2	
4	3	
⋮	⋮	

6. Titration of hydrochloric acid solution against standardised NaOH solution (Potentiometrically): Volume of HCl taken = 10 ml

Sl. No.	No. of drops of NaOH added	Observed e.m.f. (V or mV)
1	0	
2	1	
3	3	
⋮	⋮	

7. Calculation for graph plotting (for standardisation of NaOH)

n	E_{cell} (V)	ΔE (V)	Δn	$\frac{\Delta E}{\Delta n}$ V/drop
0 (n_1)	E_1	—	—	—
1 (n_2)	E_2	$E_2 - E_1$	$n_2 - n_1$	...
2 (n_3)	E_3	$E_3 - E_2$	$n_3 - n_2$	...
3 (n_4)	E_4	$E_4 - E_3$	$n_4 - n_3$	...
⋮ ⋮	⋮	⋮	⋮	...

8. Calibration of no. of drops of NaOH solution to volume: 1 cc NaOH solution = ... drops

9. Calculation for graph plotting (HCl vs. NaOH)

n	E_{cell} (V)	ΔE (V)	Δn	$\frac{\Delta E}{\Delta n}$ V/drop
0 (n_1)	E_1	—	—	—
1 (n_2)	E_2	$E_2 - E_1$	$n_2 - n_1$	...
2 (n_3)	E_3	$E_3 - E_2$	$n_3 - n_2$	...
3 (n_4)	⋮	⋮	⋮	...
⋮	⋮	⋮	⋮	...

10. Calculation of strength of HCl solution

From the graphs:

a. Volume of NaOH required for oxalic acid = ... ml

∴ strength of NaOH solution = ... (N)

b. Volume of NaOH required for HCl = ... ml

∴ strength of HCl solution = ... (N)

Compare this strength of hydrochloric acid obtained potentiometrically to the strength you have prepared analytically.

Experiment 2

Potentiometric titration of a weak acid (e.g., acetic acid) with caustic soda solution and determination of the dissociation constant of the acid using quinhydrone electrode at room temperature.

Theory: The theoretical portion is same as the previous experiment "Potentiometric titration of a strong acid (e.g., HCl) with caustic soda solution using quinhydrone electrode", i.e., e.m.f. of the experimental cell

$$\underset{(L)}{\overset{(-)}{\text{Saturated calomel electrode}}} \,\|\, \underset{(R)}{\overset{(+)}{\text{acid solution (saturated with quinhydrone)}}} \,|\, \text{Pt}$$

is given by

$$E_{cell} = (E_Q^{\circ} - E_{cal}) - 0.0591\,\text{pH} \quad [\text{at } 25\ ^{\circ}\text{C}] \tag{1}$$

If we plot E_{cell} vs n (n is the number of drops of NaOH added), then the nature of curve will be as shown in Fig. 6.11. From the equivalence point, volume of NaOH solution required can be known. Thus the strength of acetic acid solution can be calculated as the volume of acetic acid and strength of NaOH solution are known.

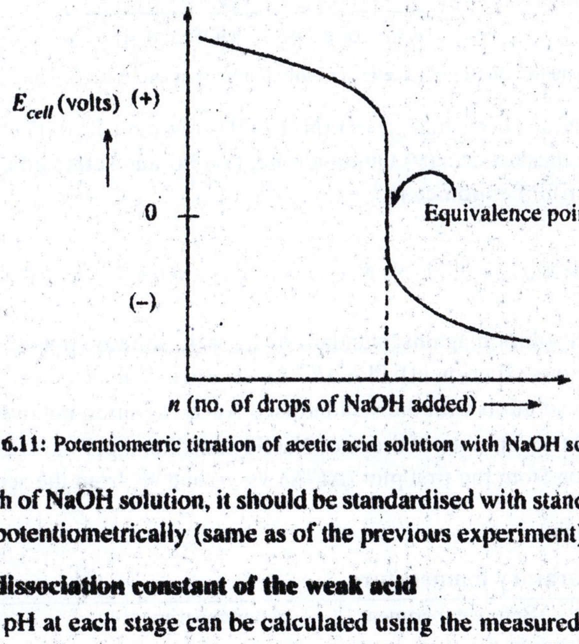

Fig. 6.11: Potentiometric titration of acetic acid solution with NaOH solution

To know the strength of NaOH solution, it should be standardised with standard oxalic acid solution. This can also be done potentiometrically (same as of the previous experiment).

Determination of dissociation constant of the weak acid

From equation (1), pH at each stage can be calculated using the measured value of e.m.f. (E_{cell}) of the cell. At 25 °C.

$$\text{pH} = \frac{E_Q^{\circ} - E_{cal} - E_{cell}}{0.0591} = \frac{0.6990 - 0.2415 - E_{cell}}{0.0591}$$

$$\text{or,}\quad \text{pH} = \frac{0.4575 - E_{cell}}{0.0591}. \tag{2}$$

Thus, if we plot pH vs n (number of drops NaOH added), then the nature of curve will be the following as shown in Fig. 6.12.

Since NaOH solution is added to acetic acid solution, it will form a buffer solution, whose pH can be calculated using Henderson's equation,

$$pH = pK_a + \log\frac{[CH_3COONa]}{[CH_3COOH]}$$. At half-equivalence point, $\frac{[CH_3COONa]}{[CH_3COOH]} = 1$,

i.e. $pH = pK_a = -\log K_a$ [K_a is dissociation constant of the acid]

In the above graph, half-equivalence point and equivalence points are clearly indicated. Thus, pK_a, i.e., K_a can be calculated.

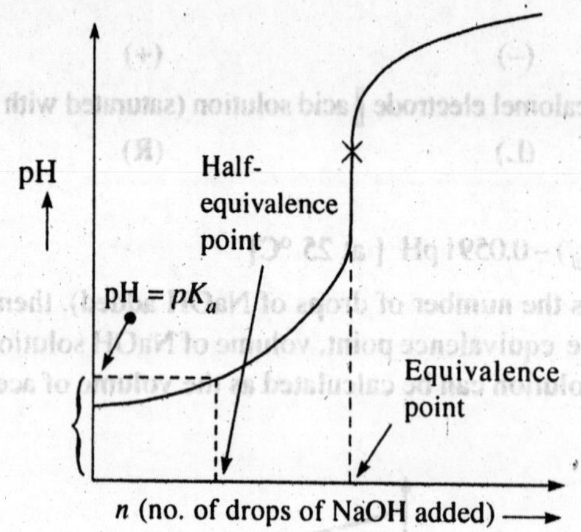

Fig. 6.12: Potentiometric titration of acetic acid solution by standard NaOH solution (pH vs. *n* plot)

Materials: (1) 0.05 (N) acetric acid, (ii) 0.5 (N) NaOH solution (iii) 0.05 (N) oxalic acid solution, (iv) Microburette, pipette, beaker, etc. (v) Quinhydrone, (vi) Pt- and saturated calomel electrodes (vii) KCl-Agar salt bridge, (viii) Potentiometer.

Procedure

i. Prepare 0.05 (N) acetic acid, 0.5 (N) NaOH solution and 0.05 (N) oxalic acid solutions (100 ml each).

ii. Standardise NaOH solution against standard oxalic acid solution potentiometrically (procedure is same as the previous experiment). Plot $\Delta E / \Delta n$ vs. n to calculate the strength of NaOH solution.

iii. Titrate CH_3COOH solution with the standardised NaOH solution potentiometrically.

iv. Plot $\Delta E_{cell}/\Delta n$ vs n (n = number of drops of NaOH added) and pH vs. n. Calculate the strength of CH_3COOH solution from the first plot ($\Delta E/\Delta n$ vs. n) and K_a from the second plot (pH vs. n).

Results and Calculations

1. Recording of laboratory temperature

	Before experiment	After experiment	Mean
Temperature ⟶	°C	°C	°C

2. Preparation of 100 ml 0.05 (N) oxalic acid solution

Initial mass (g)	Final mass (g)	Mass of oxalic acid transferred (g)	Mass of oxalic acid to be taken (g)	Strength of oxalic acid solution
...	...	...	0.315	... (N)

Relative equivalent mass of oxalic acid $\begin{array}{c} COOH \\ | \\ COOH \end{array} \cdot 2H_2O$ = 63 gm.

3. Standardisation of NaOH solution (Potentiometically)

Volume of oxalic acid (ml)	Strength of oxalic acid	No. of drops of NaOH added	Observed *e.m.f.* (E_{cell}) (volt)
10	··· (N)	0 1 2 3 ⋮	

4. Graph plotting and calculation of strength of NaOH solution

No. of drops of NaOH added (n)	E_{cell} (volt)	ΔE_{cell} (volt)	Δn	$\frac{\Delta E_{cell}}{\Delta n}$ volt/drop
0 (n_2)	E_1	—	—	—
1 (n_2)	E_2	$E_2 - E_1$	$n_2 - n_1$	···
2 (n_3)	E_3	$E_3 - E_2$	$n_3 - n_2$	···
⋮	⋮	⋮	⋮	···

From the plot of $-\frac{\Delta E}{\Delta n}$ vs. n (Fig. 6.10) strength of NaOH solution = ··· (N).

5. Graph plotting for calculation of strength of acetic acid and its dissociation constant

n	E_{cell} (volt)	ΔE_{cell} (volt)	Δn	$\frac{\Delta E}{\Delta n}$ volt/drop	pH
0 (n_1)	E_1	—	—	—	···
1 (n_2)	E_2	$E_2 - E_1$	$n_2 - n_1$	···	···
2 (n_3)	E_3	$E_3 - E_2$	$n_3 - n_2$	···	···
⋮	⋮	⋮	⋮	···	···

From **Fig. 6.10** (plot of $\frac{\Delta E}{\Delta n}$ vs. n): At equivalence point amount of NaOH required = ··· drops of NaOH = ··· ml of NaOH

∴ strength of acetic acid = ··· (N)

From **Fig. 6.12**: (Plot of pH vs. n): At half equivalance point pK_a = pH = ···

∴ $-\log K_a$ = ··· i.e., K_a = ···

Expected value of $pK_a = 4.74$ at 25 °C

Experiment 3

Determination of standard reduction potential (E^o) of quinhydrone electrode at room temperature.

Theory: In a given reversible electrode if all the substances taking part in the reaction are in their standard states, i.e., at unit activity, the potential of the electrode is called standard electrode potential, denoted by E^o.

Quinhydrone is a compound of equimolar mixture of quinone (Q) and hydroquinone (H_2Q) which in aqueous solution dissociates, giving equal concentrations of quinone (Q) and hydroquinone (H_2Q). According to the following electrochemical reaction (reduction), Q is transformed to H_2Q.

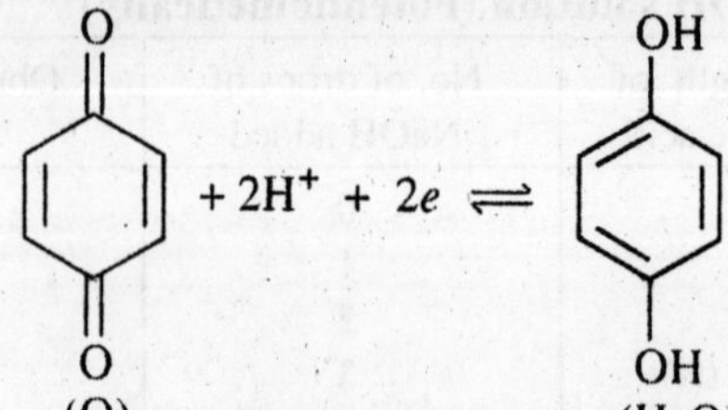

(Q) (H₂Q)

The reduction potential of the system is, therefore, given by Nernest equation:

$$E = E^{\circ}_{Q/H_2Q} + \frac{RT}{2F}\ln\frac{a_Q \cdot a^2_{H^+}}{a_{H_2Q}}$$

$$= E^{\circ}_{Q/H_2Q} + \frac{RT}{2F}\ln a_{H^+} \quad \left[\because \frac{a_Q}{a_{H_2Q}} = 1\right]$$

$$= E^{\circ}_{Q/H_2Q} + 0.0591\log a_{H^+} \quad [\text{at } 25\ {}^{\circ}\text{C}]$$

$$= E^{\circ}_{Q/H_2Q} - 0.0591\ \text{pH} \qquad (1)$$

[E°_{Q/H_2Q} is the standard reduction potential of quinhydrone electrode]

If this electrode is coupled with a saturated calomel electrode, the cell used for the experiment is given as

$$\underset{(L)}{\overset{(-)}{\text{Hg}\,|\,\text{Hg}_2\text{Cl}_2(\text{s}),\ \text{KCl (satd.)}}}\ \|\ \underset{(R)}{\overset{(+)}{\text{Solution}(\text{H}^+),\ \text{quinhydrone}\,|\,\text{Pt}}}$$

The observed e.m.f. of the cell is given by

$$E_{obs} = E_R - E_L = E^{\circ}_{Q/H_2Q} - 0.0591\ \text{pH} - E_{cal}$$

$$\text{or,} \quad E_{obs} = \left(E^{\circ}_{Q/H_2Q} - E_{cal}\right) - 0.0591\ \text{pH at } 25\ {}^{\circ}\text{C} \qquad (2)$$

E_{cal} is the reduction potential of saturated calomel electrode.

The e.m.f.'s of different cells, i.e., solutions of different pH are measured by a potentiometer. Then the plot of E_{obs} vs. pH will yield a straight line with negative slope and intercept $\left(E^{\circ}_{Q/H_2Q} - E_{cal}\right)$. Thus knowing intercept from the graph and E_{cal} at laboratory temperature, E°_{Q/H_2Q} can be found out at that temperature [Given: $E_{cal} = 0.2415 - 7 \times 10^{-4}\,(t - 25)$ volts, where t = 15 – 40 °C].

The nature of curve will be the following as shown in Fig. 6.13.

It is very important to note that to obtain the intercept of the curve pH scale along *X*-axis must start from zero, otherwise E° value will not be correct according to equation (2).

Materials

i. 0.4 (N) oxalic acid solution (100 ml)

ii. ≈ 0.5 (N) NaOH solution (250 ml)

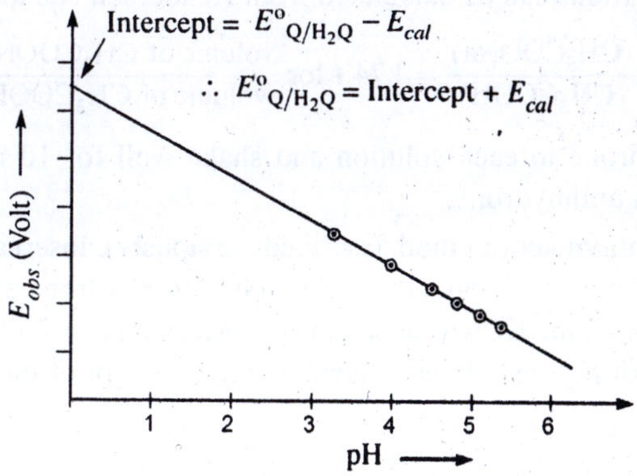

Fig. 6.13: Plot of E_{obs} vs. pH

iii. ≈ 0.5 (N) acetic acid (250 ml)
iv. 100 ml beaker (8 pcs.), burettes, pipettes, etc.
v. Platinum electrode and saturated calomel electrode
vi. Salt bridge (KCl-Agar)
vii. Exact 0.2 (N) acetic acid and 0.2 (N) sodium acetate (to prepare different buffer solutions)
viii. Potentiometer.

Procedure

i. Prepare 100 ml 0.4 (N) oxalic acid solution by accurate weighing.
ii. Prepore 250 ml ≈ 0.5 (N) NaOH solution by rough weighing (≈ 5 gm NaOH dissolved in 250 ml water)
iii. Prepare 250 ml ≈ 0.5 (N) acetic acid solution from glacial acetic acid (≈ 17 N). Take ≈ 7.5 ml glacial acetic acid in 250 ml volumetic flask and make the volume up to the mark.
iv. Standardise NaOH solution by standard oxalic acid solution and then acetic acid solution by standardised NaOH solution.
v. Preare 100 ml each of exact 0.4 (N) NaOH solution and exact 0.4 (N) CH_3COOH solution by quantitative dilution. Mix, in a seperate beaker, 50 ml 0.4 (N) NaOH solution and 50 ml 0.4 (N) CH_3COOH solution to prepare 100 ml 0.2 (N) CH_3COONa. Again, in another beaker mix 40 ml 0.4 (N) CH_3COOH solution and 40 ml distilled water to prepare 80 ml 0.2 (N) CH_3COOH solution.
vi. **Preparation of buffer solutions of different pH**

100 ml beaker no.	Volume of 0.2 (N) CH_3COOH solution (ml)	Volume of 0.2 (N) CH_3COONa solution (ml)	Total volume (ml)	Calculated pH
1	2	18	20	
2	4	16	20	
3	6	14	20	
4	8	12	20	
5	10	10	20	
6	12	8	20	
7	14	6	20	
8	16	4	20	

[pH's of the buffer solutions can be calculated from Henderson equation,

$$pH = pK_a + \log\frac{(CH_3COONa)}{(CH_3COOH)} = 4.74 + \log\frac{\text{volume of } CH_3COONa}{\text{volume of } CH_3COOH}\ ]$$

vii. Add pinch of quinhydrone to each solution and shake well for 10 minutes so that solutions become saturated with quinhydrone.

viii. Standardise your potentiometer (method described previously). Insert a clean platinum electrode to the solution (beaker no. 1)—connect this electrode (quinhydrone electrode) with a saturated calomel electrode through the KCl-Agar salt bridge. Measure e.m.f. of the cell. Repeat the same for other solutions. Plot E_{obs} vs. pH and calculate the intercept of the curve and then calculate E^o_{Q/H_2Q}.

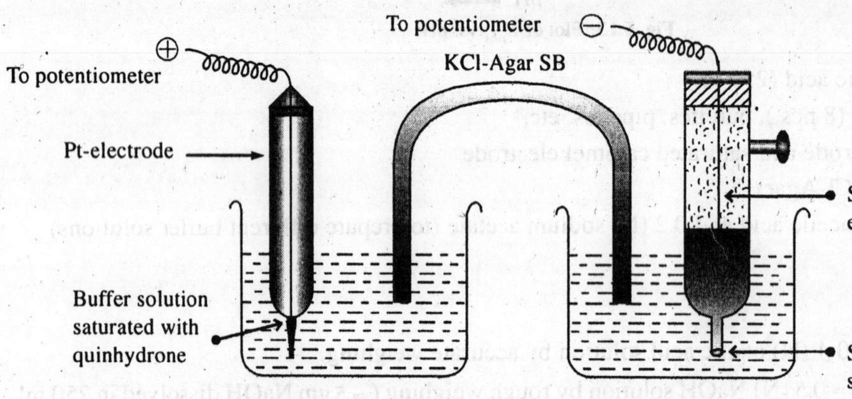

Fig. 6.14: Cell set-up for measurement of E^o of quinhydrone electrode

Results and Calculation

1. Recording of room temperature

	Before experiment	After experiment	Mean
Temperature →	°C	°C	°C

2. Preparation of 100 ml 0.4 (N) oxalic acid solution

Initial mass (g)	Final mass (g)	Mass of oxalic acid transferred (g)	Mass of oxalic acid to be taken (g)	Strength of oxalic acid solution
...	...	...	2.52	... (N)

Relative equivalent mass of oxalic acid, $\begin{matrix} COOH \\ | \\ COOH \end{matrix} \cdot 2H_2O = 63$ g.

3. Standardization of NaOH solution by standard oxalic acid solution

Observation no.	Volume of oxalic acid (ml)	Volume of NaOH required (ml)	Mean volume of NaOH required (ml)	Strength of NaOH solution
1	10	...		
2	10	...	...	... (N)
3	10	...		

To prepare 100 ml exact 0.4 (N) NaOH ... ml NaOH solution is diuted with distilled water to 100 ml (using burette).

4. Standardization of CH_3COOH solution by standardised NaOH solution

Observation no.	Volume of CH_3COOH (ml)	Volume of NaOH required (m)	Mean volume of NaOH (ml)	Strength of CH_3COOH solution
1	10	⋮		
2	10	⋮	...	... (N)
3	10	⋮		

To prepare 100 ml exact 0.4 (N) CH_3COOH solution ··· ml CH_3COOH is diluted with distilled water to 100 ml (using burette).

5. Measurement of e.m.f.'s of different buffer solutions

Solution numbers	pH	Observed e.m.f. in volt
1		⋮
2		⋮
3		⋮
4		⋮
5		⋮
6		⋮
7		⋮
8		⋮

6. Graph plotting and calculation of E^o_{Q/H_2Q}

From the plot of E_{obs} vs. pH, it is found that intercept (at pH = 0) = ··· volts.

Thus, E^o_{Q/H_2Q} = intercept (at pH = 0) + E_{cal} = ··· volt at ··· °C.

Expected value of $E^o_{Q/H_2Q} = 0.699 - 7.4 \times 10^{-3}\ (t - 25)$, t = 15–40°C.

Experiment 4

Potentiometric titration of a solution of silver nitrate with a solution of potassium chloride (i.e., standardisation of silver nitrate solution by standard potassium chloride solution potentiometrically.

Theory: Silver nitrate which are generally supplied in the laboratory sometimes contains some impurity. Again, in presence of air some of it decomposes and becomes black forming Ag_2O. So it cannot be taken as a primary standard substance for the titration of halides with silver nitrate solution. Hence, it is necessary to standardise $AgNO_3$ solution by a standard solution of KCl (to be taken as a primary standard substance).

When a clean silver electrode is immersed into a solution of silver nitrate an electrode $Ag|Ag^+$ is formed. If this electrode is coupled with a saturated calomel electrode through a salt bridge (KNO_3^- Agar salt bridge), the cell will be

$$\underset{L}{\text{Saturated calomel}} \overset{(-)\ (KNO_3\text{-Agar})\ (+)}{\Big\|} \underset{R}{Ag^+ \,|\, Ag}$$

At the right-hand electrode, the half-cell reaction is

$$Ag^+ + e \rightleftharpoons Ag$$

The e.m.f. is given by

$$E_R = E^o_{Ag^+/Ag} + \frac{RT}{F}\ln a_{Ag^+} = E^o_{Ag^+/Ag} + 0.0591\log\left[Ag^+\right] \text{ (at 25 °C)}$$

The observed cell e.m.f. is

$$E_{obs} = E_R - E_L = E^o_{Ag^+/Ag} + 0.0591\log\left[Ag^+\right] - E_{cal}$$

$$= \left(E^o_{Ag^+/Ag} - E_{cal}\right) + 0.0591\log\left[Ag^+\right]$$

$$= \text{constant} + 0.0591 \log\ [Ag^+] \text{ at 25 °C.}$$

From the above equation it is evident that if KCl solution is added to the system $Ag/AgNO_3$, then Ag^+ will be removed as AgCl and therefore there will be a drop in e.m.f. of the cell. With the progress of titration, e.m.f. will decrease and at the equivalence point there will be a sharp decrease in e.m.f. as shown in the following figures 6.15 and 6.16.

The number of drops of KCl solution is converted into volumes of KCl solution whose strength is known. Volume of $AgNO_3$ solution taken for standardisation is known to us. Therefore, strength of $AgNO_3$ solution can be determined accurately.

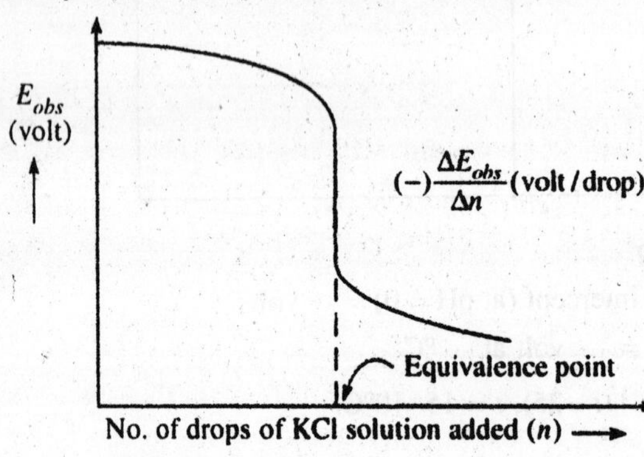

Fig. 6.15: Plot of E_{obs} vs. n (no. of drops of KCl solution added) for the potentiometric titration of $AgNO_3$ solution by standard KCl solution.

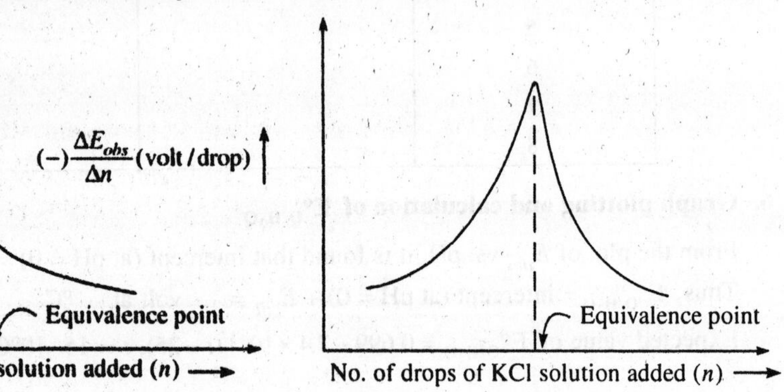

Fig. 6.16: Plot of $\left(-\frac{\Delta E_{obs}}{\Delta n}\right)$ vs. n for the potentiometric titration og $AgNO_3$ solution by standard KCl solution.

Materials required: (i) Silver nitrate, (ii) Postassium chloride, (iii) KNO_3-Agar salt bridge, (iv) Silver and saturated calomel electrodes, (v) Potentiometer, (vi) Beaker, (vii) Burettes and pipettes, (viii) Volumetric flasks (100 ml), etc.

Procedure

i. Prepare 100 ml each of 0.1 (N) $AgNO_3$ solution and KCl solution.

ii. Dilute 10 ml 0.1 (N) $AgNO_3$ solution to 100 ml to prepare 0.01 (N) $AgNO_3$ solution.

iii. Take 10 ml $AgNO_3$ solution (~0.01 N) in a 100 ml beaker and insert the clean Ag electrode into it (add some distilled water, if necessary, to immerse the electrode well). Join the saturated calomel electrode through KNO_3-Agar salt bridge and then connect the cell with the potentiometer. Note the meter reading initially.

iv. Add 1 drop of KCl solution (~0.1 N) from a burette. A white precipitate will be formed. Shake well and note the meter reading again. Continue your titration adding 1 drop KCl solution at a time. Take 5-6 more readings after sharp equivalence point.

v. Wash the silver electrode with dilute NH_4OH solution and finally with distilled water.

vi. Calibrate the number of drops of KCl solution into volume.

vii. Plot-$\Delta E_{obs}/\Delta n$ vs. n, where n is the number of drops of KCl solution added and calculate the strength of $AgNO_3$ solution (follow the theory).

Results and Calculations

1. **Recording of room temperatures**

	Before experiment	After experiment	Mean
Temperature →	... °C	... °C	... °C

2. **Preparation of 0.1 (N) KCl solution (100 ml)**

Initial mass (g)	Final mass (g)	Mass of KCl transferred (g)	Mass of KCl to be taken (g)	Strength of KCl solution
			0.7446	... (N)

3. **Preparation of 100 ml 0.1 (N) $AgNO_3$ solution:** Relative equivalent mass of $AgNO_3$ = 169.89

Initial mass (g)	Final mass (g)	Mass of $AgNO_3$ transferred (g)	Mass of $AgNO_3$ to be taken (g)	Strength of $AgNO_3$ solution
...	...	...	1.6989	... (N)

(This is diluted 10 times)

4. **Recording of titration result:** Volume of $AgNO_3$ solution taken = 10 ml

No. of observation	No. of drops of KCl solution added	E_{obs} (volt or mV)
1	0	
2	1	
3	2	
4	3	
⋮	⋮	
⋮	⋮	

5. Calibration of 1 ml of KCl solution into number of drops

No. of observation	Volume of KCl solution	No. of drops counted	Mean number of drops
1	1 ml	...	
2	1 ml	...	...
3	1 ml	...	

6. Table for graph plotting

No. of drops of KCl solution (n)	E_{obs} (volt or mV)	Δn	ΔE_{obs} (V or mV)	$-\Delta E_{obs}/\Delta n$ (volts or mV) per drop
0 (n_0)	E_0	—	—	
1 (n_1)	E_1	$n_1 - n_0$	$E_1 - E_0$	
2 (n_2)	E_2	$n_2 - n_1$	$E_2 - E_1$	
3 (n_3)	E_3	$n_3 - n_2$	$E_3 - E_2$	

7. Calculation of strength of $AgNO_3$ solution

From the plot of $-\Delta E_{obs}/\Delta n$ vs. n, it is found that at equivalence point no. of drops KCl solution required = ··· = ··· ml.

Volume of $AgNO_3$ solution taken = 10 ml and strength of KCl solution that is prepared = ··· (N)

Hence, the strength of $AgNO_3$ solution = ··· (N).

Compare the strength of $AgNO_3$ solution obtained potentiometrically and prepared by weighing.

Experiment 5

Potentiometric titration of a solution of chloride ion with a solution of silver nitrate.

Theory: If a very small amount of silver nitrate solution (i.e., Ag^+ ions) is added to a solution of potassium chloride or sodium chloride (i.e., Cl^- ions) contained in a beaker the precipitate of silver choloride which is formed, undergoes dissociation slightly on shaking forming some Ag^+ ions. If a clean silver electrode is immersed into the solution in the beaker, the electrode which is formed is $Ag|Ag^+$. If this silver electrode is coupled with a reference electrode, say, a saturated calomel electrode, the cell which is formed is given as

$$\underset{(L)}{\overset{(-)}{\text{Saturated calomel}}} \,\|\, \underset{(R)}{\overset{(+)}{Ag^+ \,|\, Ag}}$$

Here potassium nitrate-Agar agar or ammonium nitrate-Agar agar salt bridge must be used to connect the two electrodes. Pictorially it may be represented as shown in Fig. 6.17.

The two electrodes are connect to marked terminals of a potentiometer.

At the right-hand electrode the half-cell reaction is

$$Ag^+ + e \rightleftharpoons Ag$$

The e.m.f. of the electrode is given by

$$E_R = E^\circ_{Ag^+/Ag} + \frac{RT}{F}\ln\frac{a_{Ag^+}}{a_{Ag}}$$

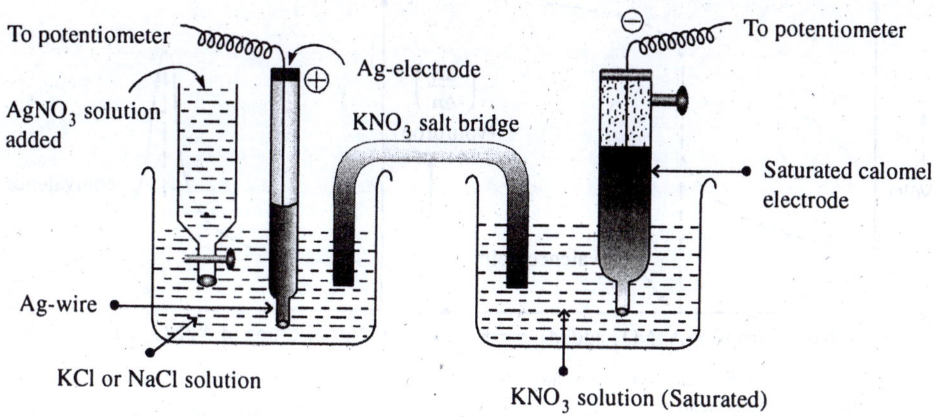

Fig. 6.17: Cell set-up for the potentimetric titration of chloride ions with $AgNO_3$ solution

$$= E^o_{Ag^+/Ag} + \frac{RT}{F}\ln a_{Ag^+} \quad [\because\ a_{Ag} = 1 \text{ as Ag is solid}]$$

So, the total e.m.f. of the cell is

$$E = E_R - E_L$$

$$= E^o_{Ag^+/Ag} + \frac{RT}{F}\ln\frac{a_{Ag^+}\cdot a_{Cl^-}}{a_{Cl^-}} - E_{cal}$$

$$= E^o_{Ag^+/Ag} - E_{cal} + \frac{RT}{F}\ln\frac{L_a}{a_{Cl^-}}$$

Here $L_a = a_{Ag^+}\cdot a_{Cl^-}$ = activity product of AgCl which is constant at room temperature. $E^o_{Ag^+/Ag}$ and E_{cal} are the standard reduction potentials of Ag^+/Ag electrode and saturated calomel electrode which are also constant at room temperature. Thus the above e.m.f. expression can be written as,

$$E = \left(E^o_{Ag^+/Ag} - E_{cal} + \frac{RT}{F}\ln L_a\right) - \frac{RT}{F}\ln a_{Cl^-}$$

$$= \text{constant} - \frac{RT}{F}\ln a_{Cl^-}$$

$$= \text{constant} - 0.0591\log a_{Cl^-} \quad [\text{at } 25\ ^oC]$$

Thus, with the progress of titration, i.e., with successive addition of $AgNO_3$ solution to the solution of chloride ions, the activity of Cl^- ions will decrease as Cl^- ions will be precipitated as AgCl. So, e.m.f. of the cell will increase gradually. At the equivalence point there will be a sharp increase in *e.m.f.* as all the Cl^- ions will be removed. After the equivalence point increase in e.m.f. of the cell will be very slow. Hence, from the sharp change in *e.m.f.* of the cell we can identify the equivalence point. If we plot E versus number of drops of $AgNO_3$ added, or $\Delta E/\Delta n$ versus number (n) of drops of $AgNO_3$ added the nature of curves, which have been found experrmentally, are shown in fig. 6.18 & 6.19.

It is more advantageous the derivative curve than the E vs. n curve, as equivalence point is more easy to understood in the $\Delta E/\Delta n$ vs. n curve.

From the graph the number of drops of $AgNO_3$ required for known volume of chloride salt can be calculated. The strength of $AgNO_3$ solution is known. Thus strength of chloride ions can be calculated. [Similar experiment may be performed to find out the strength of I^- ions].

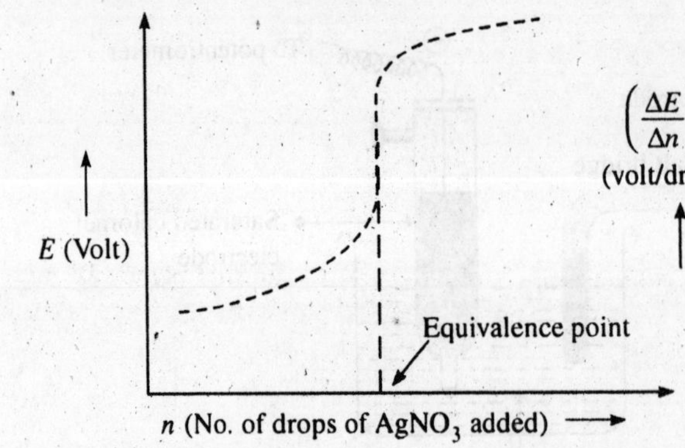

Fig. 6.18: Plot of E vs. n for the titration Cl^- ions against $AgNO_3$ solution potentiometically

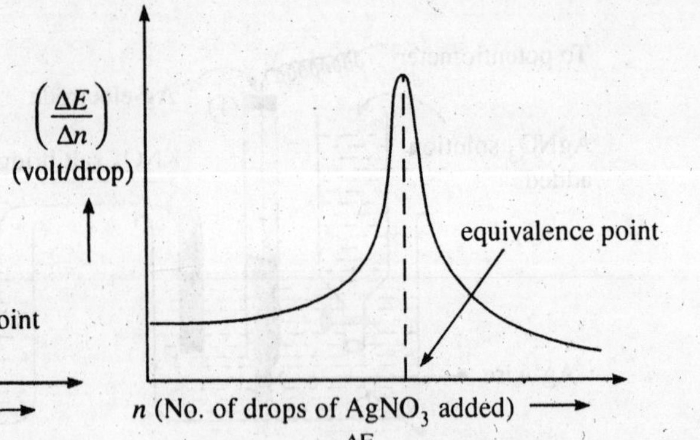

Fig. 6.19: Plot of $\frac{\Delta E}{\Delta n}$ vs. n for the titration of Cl^- ions against $AgNO_3$ solution potentiometically

Materials: (i) Standard $AgNO_3$ solution (0.1 N) (ii) NaCl or KCl solution (0.01 N) (iii) Salt bridge (KNO_3 or NH_4NO_3-Agar), (iv) Silver electrode and saturated calomel electrode, (v) Microburette (10 ml), (vi) Pipette (10 ml), (viii) Beaker (100 ml) and (viii) Potentiometer.

Procedure

i. Prepare 100 ml 0.1 (N) $AgNO_3$ solution and 100 ml 0.01 (N) KCl or NaCl solution by accurate weighing.
ii. Take 10 ml solution of chloride ions and add 1 to 2 drops of 0.1 (N) $AgNO_3$ solution to it. Shake well when white precipitate of AgCl is formed.
iii. Immerse a clean silver electrode into the solution after shaking 5 minutes. Connect the silver electrode and the saturated calomel electrode to the terminals of a potentiometer. Then connect the salt bridge (NH_4NO_3-Agar or KNO_3-Agar). On your potentiometer and adjust the scale to the reading mode. Again shake the solution of chloride ions gently and note the meter reading. [Add, if necessary, some distilled water to immerse the ends of the two electrodes in the beaker.]
iv. Continue addition of $AgNO_3$ solution (two drops or better one drop at a time), shake gently and note meter readings at each step. Take 5 to 6 more readings after passing equivalence point, i.e., after the sharp increase in e.m.f.
v. Count the number of drops of $AgNO_3$ for one ml solution using the same burette. Repeat 3 times and take average.
vi. Wash your electrodes, beaker and burette with a dilute solution of ammonium hydroxide first and then with distilled water well.

Results and Calculations

1. Recording of room temperature

	Before experiment	After experiment	Mean
Temperature →	... °C	... °C	... °C

2. Preparation of 100 ml 0.1 (N) $AgNO_3$ solution

Initial mass (g)	Final mass (g)	Mass of $AgNO_3$ transferred (g)	Mass of $AgNO_3$ to be taken (g)	Strength of $AgNO_3$ solution
...	...	...	1.699	... (N)

3. Preparation of 100 ml 0.01 (N) KCl solution

Initial mass (g)	Final mass (g)	Mass of KCl transferred (g)	Mass of KCl to be taken (g)	Strength of KCl solution
...	...	...	0.0745	... (N)

4. Titration and recording of e.m.f. at different points: Volume of KCl solution taken = 10 ml.

No. of observation	No. of drops of $AgNO_3$ solution added	Observed e.m.f. (volt or mV)
1	1	
2	2	
3	3	
⋮	⋮	
⋮	⋮	

5. Calculation of data for graph plotting

No. of drops of $AgNO_3$ added (n)	E (V or mV)	ΔE_{obs} (V or mV)	Δn	$\Delta E_{obs}/\Delta n$ (V or mV)/drop
1 (n_1)	E_1	—	—	—
2 (n_2)	E_2	$E_2 - E_1$	$n_2 - n_1$	
3 (n_3)	E_3	$E_3 - E_2$	$n_3 - n_2$	
⋮	⋮	⋮	⋮	

6. Calculation of strength of Cl^- ions

From the graph, plotting $\Delta E/\Delta n$ vs. n, number of drops of $AgNO_3$ required = ··· drops

1 ml $AgNO_3$ = ··· drops = ··· ml.

∴ strength of Cl^- ions = ··· (N).

Experiment 6

Potentiometric titration of a mixture of halides (KCl and KI)

Theory: If silver nitrate (i.e., $Ag^{(+)}$ ion) is added to a solution of $Cl^{(-)}$ or $I^{(-)}$ ions, the reaction taking place is a precipitation reaction,

$$Ag^+ + X^- \rightleftharpoons AgX(s) \qquad [X = Cl, I]$$

The AgX which is formed is a sparingly soluble salt and it forms some Ag^+ ions after shaking. Now if a clean silver electrode (Ag-foil or Ag-wire) is immersed into that solution, it will function as a Ag/Ag^+ electrode. Now, if this electrode is coupled with a standard saturated calomel electrode, a cell will be formed, represented as

$$\underset{(L)}{\overset{(-)}{\text{Calomel (saturated)}}} \; \| \; \underset{(R)}{\overset{(+)}{Ag^+ \mid Ag}}$$

The electrode potential of the half-cell reaction (right hand electrode)

$Ag^+ + e \rightleftharpoons Ag$, is given by

$$E_R = E^o_{Ag^+/Ag} + \frac{RT}{F}\ln\frac{a_{Ag^+}}{a_{Ag}}$$

$$= E^o_{Ag^+/Ag} + \frac{RT}{F}\ln a_{Ag^+} \qquad [\because \text{ in solid state } a_{Ag} = 1]$$

Now, the total e.m.f., i.e., observed e.m.f. (E_{obs}) is given by

$$E_{obs} = E_R - E_L$$

$$= E^o_{Ag^+/Ag} + \frac{RT}{F}\ln a_{Ag^+} - E_{cal}$$

$$= E^o_{Ag^+/Ag} - E_{cal} + \frac{RT}{F}\ln a_{Ag^+}$$

$$= E^o_{Ag^+/Ag} - E_{cal} + \frac{RT}{F}\ln\frac{a_{Ag^+}\cdot a_{X^-}}{a_{X^-}}$$

$$= E^o_{Ag^+/Ag} - E_{cal} + \frac{RT}{F}\ln\frac{L_{AgX}}{a_{X^-}}$$

$$= \left(E^o_{Ag^+/Ag} - E_{cal} + \frac{RT}{F}\ln L_{AgX}\right) - \frac{RT}{F}\ln a_{X^-}$$

Here, $E^o_{Ag^+/Ag}$, E_{cal} and L_{AgX} are the standard reduction potential of Ag^+/Ag electrode, standard reduction potential of saturated calomel electrode and the activity product of silver halides which are constant at the experimental temperature. Therefore,

$$E_{obs} = \text{constant} - \frac{RT}{F}\ln a_{X^-} = \text{constant} - 0.0591\log a_{X^-} \quad [\text{at } 25\ ^oC]$$

Thus, if $AgNO_3$ is added to KCl or KI solution, E_{obs} will gradually increase as $X^{(-)}$ ions will be removed from the solution as AgX. At the equivalence point there will be a sharp rise in E_{obs} from which equivalence point can be detected. If $AgNO_3$ is added to a mixture of Cl^- and I^- ions, AgI will be selectively precipilated first as AgI has much lower solubility product than AgCl. ($L_{AgCl} = 1.5 \times 10^{-10}$ and $L_{AgI} = 1.5 \times 10^{-16}$ at 25 °C). After complete removal of I^- ions, Cl^- ions will be removed. In both equivalence points there will be a sharp rise in e.m.f. If we plot $\Delta E/\Delta n$ vs. n (where n is the number of drops of $AgNO_3$ added) then the following curve (Fig. 6.20) will be obtained from which equivalence points for I^- and Cl^- ions can be detected separately.

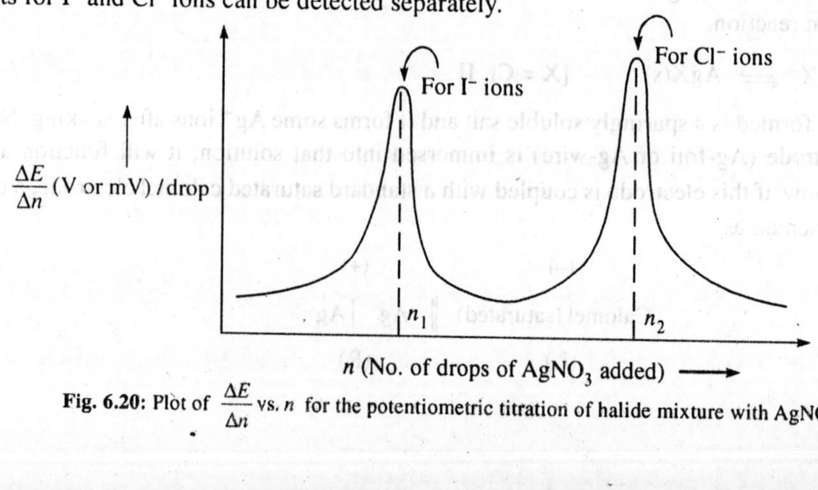

Fig. 6.20: Plot of $\frac{\Delta E}{\Delta n}$ vs. n for the potentiometric titration of halide mixture with $AgNO_3$ solution

Let n_1 be the number of drops of $AgNO_3$ required for I^- ions and n_2 be the total number of drops of $AgNO_3$ required for total halide ions. Then number of drops of $AgNO_3$ required for Cl^- ions will be $(n_2 - n_1)$ drops. The volume of mixture taken and the strength of $AgNO_3$ solution which is prepared are known. Then the strength of I^- ions and Cl^- ions can be calculated separately in their mixture.

It is very important to remember that the salt bridge which is to be used to connect the silver electrode and the calomel electrode must be KNO_3-Agar or NH_4NO_3-Agar salt bridge as we are titrating halide ions. Don't use KCl-Agar salt bridge in this type of experiment. The picture of constructed cell will be same as the potentiometric titration of chloride ions by $AgNO_3$ solution (experiment 5).

Materials: (i) 0.1 (N) $AgNO_3$ solution, (ii) 0.05 (N) KCl and 0.05 (N) KI solution (separate) (iii) Silver electrode and saturated calomel electrode. (iv) NH_4NO_3-Agar or KNO_3-Agar salt bridge, (v) Microburette (10 ml capacity), (vi) Pipette (10 ml), (vii) Beaker (100 ml), (viii) Potentiometer.

Procedure

i. Prepare 100 ml 0.1 (N) $AgNO_3$ solution by accurate weighing.
ii. Prepare 100 ml each of 0.05 (N) KCl solution and 0.05 (N) KI solution.
iii. Mix equal volumes of KCl and KI solutions in a separate beaker and take 10 ml of the mixture in an another beaker (100 ml).
iv. Add 1 to 2 drops of $AgNO_3$ solution (0.1 N) from a microburette. A yellow precipitate is formed. Shake it gently.
v. Insert the silver electrode into the solution and then connect the cell through NH_4NO_3-Agar or KNO_3-Agar salt bridge.
vi. Connect the two electrodes (Ag- and calomel) to the terminals of a potentiometer. Adjust the meter to the reading mode and note the meter reading (in V or mV) after shaking gently.
vii. Continue your titration by adding 1 drop of $AgNO_3$ at a time—shake gently and note the meter readings. There will be two sharp changes. Take 5 to 6 more readings after passing the second equivalence point.
viii. Calibrate number of drops of $AgNO_3$ in volume (ml)
ix. Wash the electrodes with a dilute solution at ammonium hydroxide.
x. Calculate $\Delta E/\Delta n$ and plot $\Delta E/\Delta n$ vs. n, from the graph calculate the volumes of $AgNO_3$ required for KI and KCl separately and calculate their strengths in the mixture separately.

Results and Calculations

1. Recording of temperature during the experiment

	Before experiment	After experiment	Mean
Temperature →	... °C	... °C	... °C

2. Preparation of 100 ml 0.1 (N) $AgNO_3$ solution

Initial mass (g)	Final mass (g)	Mass of $AgNO_3$ transferred (g)	Mass of KCl to be taken (g)	Strength of $AgNO_3$ solution
...	...	...	1.699	... (N)

3. Preparation of 100 ml 0.05 (N) KCl solution

Initial mass (g)	Final mass (g)	Mass of KCl transferred (g)	Mass of KCl to be taken (g)	Strength of KCl solution
...	...	...	0.3725	... (N)

4. Preparation of 0.05 (N) KI solution (100 ml)

Initial mass (g)	Final mass (g)	Mass of KI transferred (g)	Mass of KI to be taken (g)	Strength of KI solution
...	...	...	0.830	... (N)

5. Preparation of mixture of KCl and KI solution

Volume of KCl solution of strength S_1 (N) taken (ml)	Volume of KI solution of strength S_2 (N) (ml)	Total volume of the mixture (ml)	Strength of KCl in the mixture	Strength of KI in the mixture
25	25	50	$\frac{S_1}{2}$(N)	$\frac{S_2}{2}$(N)

[S_1 and S_2 are nearly eqaul to 0.05 (N)]

6. Titration and recording of meter readings (e.m.f): Volume of mixture taken = 10 ml.

No. of observations	No. of drops of $AgNO_3$ added	E_{obs} (V or mV)
1	1	
2	2	
3	3	
⋮	⋮	

7. Calculation for graph plotting

No. of drops of $AgNO_3$ added (n)	E_{obs} (V or mV)	ΔE_{obs} (V or mV)	Δn	$\Delta E_{obs}/\Delta n$ (volt/drop)
1 (n_1)	E_1	—	—	—
2 (n_2)	E_2	$E_2 - E_1$	$n_2 - n_1$	
3 (n_3)	E_3	$E_3 - E_2$	$n_3 - n_2$	
⋮	⋮	⋮	⋮	

8. Calibration of number of drops of $AgNO_3$ solution

No. of observations	Volume of 20 drops of $AgNO_3$ solution in the microburette (ml)	Mean volume of 20 drops of $AgNO_3$ solution in the microburette (ml)
1	V_1	
2	V_2	$\frac{V_1 + V_2 + V_3}{3} = V$ (say)
3	V_3	

i.e., 20 drops = V ml. Thus, 1 ml = $\frac{20}{V}$ drops = n' (say).

9. Calculation of strengths of Cl^- and I^- ions in their mixture separately

From the plot of $\Delta E_{obs}/\Delta n$. vs. n, it is found that

i. No. of drops of $AgNO_3$ equivalent to $I^{(-)}$ ious requied (corresponding to first peak)

$$= n_1 \text{ drops} = \frac{n_1}{n'} \text{ ml}, \quad \therefore \text{ strength of KI} = \frac{(n_1/n_2) \times S}{10} (\text{N}) \quad [S = \text{strength of } AgNO_3 \text{ solution}]$$

ii. No. of drops of $AgNO_3$ equivalent to Cl^- ions (corresponding to second peak)

$$= (n_2 - n_1) \text{ drops} = \frac{(n_2 - n_1)}{n'} \text{ ml},$$

$$\therefore \quad \text{strength of KCl in the mixture} = \frac{\frac{(n_2 - n_1)}{n'} \times S}{10} (\text{N}).$$

10. Conclusion: Verify the strengths of halides in the mixture obtained potentiometrically with the strengths of halide mixtures you have prepared.

Experiment 7

Determination of activity solubility product of silver chloride by e.m.f. measurements.

Theory: The solubility product of sparingly soluble salts like AgCl can be determined by e.m.f. measurements. If a few drops of $AgNO_3$ solution are added to a solution of KCl, a slight precipitate of AgCl is formed. If it is shaked well, it will be a solution of potassium chloride saturated with AgCl. As AgCl is a sparingly soluble salt a part of it will dissociate forming some Ag^+ ions. If a clean silver electrode (silver foil or silver wire) is inserted into the solution saturated with AgCl, then this can serve as a silver electrode Ag/Ag^+. This electrode is coupled with a saturated calomel electrode whose e.m.f. is known. The cell which is constructed is

$$\underset{(\text{L})}{\overset{(-)}{\text{Ag} \mid \text{KCl } (x\,\text{N}) \text{ (saturated with AgCl)}}} \;\|\; \underset{(\text{R})}{\overset{(+)}{\text{Calomel electrode (saturated)}}}$$

Here x is the strength of KCl solution in normality. The salt bridge which is used to connect the electrodes must be NH_4NO_3-Agar or KNO_3-Agar (instead of KCl-Agar) salt bridge. The construction of cell will be more clear to us if we represent it pictorially as follows (Fig. 6.21):

At the left-hand electrode the half-cell reaction is

$$Ag^+ + e \rightleftharpoons Ag$$

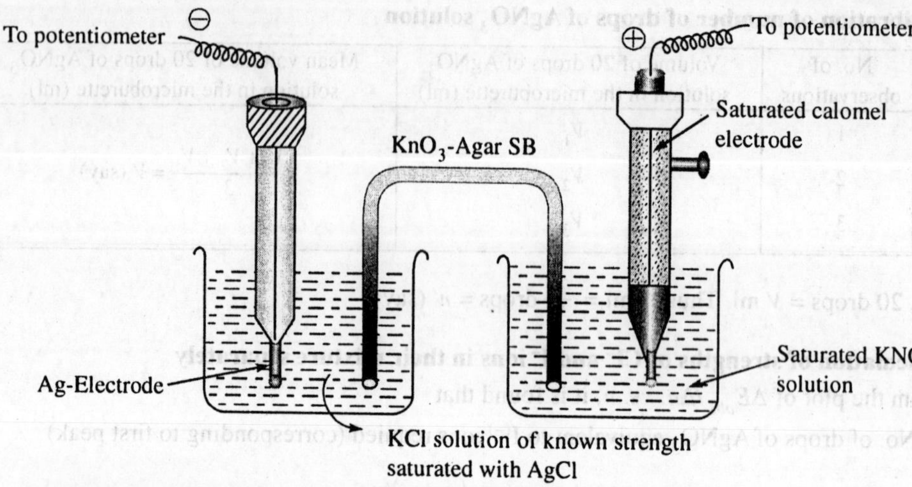

Fig. 6.21: Cell Constructed fro measurement of activity solubility product of AgCl by e.m.f. measurements.

The e.m.f. of this electrode is given by

$$E_L = E^o_{Ag^+/Ag} + \frac{RT}{F}\ln\frac{a_{Ag^+}}{a_{Ag}} = E^o_{Ag^+/Ag} + \frac{RT}{F}\ln a_{Ag^+} \quad \text{[in solid state } a_{Ag} = 1]$$

Total e.m.f. of the constructed cell is

$$E = E_R - E_L = E_{cal} - E^o_{Ag^+/Ag} - \frac{RT}{F}\ln a_{Ag^+}$$

Thus, if we measure e.m.f. (E) of the cell, a_{Ag^+} can be calculated as standard reduction potentials of calomel electrode (E_{cal}) and silver electrode $\left(E^o_{Ag^+/Ag}\right)$ are known.

For the calculation of a_{Cl^-}, the activity coefficient of Cl^- ions is to be calculated. This can be done using Debye-Hückel limiting equation as follows:

$$\log f_{Cl^-} = -A|Z_+||Z_-|\sqrt{\mu}$$

$$= -A\sqrt{\mu} \quad \text{(as KCl is uni-univalent electrolyte)}$$

$$= -0.509\sqrt{\mu} \quad \text{(at 25 °C } A \text{ is taken as 0.509)}$$

$$\therefore \quad f_{Cl^-} = 10^{-0.509\sqrt{\mu}}$$

where $\mu = \frac{1}{2}\sum CiZi^2 = \frac{1}{2}\left(C_{K^+}\cdot 1^2 + C_{Cl^-}\cdot 1^2\right)$

$= \frac{1}{2}c$ [c is concentration of KCl in nomality), $\quad \therefore \; f_{Cl^-} = 10^{-0.509\sqrt{c/2}}$

Hence activity solubility product of AgCl can be calculated by following equation:

$$K_{AgCl} = a_{Ag^+}\cdot a_{Cl^-} = a_{Ag^+}\cdot(c\cdot f_{Cl^-}).$$

Using the similar method, K_{AgCl} can be determined applying the same type of cell but different concentration of KCl solution. The mean of these is calculated.

Materials: (i) 0.1 (N) KCl solution, (ii) 0.01 (N) $AgNO_3$ solution, (iii) Silver and saturated calomel electrode, (iv) KNO_3-Agar salt bridge, (v) 100 ml beakers, (vi) Potentiometer.

Procedure

i. Prepare 250 ml 0.1 (N) KCl solution and 100 ml 0.01 N $AgNO_3$ solution.

ii. Prepae four KCl solutions of different strengths 0.075 N, 0.05 (N) 0.025 N and 0.01 N from the mother KCl solution (0.1 N) by proper dilution.

iii. Add 1 to 2 drops of $AgNO_3$ solution ($\approx$ 0.01 N) to each of these KCl solutions. A slight precipitate will form. Shake each composition well to make the solution "saturated with AgCl".

iv. Take first beaker containg the solution saturated with AgCl—insert into it a clean Ag-electrode—connect the saturated calomel electrode through KNO_3-Agar salt bridge (as shown in the Fig. 6.21. On your potentiometer—adjust the meter to reading mode and note the e.m.f. of the cell. Repeat the same for other solutions of KCl saturated with AgCl.

v. Calculate a_{Ag^+}, $\sqrt{\mu}$, a_{Cl^-} and then K_{AgCl} for different compositions. Calculate the mean K_{AgCl}.

Results and Calculations

1. Recording of temperature during the experiment

	Before experiment	After experiment	Mean
Temperature →	··· °C	··· °C	··· °C

2. Preparation of 0.1 (N) 250 ml KCl solution

Initial mass (g)	Final mass (g)	Mass of KCl transferred (g)	Mass of KCl to be taken (g)	Strength of KCl solution
···	···	···	1.8615	··· (N)

3. Preparation of 100 ml 0.01 (N) $AgNO_3$ solution

Initial mass (g)	Final mass (g)	Mass of $AgNO_3$ transferred (g)	Mass of $AgNO_3$ to be taken (g)	Strength of $AgNO_3$ solution
···	···	···	0.1699	··· (N)

4. Preparation of KCl of different compositions

100 ml beaker number	Volume of 0.1 (N) KCl solution (ml)	Volume of distilled water (ml)	Concentration of KCl solution
1	50	0	0.1 (N)
2	37.5	12.5	0.075 (N)
3	25	25	0.05 (N)
4	12.5	37.5	0.025 (N)
5	5.0	45.0	0.01 (N)

(Use burette to measure the above volumes)

5. Measurement of *e.m.f.* of different sets

Set Number	Observed e.m.f. in volt
1	⋮
2	⋮
3	⋮
4	⋮
5	⋮

6. Calculation of K_{AgCl} for different sets

Set No.	Observed e.m.f. (volts)	a_{Ag^+}	Concentration of KCl Soln. (c)	$\sqrt{\mu} = \sqrt{c/2}$	f_{Cl^-}	a_{Cl^-}	K_{AgCl}
1			0.1 (N)	⋮			
2			0.075 (N)	⋮			
3			0.05 (N)	⋮			
4			0.025 (N)	⋮			
5			0.01 (N)	⋮			

7. **Interpretation of results:** It is observed that solubility of AgCl, i.e., a_{Ag^+} decreases with increase in KCl concentration due to common ion effect but solubility product i.e. activity product remains almost constant. The mean value of K_{AgCl} = ··· at ··· °C.

Experiment 8

Determination of solubility and solubility product of silver chloride by e.m.f. measurement of a concentration cell.

Theory: e.m.f. measurements may be made for determining the solubility product K_s of a sparingly soluble salt like AgCl. This has already been discussed in the previous experiment using saturated calomel electrode. However, we can determine also the solubility product of AgCl by considering the cell.

$$\underset{(L)}{\text{Ag} \mid \text{AgCl, 0.01 (N) KCl}} \parallel \underset{(R)}{\text{0.01 (N) AgNO}_3 \mid \text{Ag}}$$

and treating the cell as a concentration cell with the concentration of Ag^+ different in the L.H.E. and R.H.E.

The cell can be set up by dipping one of the clean silver electrodes into 0.01 (N) $AgNO_3$ solution and the other into 0.01 (N) KCl solution, the two solutions being connected through a salt bridge (ammonium nitrate-Agar salt bridge). A drop of $AgNO_3$ solution is added to the beaker containing 0.01 (N) KCl solution and shaken well. A small amount of AgCl which is formed is sufficient to give its saturated solution. At the R.H.E. the concentration of Ag^+ is known (0.01 N, assuming $AgNO_3$ to be completely dissociated) while at the L.H.E. concentration of Ag^+ is much lower (furnished by ionisation of sparingly soluble AgCl) which is unknown to us and is to be determined.

The e.m.f. of the above concentration cell is given by the following expression:

$$E = \frac{RT}{F} \ln \frac{\left(a_{Ag^+}\right)_{RHE}}{\left(a_{Ag^+}\right)_{LHE}}$$

or, $E = \frac{RT}{F} \ln \frac{0.01}{c}$ [assuming the solutions are dilute]

or, $E = 0.0591 \log \frac{0.01}{C}$ (at 25 °C) (1)

where c is the unknown concentration of Ag^+ furnished by sparingly soluble AgCl at LHE. If E is measured by a potentiometer, c can be evaluated from the above expression. Since the concentration of Cl^- ions is 0.01 (M), so the solubility product of AgCl is given by

$$K_s = c \times 0.01 \quad (\text{moles lit}^{-1})^2. \quad (2)$$

Thus, solubility of AgCl, $S = \sqrt{K_S}$ moles/lit. (3)

Materials required: (i) Potassium chloride and silver nitrate (G.R. Grade), (ii) KNO_3-Agar or NH_4NO_3-Agar salt bridge, (iii) Silver electrodes, (iv) Potentiometer, (v) Volumetric flasks (100 ml), (vi) Pipettes, (viii) Burettes, (viii) Beakers (100 ml), (ix) Tissue papers, etc.

Short Procedure

i. Prepare 100 ml each of 0.1 (N) KCl solution and 0.1 (N) $AgNO_3$ solution.

ii. Prepare 100 ml each of exact 0.01 (N) KCl solution and exact 0.01 (N) $AgNO_3$ solution by quantitative dilution.

iii. Take two dry and clean beakers (100 ml). In one beaker take 50 ml 0.01 (N) KCl solution and in another take 50 ml 0.01 (N) $AgNO_3$ solution. Add 1 to 2 drops of 0.01 (N) $AgNO_3$ solution to the beaker containing 50 ml 0.01 (N) KCl solution and shake it well. Immerse the clean silver electrode into the solutions and couple the two solutions through a KNO_3-Agar or NH_4NO_3-Agar salt bridge as described in the theory. A complete cell is thus constructed.

iv. Connect the electrodes (cell) to the potentiometer which is already standardished (using the method described at the beginnings of this chapter) and measure the e.m.f. of the cell.

v. Calculate concentration of Ag^+ ions furnished by sparingly soluble AgCl, i.e., c from equation (1).

vi. Calculate the solubility product K_S and solubility (s) of AgCl using equation (2) and (3) respectively.

Results and Calculations

1. Recording of laboratory temperature:

	Before experiment	After experiment	Mean
Temperature →	... °C	... °C	... °C

2. Preparation of 100 ml 0.1 (N) KCl solution:

Initial mass (g)	Final mass (g)	Mass of KCl transferred (g)	Mass of KCl to be taken (g)	Strength of KCl solution
...	...	...	0.7446	... (N)

3. Preparation of 100 ml 0.1 (N) $AgNO_3$ solution

Initial mass (gm)	Final mass (gm)	Mass of $AgNO_3$ transferred (gm)	Mass of $AgNO_3$ to be taken (gm)	Strength of $AgNO_3$ solution
...	...	...	1.6989	... (N)

4. Preparation of exact 0.01 (N) KCl solution and 0.01 (N) $AgNO_3$ solution (100 ml each)

a. Dilule ... ml of the prepared KCl solution to 100 ml with distilled water.

b. Dilute ... ml of the prepared (nearly 0.1 N) $AgNO_3$ solution to 100 ml with distilled water.

5. Recording of e.m.f. of the cell and calculation of solubility product and solubility of AgCl

Observed e.m.f. of the cell E (volts)	Concentration of Ag^+ ions (c) moles/lit From eq. (1)	Concentration of Cl^- ions moles/lit	Solubility products (K_S) of AgCl From eq. (2)	Solubility (S) of AgCl from From eq. (3) moles/lit
...	...	...	...	...

Literature value of solubility product of AgCl = 1.6×10^{-10} at 25°C.

Experiment 9

Potentiometric estimation of Mohr salt solution with standard potassium dichromate solution and also determination of formal potential (reduction) of ferric-ferrous system.

Theory: The reduction potential for the redox system $Fe^{3+} + e \rightleftharpoons Fe^{2+}$ is given by the Nernst equation as $E = E^{o}_{Fe^{3+}/Fe^{2+}} + \frac{RT}{F}\ln\frac{[Fe^{3+}]}{[Fe^{2+}]}$.

If a clean platinum electrode (Pt-wire or Pt-foil) is dipped into the $Fe^{2+} - Fe^{3+}$ system, it will serve as an electrode of the type $Fe^{3} - Fe^{2}/(Pt)$. When this electrode is coupled with a reference electrode such as saturated calomel electrode and the two electrodes are connected through a KCl-Agar salt bridge, the experimental cell will be

$$\underset{(-)}{\text{Saturated calomel electrode}} \,\|\, \underset{(+)}{Fe^{3+} - Fe^{2+}} \mid Pt$$

The e.m.f. of the experimental cell will be

$$E_{cell} = E_R - E_L = E_{Fe^{3+}/Fe^{2+}} + \frac{RT}{F}\ln\frac{[Fe^{3+}]}{[Fe^{2+}]} - E_{cal}$$

$$\text{or,}\quad E_{cell} = \left(E^{o}_{Fe^{3+}/Fe^{2+}} - E_{cal}\right) + 0.0591\log\frac{[Fe^{3+}]}{[Fe^{2+}]}\quad \text{(at 25 °C)} \tag{1}$$

Here, E_{cal} is the standard reduction potential of saturated calomel electrode and $E^{o}_{Fe^{3+}/Fe^{2+}}$ is called formal (reduction) potential of ferric-ferrous system.

[*Definition of formal potential*: The formal potential is the potential observed experimentally in a solution containing one mole each of the oxidised and reduced substances together with other specified substances at specified concentrations. It is found that formal potentials vary appreciably, for example,

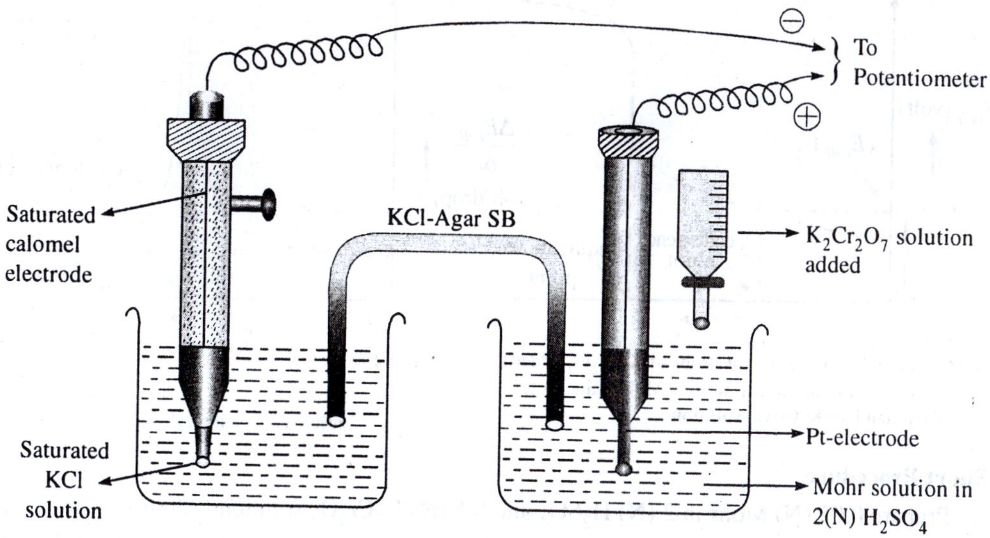

Fig. 6.22: Cell set-up for potentiometric titration of Mohr salt solution with standard potassium dichromate solution

with the nature and concentration of the acid that is present. The formal potential incorporates in one value, the effects resulting from variation of activity coefficients with ionic strength, acid-base dissociation, complexation, liquid-juction-potentials, etc., and thus has a real practical value. Formal potentials do not have the theoretical significance of standard potentials, but they are observed values in actual potentiometric measurements].

When $K_2Cr_2O_7$ solution is added to the system, Fe^{2+} ion is removed and the concentration of Fe^{3+} increases. From equation (1), it is evident that with the progressive addition of $K_2Cr_2O_7$ solution, the observed e.m.f. (E_{cell}) will gradually increase. At the equivalence point there will be a sharp increase in *e.m.f.* of the cell due to the removal of all Fe^{2+} ions. (Here, actually just after the equivalence point Fe^{3+}/Fe^{2+} system changes to $Cr_2O_7^{2-}/Cr^{3+}$ system which has much higher E^o value than that the Fe^{3+}/Fe^{2+} system. Hence, the sharp increase in e.m.f. of the cell at equivalence point).

From equation (1), at half equivalence point $[Fe^{2+}] = [Fe^{3+}]$; thus, $\left(E_{cell}\right)_{1/2} = E^o_{Fe^{3+}/Fe^{2+}} - E_{cal}$

$$\therefore \; E^o_{Fe^{3+}/Fe^{2+}} = \text{formal reduction potential} = E_{cal} + (E_{cell})_{1/2}.$$

If we plot E_{cell} vs. n (n is number of $K_2Cr_2O_7$ solution added), then the following nature of curve (Fig. 6.23) will be obtained.

Thus, $\left(E_{cell}\right)_{1/2}$ will be obtained from the graph, E_{cal} is known [$E_{cal} = 0.2415\text{ V} - 7 \times 10^{-4}\,(t - 25)$, t = 15 - 40 °C]. So the formal reduction potential of ferric-ferrous system $E^o_{Fe^{3+}/Fe^{2+}}$ can be calculated.

Again, if we plot $\Delta E_{cell}/\Delta n$ vs. n, the nature of curve will be obtained as shown in Fig. 6.24:

Let the number of drops of $K_2Cr_2O_7$ added at equivalence point be $n'(= \text{V ml})$. Volume of Mohr solution is known to us, strength of $K_2Cr_2O_7$ solution is also known. Hence, strength of Mohr solution can be calculated.

Materials: (1) 0.05 (N) Mohr solution in 2 (N) H_2SO_4, (ii) 0.5 (N) $K_2Cr_2O_7$ solution, (iii) Pt- and saturated calomel electrode, (iv) KCl-Agar salt bridge, (v) Beaker, micro-burette, pipette, etc. (vi) Potentiometer.

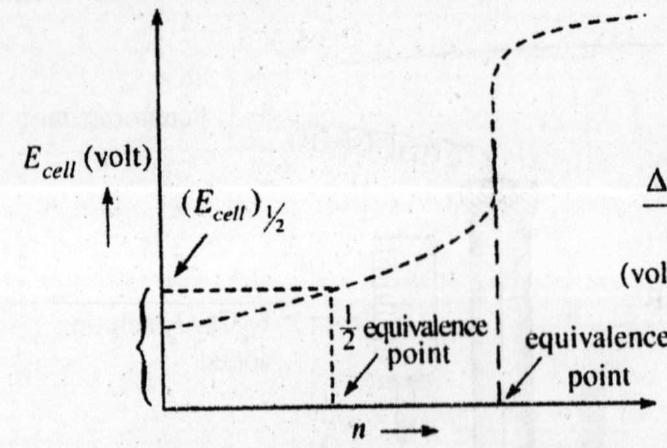

Fig. 6.23: Plot of E_{cell} vs. n (no. of drops of $K_2Cr_2O_7$ added) for potentiometric titration of Mohr salt solution by standard $K_2Cr_2O_7$ solution.

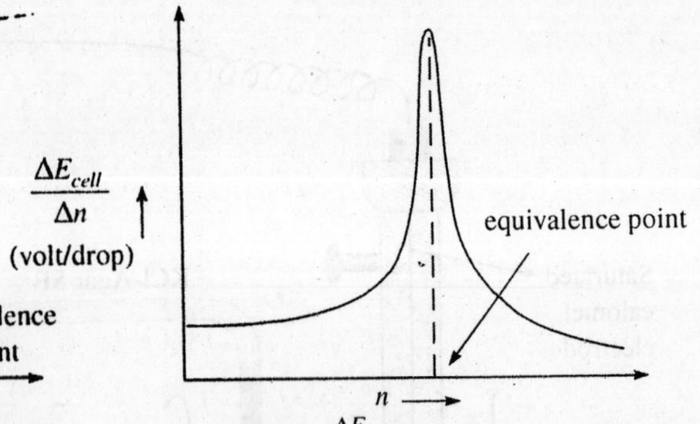

Fig. 6.24: Plot of $\frac{\Delta E_{cell}}{\Delta n}$ vs. n (no. of $K_2Cr_2O_7$ drops added) for potenttiometric titration of Mohr salt solution by standard $K_2Cr_2O_7$ solution.

Short Procedure

i. Prepare 0.05 (N) Mohr in 2 (N) H_2SO_4 and 0.5 (N) $K_2Cr_2O_7$ solution (100 ml each).

ii. Take 25 ml Mohr solution in a 100 ml clean beaker, insert the electrodes into the solution. It will be better if pinch of $FeCl_3$ be added to the Mohr solution. Calibrate your potentiomter to the reading mode and take initial reading of the meter before addition of $K_2Cr_2O_7$ into the Mohr solution.

iii. Add 2 drops of $K_2Cr_2O_7$ solution—shake gently and take the meter reading. Continue your addition and note meter reading each time. Take 5 to 6 more readings after passing sharp change in e.m.f. i.e., equivalence point.

iv. Clean the electrodes with distilled water and keep immersed them into distilled water.

v. Plot two graphs: (a) E_{cell} vs. n, and (b) $\Delta E_{cell}/\Delta n$ vs. n.

Calculate the strength of Mohr solution and the formal potential of the ferric-ferrous system.

Results and Calculations

1. Recording of laboratory temperature

	Before experiment	After experiment	Mean
Temperature →	... °C	... °C	... °C

2. Preparation of 0.5 (N) $K_2Cr_2O_7$ solution (100 ml)

Equivalent relative mass of $K_2Cr_2O_7 = \frac{249.22}{6} = 49.037$ (in acid medium)

Initial mass (g)	Final mass (g)	Mass of $K_2Cr_2O_7$ transferred (g)	Mass of $K_2Cr_2O_7$ to be taken (g)	Strength of $K_2Cr_2O_7$ solution
...	...	...	2.4518	... (N)

3. Preparation of 0.05 (N) Mohr solution in 2 (N) H_2SO_4 (100 ml)

Relative equivalent mass of $FeSO_4 \cdot (NH_4)_2SO_4 \cdot 6H_2O = 392.16$

Initial mass (g)	Final mass (g)	Mass of Mohr salt transferred (g)	Mass of Mohr salt to be taken (g)	Strength of Mohr solution
...	...	...	1.96	... (N)

[Dissolve the required amount of Mohr salt in 100 ml 2 (N) H_2SO_4 which is prepared by diluting 5.6 ml concentrated H_2SO_4 (36 N) to 100 ml carefully and then cooling.]

4. Titration and recording of e.m.f.: Volume of Mohr solution taken = 25 ml

No. of observations	No. of drops of $K_2Cr_2O_7$ (n) added	Observed e.m.f. (E_{cell}) volt or mV
1	0	
2	2	
3	4	
⋮	⋮	
⋮	⋮	
⋮	⋮	

5. Calibration of 1 cc of $K_2Cr_2O_7$ solution to number of drops : 1 cc = ... n' (drops say)

6. Graph plotting and Calculations

n	E_{cell} (volts)	ΔE_{cell} (volts)	Δn	$\Delta E_{cell}/\Delta n$ volt/drop
0 (n_1)	E_1	—	—	—
2 (n_2)	E_2	E_2-E_1	n_2-n_1	⋮
4 (n_3)	E_3	E_3-E_2	n_3-n_2	⋮
⋮	⋮	⋮	⋮	⋮

7. Calculation of strength of Mohr solution and formal potential of Fe (III)-Fe(II)—system

i. From the plot of E_{cell} vs. n, $E^{o}_{Fe^{3+}/Fe^{2+}} = E_{cal} + (E_{cell})_{1/2}$ = ... volt.

Expected value of $E^{o}_{Fe^{3+}/Fe^{2+}} = +0.68$ volt in (M) H_2SO_4 medium

ii. From the plot of $\Delta E/\Delta n$ vs. n, number of drops of $K_2Cr_2O_7$ solution required = ... = ... ml

Hence strength of Mohr solution = ... (N).

Experiment 10

Potentiometic titration of Zn(II) *solution by potassium ferrocyanide solution and also determination of the composition of* Zn(II)*-ferrocyanide complex.*

Theory: The electrode potential for the redox system

$[Fe(CN)_6]^{3-} + e \rightleftharpoons [Fe(CN)_6]^{4-}$ is given by Nernst equation as

$$E = E^{\circ} + \frac{RT}{F}\ln\frac{C_{[Fe(CN)_6]^{3-}}}{C_{[Fe(CN)_6]^{4-}}}$$

where E° is the standard reduction potential of $[Fe(CN)_6]^{3-} - [Fe(CN)_6]^{4-}$ system.

If a clean platinum electrode is inserted into the solution of a mixture of $K_4[Fe(CN)_6]$ and $K_3[Fe(CN)_6]$, it will serve as an electrode. If we couple this electrode with a standard calomel (saturated) electrode, then the cell which is constructed for the experiment is

$$\underset{(L)}{\overset{(-)}{\text{Saturated calomel electrode}}} \,\|\, \underset{\text{(Mixture of solution) (R)}}{[Fe(CN)_6]^{4-} - [Fe(CN)_6]^{3-}} \overset{(+)}{/\text{Pt}}$$

Pictorially it is represented as below:

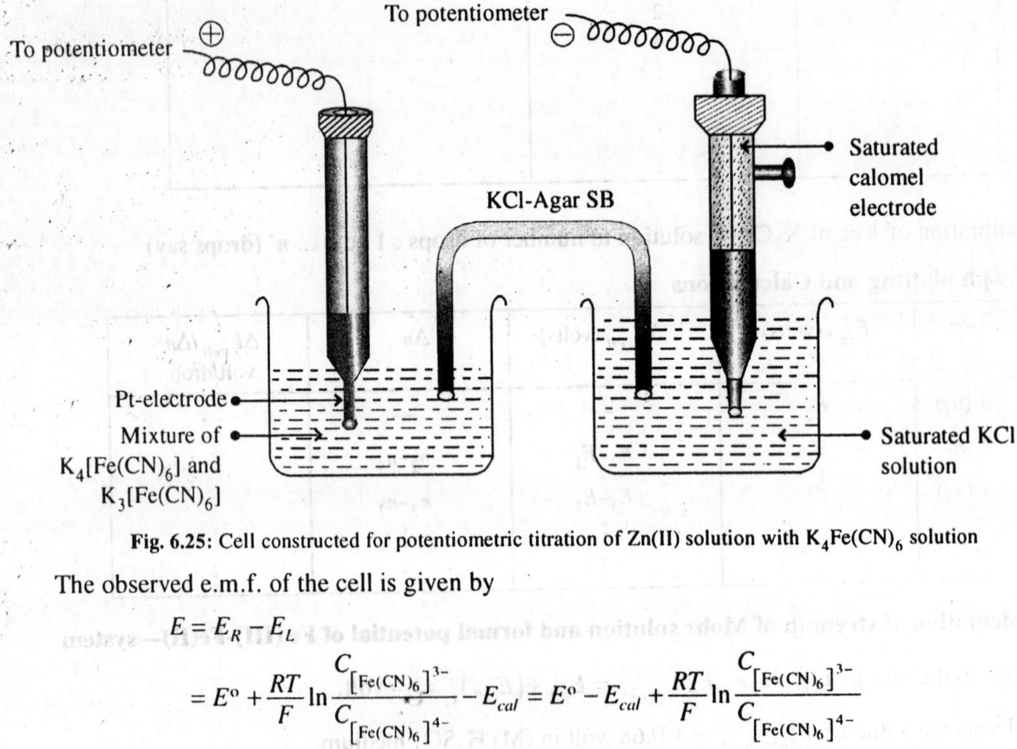

Fig. 6.25: Cell constructed for potentiometric titration of Zn(II) solution with $K_4Fe(CN)_6$ solution

The observed e.m.f. of the cell is given by

$$E = E_R - E_L$$

$$= E^{\circ} + \frac{RT}{F}\ln\frac{C_{[Fe(CN)_6]^{3-}}}{C_{[Fe(CN)_6]^{4-}}} - E_{cal} = E^{\circ} - E_{cal} + \frac{RT}{F}\ln\frac{C_{[Fe(CN)_6]^{3-}}}{C_{[Fe(CN)_6]^{4-}}}$$

$$= E^{\circ} - E_{cal} + 0.0591\log\frac{C_{[Fe(CN)_6]^{3-}}}{C_{[Fe(CN)_6]^{4-}}} \text{ [at 25 °C].} \qquad (1)$$

Here E_{cal} is the reduction potential of saturated calomel electrode.

When Zn(II) solution is added to the system, $[Fe(CN)_6]^{4-}$ is removed and the proportions of $[Fe(CN)_6]^{3-}$ increases. From equation (1) it is evident that with the progressive addition of Zn(II), the observed e.m.f. will gradually increase. At the equivalence point, there will be a sharp increase in e.m.f. due to sudden removal of all $[Fe(CN)_6]^{4-}$.

If we plot $\Delta E/\Delta n$ vs. n (n is the number of drops of $ZnSO_4$ added) the nature of curve will be of the following type (Fig. 6.26).

From the equivalence point, corresponding to the peak, volume of $ZnSO_4$ required can be known. Volume of $K_4[Fe(CN)_6]$ taken for estimation is also known. Therefore, strength of $K_4[Fe(CN)_6]$ can be calculated as strength of $ZnSO_4$ is known (prepared).

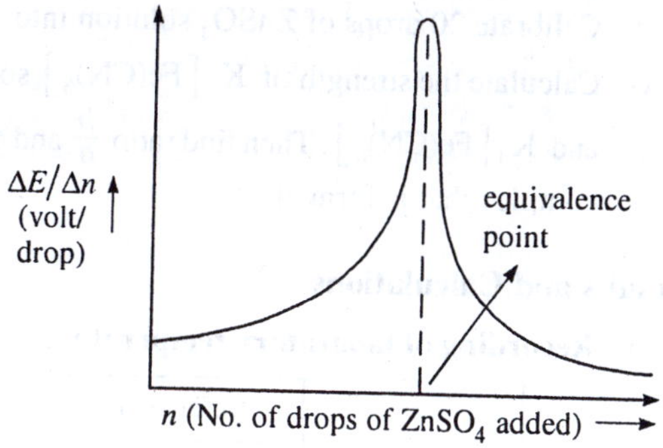

Fig. 6.26: Plot of $\Delta E/\Delta n$ vs. n (no. of drops of $ZnSO_4$ solution added for the titration of $K_4Fe(CN)_6$ solution with $ZnSO_4$ solution.

Determination of composion of the complex formed

Let 10 ml $K_4[Fe(CN)_6]$ is taken for titration.

Let the number of moles of $K_4\left[Fe(CN)_6\right]$ present in 10 ml solution = a.

If at the equivalence point V ml $ZnSO_4$ solution is required, then V ml of $ZnSO_4$ solution combines with 10 ml $K_4\left[Fe(CN)_6\right]$ solution.

Let the number of moles of $ZnSO_4$ present in V ml solution = b.

Then find the ratio, $\dfrac{\text{No. of moles of } ZnSO_4}{\text{No. of moles of } K_4\left[Fe(CN)_6\right]} = \dfrac{b}{a}$.

From this ratio we can conclude about the composition of the complex formed.

Materials

i. 100 ml $\frac{M}{6}$ $ZnSO_4$ solution

ii. 100 ml $\frac{M}{100}$ $K_4\left[Fe(CN)_6\right]$ solution

iii. 100 ml $\approx \frac{M}{100}$ $K_3\left[Fe(CN)_6\right]$ solution

iv. Pt-electrode and saturated calomel electrode

v. KCl-Agar salt bridge

vi. Potentiomter, (vii) Microburrette, 100 ml beaker, pipette (10 ml), etc.

Procedure

i. Prepare 100 ml $\frac{M}{6}$ $ZnSO_4$ solution, 100 ml $\frac{M}{100}$ $K_4\left[Fe(CN)_6\right]$ solution and 100 ml $\frac{M}{100}$ $K_3\left[Fe(CN)_6\right]$ solution.

ii. Take 10 ml $K_4\left[Fe(CN)_6\right]$ solution and 10 ml $K_3\left[Fe(CN)_6\right]$ solution (of order ~ 0.01 M) in a clean 100 ml beaker (use pipette) and insert the clean and dry Pt-electrode into the solution. Connect the saturated calomel electrode through KCl-Agar salt bridge as shown inFig. 6.25. Then connect the cell to the terminals indicated in the potentiometer.

iii. Add one drop of $ZnSO_4$ solution from a microburette into the solution in the beaker—a white precipitate will be formed. Shake well and note the meter reading. Continue the titration adding 1 drop of $ZnSO_4$ solution at each time. Take 5 to 6 more readings after sharp change (equivalence point).

iv. Calculate $\Delta E/\Delta n$ and plot it against n. Detect the equivalence point in the graph and note the number of drops of $ZnSO_4$ required.

v. Calibrate 20 drops of $ZnSO_4$ solution into volume (ml).

vi. Calculate the strength of $K_4[Fe(CN)_6]$ solution. Also calculate the number of moles of $ZnSO_4$ and $K_4[Fe(CN)_6]$. Then find ratio $\frac{b}{a}$ and conclude about the composition of Zn(II) ferrocyanide complex that is formed.

Results and Calculations

1. Recording of laboratory temperature

	Before experiment	After experiment	Mean
Temperature →	... °C	... °C	... °C

2. Preparation of 100 ml $\frac{M}{6}$ $ZnSO_4$ solution: Relative molar mass of $ZnSO_4 \cdot 7H_2O = 287.35$

Initial mass (g)	Final mass (g)	Mass of $ZnSO_4$ transferred (g)	Mass of $ZnSO_4$ to be taken (g)	Strength of $ZnSO_4$ solution
...	...	...	4.789	... (M)

3. Preparation of100 ml 0.01 (M) $K_4[Fe(CN)_6]$ solution: Relative molar mass of $K_4[Fe(CN)_6] \cdot 3H_2O = 422.26$.

Initial mass (g)	Final mass (g)	Mass of $K_4[Fe(CN)_6]$ transferred (g)	Mass of $K_4[Fe(CN)_6]$ to be taken (g)	Strength of $K_4[Fe(CN)_6)]$ solution
...	...	...	0.4222	... (M)

4. Preparation of 100 ml 0.01 (M) $K_3[Fe(CN)_6]$ solution: Relative molar mass of $K_3[Fe(CN)_6]$ = 329.16

Initial mass (g)	Final mass (g)	Mass of $K_3[Fe(CN)_6]$ transferred (g)	Mass of$K_3[Fe(CN)_6]$ to be taken (g)	Strength of $K_3[Fe(CN)_6]$ solution
...	...	...	0.3291	... (M)

5. Titration and recording of e.m.f.'s

Volume of $K_4[Fe(CN)_6]$ solution taken = 10 ml
Volume of $K_3[Fe(CN)_6]$ solution taken = 10 ml } in a 100 ml beaker

Observation number	No. of drops of $ZnSO_4$ solution added	Observed e.m.f. (in V or mV)
1	1	...
2	2	...
3	3	...
4	4	...
⋮	⋮	...

6. Calibration of number of drops of $ZnSO_4$ solution to volume (ml)

Observation No.	No. of drops of $ZnSO_4$ solution (from microburette)	Volume of 20 drops of $ZnSO_4$ solution (ml)	Mean volume of of 20 drops (ml) $ZnSO_4$ solution	No. of drops for 1 ml solution
1	20	...	V ml (say)	$\frac{20}{V} = n'$ (say)
2	20	...		
3	20	...		

7. Graph plotting and calculation of strength of $K_4[Fe(CN)_6]$ solution

No. of drops of $ZnSO_4$ solution added	Observed e.m.f. E (volt)	ΔE(volt)	Δn	$\frac{\Delta E}{\Delta n}$ volt/drop
1 (n_1)	E_1	—	—	—
2 (n_2)	E_2	E_2-E_1	n_2-n_1	⋮
3 (n_3)	E_3	E_3-E_2	n_3-n_2	⋮
⋮ ⋮	⋮	⋮	⋮	⋮
⋮ ⋮	⋮	⋮	⋮	⋮

From the graph, obtained by plotting $\Delta E/\Delta n$ vs. n, it is found that the number of drops of $ZnSO_4$ solution, at equivalence point (corresponding to the peak) = n'' (say). So, the volume of $ZnSO_4$ required for 10 ml $K_4[Fe(CN)_6]$ solution = $\frac{n''}{n'}$ ml = x ml (say).

Therefore, the strength of $K_4[Fe(CN)_6]$ solution

$$= \frac{x \times S}{10}(\text{M}) = y(\text{M}) \text{ (say), where } S \text{ is the strength of } ZnSO_4 \text{ solution in molarity.}$$

Compare this stength, obtained potentiometrically, with the strength of $K_4[Fe(CN)]$ solution you have prepared by accurate weighing.

Determination of Composition of Zn(II) – $Fe(CN)_6^{4-}$ complex

$$\text{Number of moles of } ZnSO_4 = \frac{x \times \text{molar concentration of } ZnSO_4 \text{ solution}}{1000} = b$$

Again, number of moles of $K_4[Fe(CN)_6]$ that combine with $ZnSO_4$

$$= \frac{10 \times \text{molar concentron of } K_4[Fe(CN)}{1000} = \frac{y}{100} = a.$$

$$\text{Thus, the ratio} = \frac{\text{No. of moles of } ZnSO_4}{\text{No. of moles of } K_4[Fe(CN)_6]} = \frac{b}{a}.$$

Expected value of the ratio is $\frac{3}{2}$.

Hence, the precipitation reaction that takes place in neutral medium according to the following equation.

$$3ZnSO_4 + 2K_4[Fe(CN)_6] = K_2Zn_3[Fe(CN)_6]_2 \downarrow + 3K_2SO_4$$

[**N.B.:** $K_4[Fe(CN)_6]$ and $K_3[Fe(CN)_6]$ must be fresthly prepared. Otherwise their concentrations will change in presence of air and the result will be inaccurate].

Suggestions for doing similar type of experiments

Experiment 10A

Determination of composition of Pb(*II*)*-ferrocyanide complex potentiometrically.*

Here, aqueous solution of $Pb(NO_3)_2$ may be used as Pb(II) instead of $ZnSO_4$ solution. Other set-up, procedure, etc will remain same.

Experiment 10B

Determination of composition of Cd(II)*-ferrocyanide complex potentiometrically*.

Here, aqueous solution of $CdSO_4$ may be used as titrant instead of $ZnSO_4$ solution. Other experimental set-up, procedure, etc. will remain same.

In both the above experiments, the expected value of moles ratio is 1 : 2, i.e., the reactions will be of the type: $M^{2+} + 2K_4[Fe(CN)_6] = K_6M[Fe(CN)_6]_2 + 2K^+$, M = Pb, Cd.

Experiment 11

Determination of standard reduction potential of ferricyanide-ferrocyanide system at room temperature.

Theory: Ferrocyanide-ferricyanide is a redox system of the type:

$$[Fe(CN)_6]^{3-} + e \rightleftharpoons [Fe(CN)_6]^{4-}$$

If a clean platinum electrode is immersed into the system, it will serve as an electrode whose potential is given by the Nernst equation as below:

$$E = E^\circ + \frac{RT}{F}\ln\frac{a_{[Fe(CN)_6]^{-3}}}{a_{[Fe(CN)_6]^{4-}}} \quad [a \text{ represents activity}]$$

$$= E^\circ + \frac{RT \times 2.303}{F}\log\frac{C_{[Fe(CN)_6]^{-3}}}{C_{[Fe(CN)_6]^{4-}}} + \frac{2.303\,RT}{F}\log\frac{f_{[Fe(CN)_6]^{3-}}}{f_{[Fe(CN)_6]^{4-}}} \quad (1)$$

[*C*, *f* represent molar concentration and activity coefficient respectively]

From Debye-Hückel limiting law as below activity coefficient (*f*) is related with ionic strength (μ) of the solutions:

$$\log f = -AZ_i^2\sqrt{\mu}\text{, where}$$

$$\mu = \frac{1}{2}\sum C_iZ_i^2\text{, } A \text{ is a constant, called Debye-Hückel constant.}$$

$$\therefore \log\frac{f_{[Fe(CN)_6]^{3-}}}{f[Fe(CN)_6]^{4-}} = A\left(Z^2_{[Fe(CN)_6]^{4-}} - Z^2_{[Fe(CN)_6]^{3-}}\right)\sqrt{\mu}\,.$$

If an equimolar mixture of potassium ferrocyanide and potassium ferricynaide be used, i.e.,

$$\log\frac{C_{[Fe(CN)_6]^{3-}}}{C_{[Fe(CN)_6]^{4-}}} = 0$$

then equation (1) becomes

$$E = E^\circ + \frac{2.303\,RT}{F}A\left(Z^2_{[Fe(CN)_6]^{4-}} - Z^2_{[Fe(CN)_6]^{3-}}\right)\sqrt{\mu}$$

$$= E^o + K\sqrt{\mu}\ . \qquad (2),$$

where $K = \dfrac{2.303\,RT}{F} A\left(Z^2_{[Fe(CN)_6]^{4-}} - Z^2_{[Fe(CN)_6]^{3-}}\right)$.

Our aim of the present experiment is to find out E^o.

The cell used for the experiment is

$$\underset{(L)}{\overset{(-)}{\text{Saturated calomel electrode}}} \parallel \underset{\text{(Solution of equimolar mixture)}\ (R)}{\overset{(+)}{\left[Fe(CN)_6\right]^{3-} - \left[Fe(CN)_6\right]^{4-} / Pt}}$$

Or, pictorially,

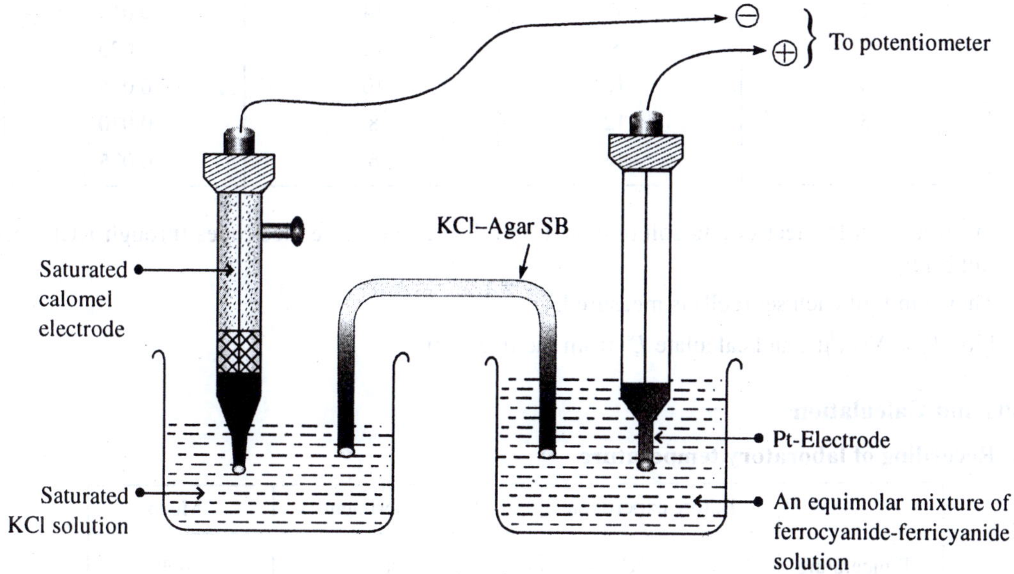

Fig. 6.27: Cell constructed for determination of E^o of ferricyanide-ferrocyanide system

The observed e.m.f. of the cell is given by

$$E_{cell} = E_R - E_L$$

$$= E^o + K\sqrt{\mu} - E_{cal} = (E^o - E_{cal}) + K\sqrt{\mu}$$

Plotting of E_{cell} vs. $\sqrt{\mu}$ for different solutions will give a straight line with positive slope, intercept of which will be $(E^o - E_{cal})$.

Thus, knowing E_{cal} at the laboratory temperature, E^o, the standard reduction potential of ferricyanide-ferrocyanide system, can be calculated at that temperature.

[Given: $E_{cal} = [\{0.2415 - 7 \times 10^{-4}(t - 25)\}$ volts, t = 15-40 °C].

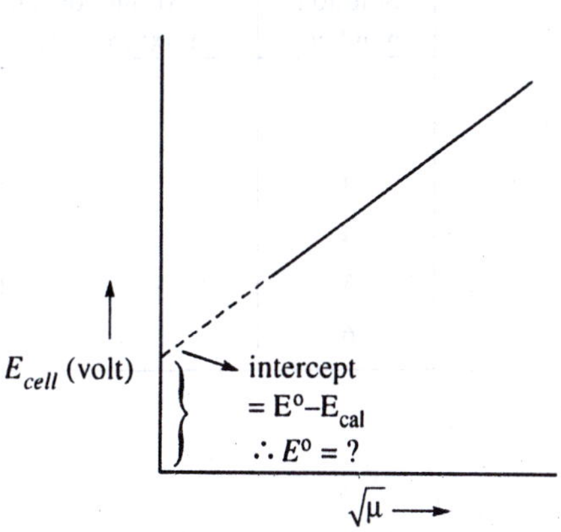

Fig. 6.28: Plot of E_{cell} vs. $\sqrt{\mu}$

Materials: (i) 0.1(M) $K_4[Fe(CN)_6]$ solution and 0.1(M) $K_3[Fe(CN)_6]$ solution (ii) Pt-electrode and saturated calomel electrode. (iii) KCl-Agar salt bridge, (iv) Potentiometer, (v) 100 ml beaker, Pipette, 50 ml burettle, Micro burette, etc.

Procedure

i. Preapare exact 0.1 (M) $K_4[Fe(CN)_6]$ solution (100 ml) and exact 0.1 (M) 100 ml $K_3[Fe(CN)_6]$ solution by accurate weighing and quantitative dilution, if necessary.

Then mix 50 ml 0.1 (M) ferrocyanide solution with 50 ml 0.1 (M) ferricyanide solution to prepare 100 ml 0.05 (M) ferricyanide-ferrocyanide mixture (i.e., equimolar mixture). From the aliquots the following sets of solutions are prepared (use burette to measure the volumes):

100 ml dry beaker No.	Volume of 0.05 (M) mixture (ml)	Volume of distilled water (ml)	Concentration of resultant mixture (M)
1	4	16	0.010
2	6	14	0.015
3	8	12	0.020
4	10	10	0.025
5	12	8	0.030
6	14	6	0.035

ii. Insert a clean Pt-electrode to solution no.1 and then connect the electrodes through KCl-Agar salt bridge.

iii. Then e.m.f. of each set (cell) is measured.

Plot E_{cell} Vs. $\sqrt{\mu}$, and calculate E^o from the intercept.

Results and Calculations

1. Recording of laboratory temperature

	Before experiment	After experiment	Mean
Temperature →	... °C	... °C	... °C

2. Recording of e.m.f. of different sets

Solution numbers	Molar concentration of equimolar mixture of ferricyaride-ferrocyanide	Observed e.m.f. (E_{cell}) (volts or mV)
1	0.010	
2	0.015	
3	0.020	
4	0.025	
5	0.030	
6	0.035	

3. **Graph ploting and calculation of E^o:**

Solution No.	Molar concentration of the equimolar solution	$\mu = \frac{1}{2}\sum C_i Z_i^2$	$\sqrt{\mu}$	E_{cell} (V or mV)
1	0.010			
2	0.015			
3	0.020			
4	0.025			
5	0.030			
6	0.035			

From the graph (plot of E_{Cell} vs. $\sqrt{\mu}$), it is found that intercept = ... volt.

$\therefore$ E^o = (intercept + E_{cal}) volt.

The expected value of E^o = +0.36 volt at 25 °C.

Experiment 12

Determination of strengths of hydrochloric acid and acetic acid in their mixture pH-*metrically.*

Theory: The negative logarithm of H^+ ion activity is defined as pH, i.e.,

$$pH = -\log_{10} a_{H^+}, \text{ when solution is very dilute, } a_{H^+} = [H^+].$$

$$\therefore \ pH = -\log[H^+].$$

pH of a solution can be measured accurately using glass electrode which consists of a thin membrane of a soft glass globe (soda lime glass) containing dilute solution of hydrochloric acid in which Ag-AgCl electrode is immersed. When this electrode is coupled with a reference electrode, say, a calomel electrode, cell obtained is

Ag-AgCl | 0.1N HCl | Glass | Unknown solution | Calomel (saturated) electrode

The *e.m.f.* of cell is $E = E_G - E_{cal} = E_G^o - E_{cal} + \frac{2.303\,RT}{F}\text{pH}$.

The *e.m.f.* of the cell is a linear function of pH of the solution in which the glass electrode is kept immersed. In view of the high resistance of the glass membrane (1 to 100 meg ohms), the e.m.f. of this cell cannot be measured with an oridnary potentiometer circuit. This necessitates the use of an electronic voltameter (as in pH-meter). The commonly used pH-meters directly read pH of the solution.

The experiment can be divided into two parts:

a. **Standardisation of NaOH solution by standard oxalic acid solution**

When oxalic acid solution is titrated with a solution of NaOH, then with progressive addition of NaOH solution pH of the solution increases slowly first and then sharply at equivalence points. The nature of curve that is found experimentally is shown in Fig. 6.29.

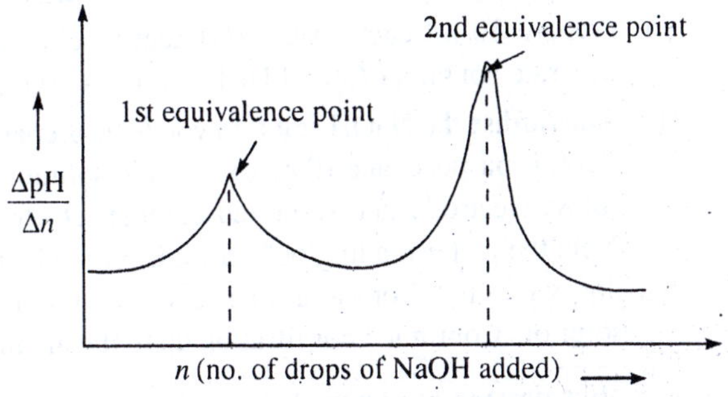

Fig. 6.29: Curve for standardisation of NaOH solution against standard oxalic acid solution (pH-metrically)

Thus, the strength of NaOH solution can

be found out from the second equivalence point as strength and volume of oxalic acid solution are known to us.

b. Titration of acid mixture

The titration of acid mixture ($HCl + CH_3COOH$) by the above standardised NaOH solution is also an acid-base titration. So, during the titration, pH of the solution increases with addition of NaOH solution. During the progress of titration of acid mixture vs. NaOH solution, first HCl will be neutralised followed by acetic acid. The titration is continued upto the alkaline range (pH = 11 to 12).

If we plot $\frac{\Delta pH}{\Delta n}$ vs. n, then the nature of graph will be, as shown in Fig. 6.30.

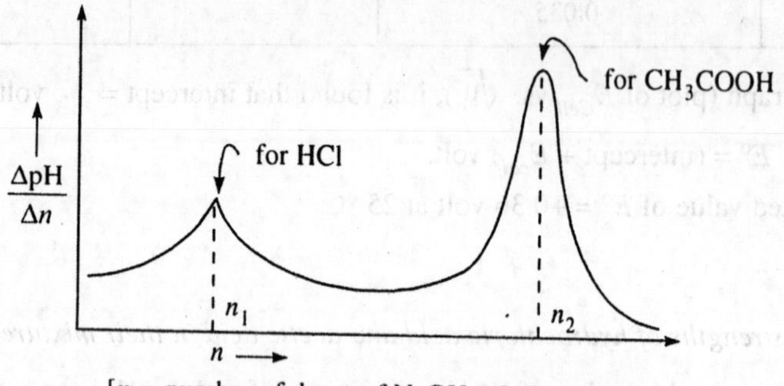

[n = number of drops of NaOH solution added]

Fig. 6.30: Curve for pH-metric titration of acid mixture by NaOH solution

So, knowing the number of drops of NaOH solution (i.e., the volumes of NaOH solution) required for neutralisation of HCl and CH_3COOH, we can calculate the individual strengths of HCl and CH_3COOH in their mixture.

Let n_1 be the number of drops of NaOH solution required for HCl and n_2 be the same for total acid mixture. Then the number of drops of NaOH solution required for neutralisation of CH_3COOH will be $(n_2 - n_1)$ drops.

Materials: (i) ~0.5N NaOH solution. (ii) ~0.05N oxalic acid solution. (iii) Acid mixture (of order ~0.05 N each). (iv) pH-electrodes and pH-meter. (v) Beaker, volumetric flasks, pipettes, microburettes, etc. (vi) Buffer solutions (pH 7 and pH 4 or 9.2).

Procedure

i. Prepare 100 ml ~0.5(N) NaOH solution (dissolve ≈ 2gm of solid NaOH beads in 100 ml water).
ii. Prepare 100 ml 0.05 (N) oxalic acid solution by accurate weighing.
iii. Prepare 50 ml each of 0.1 (N) acetic acid and 0.1 (N) hydrochloric acid and mix them in a seperate container (50 ml HCl solution +50 ml CH_3COOH, each of strength 0.1 (N).)
iv. Standardise the NaOH solution you have prepared by standard oxalic acid solution pH-metrically. For this purpose take 10 ml oxalic acid solution in a 50 ml beaker and immerse the pH-electrodes (glass electrode and saturated calomel electrode or a combined glass-calomel pH-electrode). Note the meter reading before addition of alkali. Then add NaOH solution from a micro-burette dropwise (1-2 drops at a time) and note the meter readings after each addition and shaking the solution. Continue your titration up to the alkaline range (i.e., pH > 7). Take 5 to 6 more readings after passing the equivalence point (i.e., after sharp increase in pH). Plot $\frac{\Delta pH}{\Delta n}$ vs. n (n is the number of drops of NaOH solution added) and calculate the strength of NaOH solution using the curve.

[Before starting titration standardise your pH-meter using buffer solutions of pH = 7.0 and pH = 4.2 or 9.2 following the method of standardisation of pH-meters described at the beginning of this chapter].

v. Wash the electrodes and take 10 ml of the acid mixture you have prepared (or supplied) in a beaker of 50 ml capacity and continue titration as before. Plot $\frac{\Delta pH}{\Delta n}$ vs. n and claculate the strengths of HCl and CH_3COOH separately from Fig. 6.30, as strength of NaOH solution is already known. Compare the strengths of HCl and CH_3COOH that you have already prepared.

vi. Wash and clean the pH-electrodes and immerse them into fresh distilled water or 4 (M) KCl solution.

Results and Calculations

1. Recording of laboratory temperature

	Before experiment	After experiment	Mean
Temperature →	··· °C	··· °C	··· °C

2. Preparation of 100 ml 0.05N oxalic acid solution

Relative equivalent mass of $(COOH)_2 \cdot 2H_2O = 63$.

Initial mass (gm)	Final mass (gm)	Mass of Oxalic acid transferred (gm)	Mass to be taken (gm)	Strength of oxalic acid
			0.315	··· (N)

3. Standardisation of NaOH solution (pH-metrically): Volume of oxalic acid solution taken = 10 ml.

No. of drops of NaOH solution added (n)	Observed pH	ΔpH	Δn	$\frac{\Delta pH}{\Delta n}$
0 (n_0)	$(pH)_0$	—	—	—
1 (n_1)	$(pH)_1$	$(pH)_1 - (pH)_0$	$n_1 - n_0$	⋮
2 (n_2)	$(pH)_2$	$(pH)_2 - (pH)_1$	$n_2 - n_1$	⋮
3 (n_3)	$(pH)_3$	$(pH)_3 - (pH)_2$	$n_3 - n_2$	⋮

From the plot of $\frac{\Delta pH}{\Delta n}$ vs. n (Fig. 6.29), the number of drops of NaOH solution required = ··· drops = ··· ml. [∵ 1 ml = ··· drops].

∴ strength of NaOH solution = ··· (N).

4. Titration of acid mixture: Volume of acid mixture taken = 10 ml.

No. of drops of NaOH solution added (n)	Observed pH	ΔpH	Δn	$\frac{\Delta pH}{\Delta n}$
0				
1				
2				
3				
⋮				

Calculate $\frac{\Delta pH}{\Delta n}$ using similar method given in the above table.

Calculation of strengths of acid mixture: From the plotting of $\frac{\Delta pH}{\Delta n}$ vs. *n* Fig. 6.30 it is found that for first equivalence point corresponding to HCl, the number of drops of NaOH solution required = ... drops = ... ml and for CH_3COOH (corresponding to second equivalence point) no. of drops of NaOH required = ... drops = ... ml.

$$\left.\begin{aligned} \therefore \quad &\text{strength of HCl} &= \ldots \text{ (N)} \\ \text{and} \quad &\text{strength of } CH_3COOH &= \ldots \text{ (N)} \end{aligned}\right\}$$

Experiment 13

Determination of ionic product (K_w) of water at room temperature pH-metrically.

Theory: In pure state, water dissociates to a very small extent and behaves as a weak electrolyte. The equilibrium constant (*K*) of the dissociation,

$$H_2O \rightleftharpoons H^+ + OH^- \text{ is given by } K = \frac{a_{H^+} \cdot a_{OH^-}}{a_{H_2O}}.$$

In pure water or in dilute solution, the activity of H_2O is constant and is taken as unity. Hence, $a_{H^+} \cdot a_{OH^-} = K_w$, which is called the ionisation constant of water. Replacing activities with concentration and activity coefficient,

$$K_w = C_H \cdot C_{OH^-} \cdot f_{H^+} \cdot f_{OH^-} = K'_w \cdot f_{H^+} \cdot f_{OH^-},$$

where K'_w is called ionic product of water. In pure water or in very dilute solution, the activity coefficients f_{H^+}, f_{OH^-} are almost unity and so $K_w = K'_w$.

The above experiment can be done with the help of a pH-meter. The required cell is

$$\underset{(L)}{\text{Glass electrode}} \,\Big|\, \text{HCl (0.01 N)} \,\Big\|\, \text{Saturated calomel electrode} \,\Big\|\, \text{KOH (0.01N)} \,\Big|\, \underset{(R)}{\text{Glass electrode}}$$

Now, $H^+ + e \rightleftharpoons \frac{1}{2} H_2$

$$\therefore \quad E_L = E_H^o - 0.0591 \log \frac{a_{H_2}^{1/2}}{a_{H^+}} \text{ [at 25 °C]}$$

$$= -0.0591 \log \frac{1}{a_{H^+}} \quad [E_H^o = 0 \text{ and } a_{H_2} = 1]$$

$$= -0.0591 (-\log a_{H^+})$$

$$= -0.0591 \text{ pH (1)}. \quad (1)$$

Similarly, $E_R = -0.0591 \text{ pH (2)}$. (2)

$$\therefore \quad E_{cell} = E_R - E_L = -0.0591 \text{ [pH (2)} - \text{pH (1)]} \quad (3)$$

Again, for a concentration cell,

$$E_{cell} = E_R - E_L = \frac{RT}{F} \ln \frac{a_2}{a_1}$$

$$= 0.0591 \log \frac{a_{H^+}(2)}{a_{H^+}(1)} \quad \text{[at 25 °C]}$$

$$= 0.0591 \log \frac{a_{H^+}(2) \cdot a_{OH^-}(2)}{a_{H^+}(1) \cdot a_{OH^-}(2)}$$

$$= 0.0591 \log \frac{K_w}{a_{H^+}(1) \cdot a_{OH^-}(2)} \tag{4}$$

Comparing (3) and (4), we get

$$pH(2) - pH(1) = -\log \frac{K_w}{a_{H^+}(1) \cdot a_{OH^-}(2)}$$

or, $$pH(1) - pH(2) = \log \frac{K_w}{a_{H^+}(1) \cdot a_{OH^-}(2)}$$

Without activity correction,

$$pH(1) - pH(2) = \log K_w - \log(C_{H^+} \cdot C_{OH^-}) \tag{5}$$

Here pH (1) is the measured pH of HCl solution, when $C_{H^+} = 0.01(N)$ and pH(2) is the measured pH of KOH solution, when $C_{OH^-} = 0.01(N)$. Thus, we can determine K_w, the ionic product of water at the experimental temperature, if pH's of dilute HCl and dilute KOH solutions (of 0.01 N) are measured.

Materials: (i) Pure double distilled water. (ii) KOH solution and HCl solution (0.01 N). (iii) pH-meter and its electrodes. (iii) 0.1(N) oxalic acid solution. (iv) Burettes, pipettes, beakers, conical flasks, etc.

Short Procedure

i. Preparare 100 ml 0.1 (N) oxalic acid solution by accrate weighing.
ii. Prepare 250 ml each of ≈ 0.1 (N) HCl solution (base normality 11.6 N) and ≈ 0.1 (N) KOH solution (equivalent weight 56, i.e., mass of KOH to be dissolved in 250 ml is ≈ 1.5 g].
iii. Standardise KOH solution against standard oxalic acid solution and then HCl solution against KOH solution using prenolphthalein as indicator.
iv. Then prepare 100 ml each of 0.01 N HCl solution and 0.01 (N) KOH solution by proper dilution.
v. Calibrate your pH-meter using buffer solutions of pH 7 and pH 4 following usual method described previously. Measure the pH of 0.01 (N) HCl solution and 0.01 (N) KOH solution using pH-meter.
vi. Calculate K_w from equation (5).

Results and Calculations

1. Recording of laboratory temperature

	Before experiment	After experiment	Mean
Temperature →	... °C	... °C	... °C

2. Preparation of 100 ml 0.1 (N) oxalic acid solution

Initial mass (g)	Final mass (g)	Mass of oxalic acid transferred (g)	Mass of oxalic acid to be taken (g)	Strength of oxalic acid soln.
...	...	...	0.63	... (N)

3. Standardisation of KOH solution by standard oxalic acid solution

Volume of oxalic acid taken	Volume of KOH required (ml)	Mean volume of KOH solution required (ml)	Strength of KOH solution
10 ml	(a) (b) (c)	...	... (N)

4. Standardisation of HCl solution by standardised KOH solution

Volume of HCl solution taken	Volume of KOH solution required (ml)	Mean volume of KOH solution (ml)	Strength of HCl solution
10 ml	(a) (b) (c)	...	... (N)

5. (a) 100 ml exact 0.01 (N) KOH is prepared by diluting ... ml KOH solution to 100 ml.

(b) 100 ml exact 0.01 (N) HCl solution is prepared by diluting .... ml HCl to 100 ml.

6. Recording of pH-values

Substances	Concentration	Observed pH	Calculated K_w
HCl solution	0.01 (N)	pH (1) =	
KOH solution	0.01 (N)	pH (2) =	

7. Calculation: Use equation (5).

8. Conclusion: The ionic product of water that is determined ··· at ··· °C.

Expected value is 1.05×10^{-14} at 25 °C.

[**N.B.**: The water which is to be used for this experiment must be extremely pure. Double or triple distilled water must be used. Otherwise the value of K_w will not be accurate.]

Experiment 14

Determination of hydrolytic constant (K_h) of ammonium chloride solution pH-*metrically.*

Theory: Ammonium chloride (NH_4Cl) is a salt of strong acid (HCl) and weak base (NH_4OH), which hydrolyses partly in water as below:

$$NH_4Cl + H_2O \rightleftharpoons NH_4OH + HCl$$

i.e., $NH_4^+ + Cl^- + H_2O \rightleftharpoons NH_4OH + H^+ + Cl^-$.

Suppose, the degree of hydrolysis is α (which is very small) and initial concentration of the salt is C mole/lit. Now the equilibrium constant, called hydrolytic constant (K_h), is given by

$$K_h = \frac{a_{NH_4OH} \cdot a_{H^+} \cdot a_{Cl^-}}{a_{NH_4^+} \cdot a_{Cl^-} \cdot a_{H_2O}} = \frac{a_{NH_4OH} \cdot a_{H^+}}{a_{NH_4^+}}$$

[a terms denote activity of the species] $[\because a_{H_2O} = 1]$

$$\text{or,} \quad K_h = \frac{C_{NH_4OH} \cdot C_{H^+}}{C_{NH_4^+}} \frac{f_{NH_4OH} \cdot f_{H^+}}{f_{NH_4^+}}$$

[f terms denote activity coefficient of the species]

If ionic strength of the solution is very low, then

$$f_{NH_4OH} = 1, \quad \frac{f_{H^+}}{f_{NH_4^+}} \simeq 1, \quad \therefore \; K_h = \frac{C_{NH_4OH} \cdot C_{H^+}}{C_{NH_4^+}}.$$

Now, $C_{NH_4OH} = C\alpha$, $C_{H^+} = C\alpha$ and $C_{NH_4^+} = C(1-\alpha)$.

$$\therefore \; K_h = \frac{C\alpha \cdot C\alpha}{C(1-\alpha)} \approx C\alpha^2 \quad [\because \; \alpha << 1].$$

$$\therefore \; \alpha = \sqrt{K_h/C}, \qquad \therefore \; C_{H^+} = C\alpha = \sqrt{K_h \cdot C}$$

$$\text{or,} \quad \log C_{H^+} = \frac{1}{2}\log K_h + \frac{1}{2}\log C$$

$$\text{or,} \quad \text{pH} = -\frac{1}{2}\log K_h - \frac{1}{2}\log C.$$

Thus, if we plot pH vs. $\log C$, a straight line with slope $\left(-\frac{1}{2}\right)$ and intercept $\left(-\frac{1}{2}\log K_h\right)$ will be obtained (Fig. 6.31). From the intercept, K_h of NH_4Cl can be calculated.

The principle of pH-meter and its use are same as previous experiment.

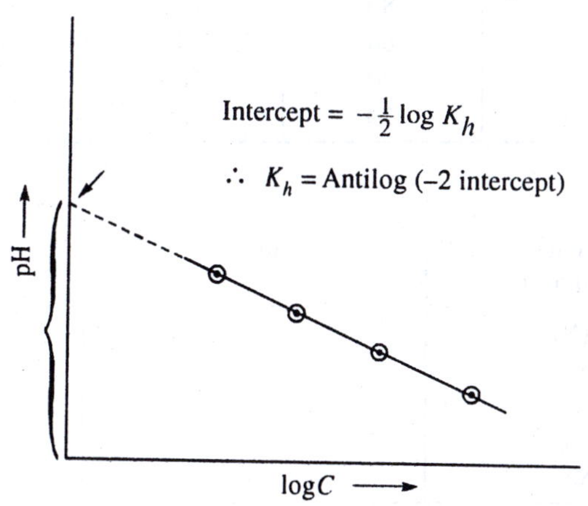

Fig. 6.31: Plot of pH vs. $\log C$ for determination of K_h

Materials: (i) 0.5 (N) NH_4Cl solution. (ii) pH-meter, (iii) Buffer solutions (pH 4 and 7) for calibration of pH-meter. (iv) pH-electrodes, (v) Beaker, volumetric flasks, burettes, pipettes, etc.

Short Procedure

i. Prepare 250 ml exact 0.5 (N) NH_4Cl solution and prepare 100 ml each of 0.4 (N), 0.3 (N), 0.2 (N), 0.1 (N) and 0.05 (N) from the 0.5 (N) solution by proper dilution.

ii. Standardise your pH-meter using buffer solutions of pH 7 and 4 (method of calibration is given at the begining of this chapter).

iii. Measure the pH's of NH_4Cl solutions of different concentrations starting from lower concentration [0.05 (N) NH_4Cl].

iv. Plot pH vs log C, and calculate the K_h value from the intercept.

Results and Calculations

1. Recording of laboratory temperature

	Before experiment	After experiment	Mean
Temperature →	... °C	... °C	... °C

2. Preparation of 250 ml 0.5 (N) NH_4Cl solution: Relative equivalent mass of NH_4Cl = 53.46.

Initial mass (g)	Final mass (g)	Mass of NH_4Cl transferred (g)	Mass of NH_4Cl to be taken (g)	Concentration of NH_4Cl solution
...	...	...	6.6825	... (N)

3. Preparation of NH_4Cl solutions of different concentration and recordings their pH-values using pH-meter

Solution Number	Volume of 0.5 (N) NH_4Cl solution (ml)	Volume of water added (ml)	Total volume (ml)	Concentration of NH_4Cl solutions	Observed pH
1	10	90	100	0.05 (N)	...
2	20	80	100	0.10 (N)	...
3	40	60	100	0.20 (N)	...
4	60	40	100	0.30 (N)	...
5	80	20	100	0.40 (N)	...

4. Graph plotting

Concentration of NH_4Cl solution (C)	log C	Observed pH
1 0.05 (N)	...	...
2 0.1 (N)	...	...
3 0.2 (N)	...	...
4 0.3 (N)	...	...
5 0.4 (N)	...	...

5. Calculation of K_h: From Fig. 6.31, the plot of pH vs. log$C$, intercept $= -\frac{1}{2}\log K_h$.

$\therefore$ K_h = Antilog of (–2 intercept).

Expected value of hydrolytic constant (K_h) of NH_4Cl is 5×10^{-10} at 25 °C.

[**N.B.** Ammonium chloride which is to be used must be pure and moisture free. Otherwise NH_4Cl absorbs moisture and undergoes hydrolysis to some extent before preperation of solution, and the result obtained will be inaccurate].

Experiment 15

Determination of the hydrolytic constant (K_h) of aniline hydrochloride using a pH-meter.

Theory: Aniline hydrochloride, the salt of a weak base and strong acid, undergoes hydrolysis according to the following equation.

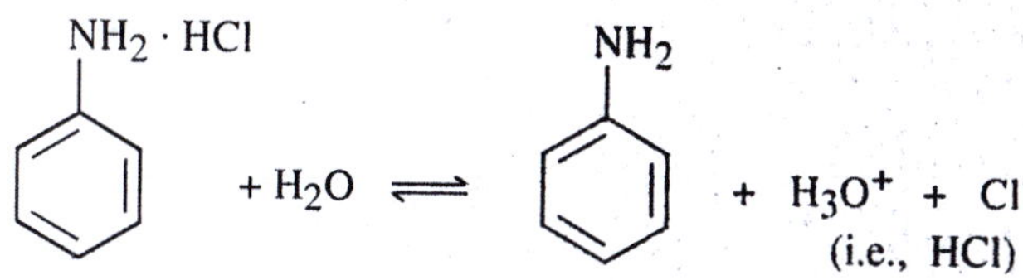

$C(1-\alpha)$ $\qquad$ $C\alpha$ $\qquad$ $C\alpha$

The hydrolytic constant (K_h) is given by

$$K_h = \frac{C\alpha \cdot C\alpha}{C(1-\alpha)}, \qquad (1)$$

where α is the degree of hydrolysis and C is the total concentration of the salt in moles per litre. $C\alpha$, the concentration of free acid, may be taken as equal to $[H^+]$ (as HCl is completely ionised), where $[H^+]$ is the molar concentration of hydrogen ions in solution. The hydrolytic constant then may be written from equation (1) as below:

$$K_h = \frac{[H^+][H^+]}{C-[H^+]} = \frac{[H^+]^2}{C-[H^+]}. \qquad (2)$$

From equation (2), it follows that by measuring the pH, i.e., $[H^+]$ in a salt solution of known concentration, the hydrolytic constant (K_h) can be found out. Method for measurement of pH of the salt solution using glass electrode is as usual.

Materials: (i) Aniline hydrochloride crystals, (ii) pH-meter and its electrodes, (iii) Beakers, (iv) Volumetric flasks, (v) Burettes and pipettes, (vi) Tissue paper, (vii) Buffer solutions of pH 7 and 4.

Short Procedure

i. Prepare 250 ml $\frac{M}{10}$ aniline hydrochloride solution by accurate weighing and then prepare 100 ml each of $\frac{M}{50}$ and $\frac{M}{250}$ solutions from $\frac{M}{10}$ solution by quantitative dilution.

ii. Standardise the pH-meter using buffer solutions of pH 7 and pH 4 using usual method.

iii. Measure the pH's of the three aniline hydrochloride solutions $\frac{M}{250}$, $\frac{M}{50}$ and $\frac{M}{10}$ starting from lower concentration.

iv. Calculate $[H^+]$ of the three solutions and then K_h from equation (2). Calculate the mean K_h.

Results and Calculations

1. Recording of laboratory temperature

	Before experiment	After experiment	Mean
Temperature →	... °C	... °C	... °C

2. Preparation of 250 ml $\frac{M}{10}$ aniline hydrochloride solution

Initial mass (g)	Final mass (g)	Mass of the salt transferred (g)	Mass of the salt to be taken (g)	Concentration of the salt
...	...	...	...	... (M)

Prepare exact $\frac{M}{10}$ salt solution, if necessary, by proper dilution.

3. Preparation of solutions of aniline hydrochloride of different concentrations

Solution number	Volume of $\frac{M}{10}$ salt solution (ml)	Volume of distilled water added (ml)	Total volume (ml)	Concentration of the prepared salt solution
1	4	96	100	$\frac{M}{250}$
2	20	80	100	$\frac{M}{50}$
3	100	0	100	$\frac{M}{10}$

4. Recording of pH of the solutions

Solution no.	Concentation	Observed pH	Concentration of hydrogen ions $[H^+]$ (gm-ions/lit)
1	$\frac{M}{250}$	...	pH = $-\log [H^+]$
2	$\frac{M}{50}$	...	$\therefore$ $[H^+]$ = ...
3	$\frac{M}{10}$	...	

5. Determination of hydrolytic constants (K_h)

Solution number	Concentration of the salt (C)	$[H^+]$ gm-ions/lit	K_h (moles/lit)	Mean K_h (moles/lit)
1	$\frac{M}{250}$	...	...	
2	$\frac{M}{50}$	...	...	...
3	$\frac{M}{10}$	...	...	

6. **Conclusion:** Hence hydrolytic constant (K_h) of aniline hydrochloride is moles/lit at °C.
Expected value = moles/lit.

Experiment 16

Determination of pK_a-values of dibasic acid (e.g., oxalic acid) pH-metrically at room temperature.

Theory: Oxalic acid is a dibasic acid and it is neutralised in two steps as follows:

$$\underset{\text{(Oxalic acid)}}{\begin{array}{c} COOH \\ | \\ COOH \end{array}} + H_2O \rightleftharpoons H_3O^+ + \underset{\text{(mono-oxalate)}}{\begin{array}{c} COO^- \\ | \\ COOH \end{array}}$$

$$\begin{array}{c} COO^- \\ | \\ COOH \end{array} + H_2O \rightleftharpoons H_3O^+ + \underset{\text{(Oxalation)}}{\begin{array}{c} COO^- \\ | \\ COO^- \end{array}}$$

The second dissociation is approximately 10^3 times slower than the first one. Therefore, if we titrate it with a standard solution of caustic soda pH-metrically, the two neutratisation points will be distinct. The pH of the solution at half neutralisation points of the first and half neutralisation point of the second are equal to pK_{a_1} and pK_{a_2} respectively.

If a known volume of oxalic acid solution is titrated with a solution of standard caustic soda pH-metrically, the nature of graph, when pH-values at each step of addition of NaOH solution are plotted, will be as follows (Fig. 6.32). The use of glass electrode and calomel electrode or combined pH-electrode in the pH-meter and its calibration is as usual.

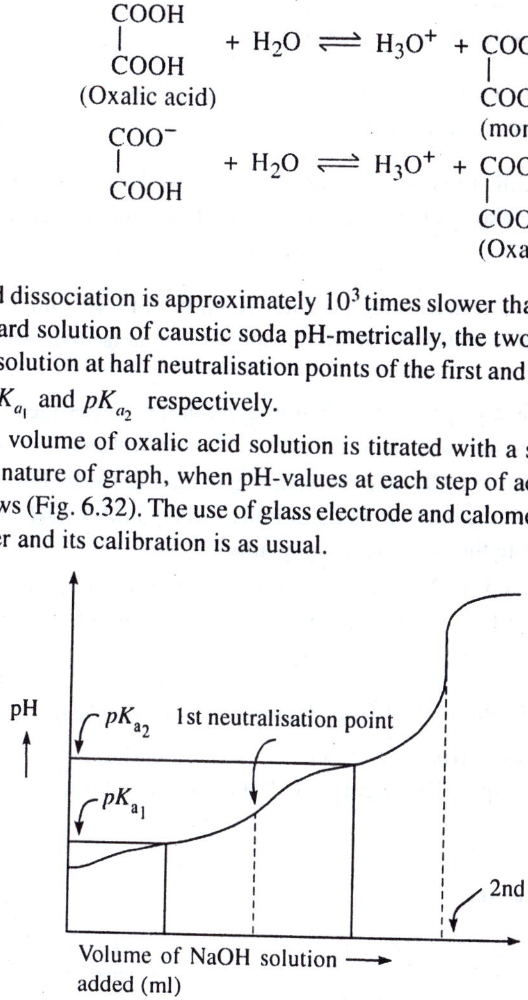

Fig. 6.32: Plot of observed pH values vs. volume of NaOH solution added for titration of oxalic acid salution against standard NaOH solution

pK_a-values thus can be determined from the pH-values corresponding to the half neutralisation points f the first and second neutralisation points respectively as shown in the above figure.

Also the strength of oxalic acid solution can be determined from the volume of NaOH solution equired corresponding to the second neutralisation point, if the strength of NaOH solution is known. owever, the strength of oxalic acid solution will be more accurate if we plot (volume of NaOH solution dded) as shown in the following figure 6.33.

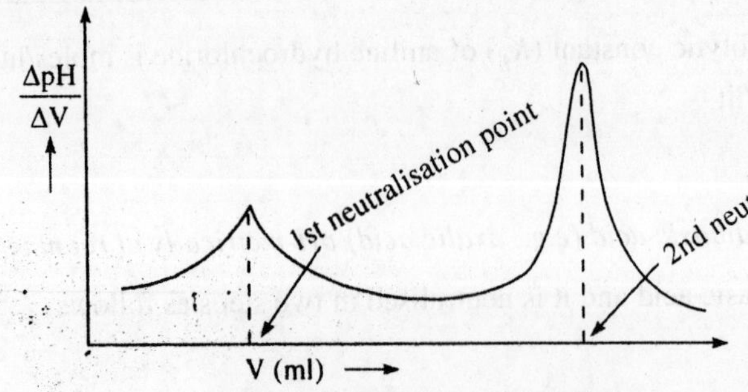

Fig. 6.33: Plot of $\frac{\Delta pH}{\Delta V}$ vs. V (volume of NaOH solution added) for titration of oxalic acid solution against standard NaOH solution.

Materials: (i) Oxalic acid crystals (G.R. grade), (ii) Caustic soda, (iii) pH-meter with electrodes, (iii) Buffer solutions of pH 4 and 7, (v) Beaker, volumetric flaskes, burettes, pipettes, tissue paper, etc.

Short Procedure

i. Prepare 100 ml 0.4 (N) oxalic acid solution by accurate weighing and 100 ml 1 (N) NaOH solution ($\approx$ 4 g NaOH pallets dissolve in 100 ml distilled water) and standardise NaOH solution by oxalic acid analytically.

ii. Prepare buffer solutions 100 ml each of pH 7 and pH 4 and calibrate the pH-meter to be used following usual method or instrumental manual supplied by the manufacturer.

iii. Take 25 ml of the oxalic acid solution in a 100 ml beaker, immerse the electrodes into the solution and connect with the pH-meter. Note the meter reading initially. Add 0.5 ml NaOH solution from a micro-burette and shake well. Note the meter reading again.

iv. Continue your titration by adding 0.5 ml NaOH solution at a time and note meter readings at each step. Take 5-6 more readings after complete neutralisation point, i.e., above pH 7.

v. Plot pH vs. V and determine the pK_a values from the graph as described in the theory.

vi. Plot $\frac{\Delta pH}{\Delta V}$ vs. V (volume of NaOH solution added) and determine the volume of NaOH solution required for 25 ml oxalic acid solution from the graph as described in the theory. Calculate the strength of oxalic acid solution. Compare the strength with that prepared by weighing.

Results and Calculations

1. Recording of laboratory temperature

	Before experiment	After experiment	Mean
Temperature →	... °C	... °C	... °C

2. Preparation of 100 ml 0.4 N oxalic acid solution

Relative equivalence mass of $(COOH) \cdot 2H_2O = 63$

Initial mass (g)	Final mass (g)	Mass of oxalic acid transferred (g)	Mass of oxalic acid to be taken (g)	Strength of oxalic acid soln.
...	...	...	2.52	... (N)

3. Titration of oxalic acid solution pH-metrically

No. of observation	Volume of oxalic acid solution taken (ml)	Volume of NaOH solution added (ml)	Observed pH
1		0	
2		0.5 *	
3	25	1.0	
⋮		⋮	
⋮		⋮	
⋮		⋮	

*For accurate resullts add 0.2 ml NaOH solution at each step of titration.

4. Table of graph plotting

Volume of NaOH solution added (ml) V	pH observed	ΔpH	ΔV ml	$\frac{\Delta pH}{\Delta V}$
0 (V_0)	$(pH)_0$	—	—	—
0.5 (V_1)	$(pH)_1$	$(pH)_1 - (pH)_0$	$V_1 - V_0$	⋮
1.0 (V_2)	$(pH)_2$	$(pH)_2 - (pH)_1$	$V_2 - V_1$	⋮
⋮ ⋮	⋮	⋮	⋮	⋮
⋮ ⋮	⋮	⋮	⋮	⋮

5. Calculations

a. From Fig. 6.32 (plot of pH vs *V*)

$$\left.\begin{aligned} pK_{a_1} &= \cdots \\ pK_{a_2} &= \cdots \end{aligned}\right\} \text{ at } \cdots \ ^{\circ}\text{C}$$

b. From Fig. 6.33 (Plot of $\frac{\Delta pH}{\Delta V}$ vs. *V*)

Volume of NaOH solution required = ... ml

∴ strength of oxalic acid solution = ... (N).

6. Comparison of the strength of the oxalic acid solution

Method	Strength of oxalic acid solution	Ratio
pH-metric	... (N)	...
By weight	... (N)	

Suggestions of performing similar type of experiment.

Experiment 16A

Determination of pK_a values of phosphoric acid pH-metrically at laboratory temperature.

The titration selected is that of phosphoric acid with sodium hydroxide solution. Phosphoric acid is a tribasic acid with three successive ionisations:

$H_3PO_4 \rightleftharpoons H_2PO_4^- + H^+$

$H_2PO_4^- \rightleftharpoons HPO_4^{2-} + H^+$

$HPO_4^{2-} \rightleftharpoons PO_4^{3-} + H^+$

10 ml of 0.1 (N) H_3PO_4 is taken in a 100 ml beaker and add water such that the pH-electrodes (electrode assembly) is just immersed into the solution. Using usual method titrate it dropwise with NaOH solution (say, 0.5 N) pH-metrically. The solution should be stirred after each addition of alkali before the pH-measurement. On completion of the titration the electrode system is immersed in distilled water. The results are recorded in tabular form under the headings:

Observation no.	No. of drops of NaOH added (n)	Observed pH	$\frac{\Delta pH}{\Delta V}$
1	2	⋮	
2	4	⋮	
3	6	⋮	
⋮	⋮	⋮	

Two graphs are plotted:

a. pH vs. n (no. of drops of NaOH)

b. $\frac{\Delta pH}{\Delta V}$ vs. V

The points of inflection on the first graph are correlated with the sharp peaks on the second graph.

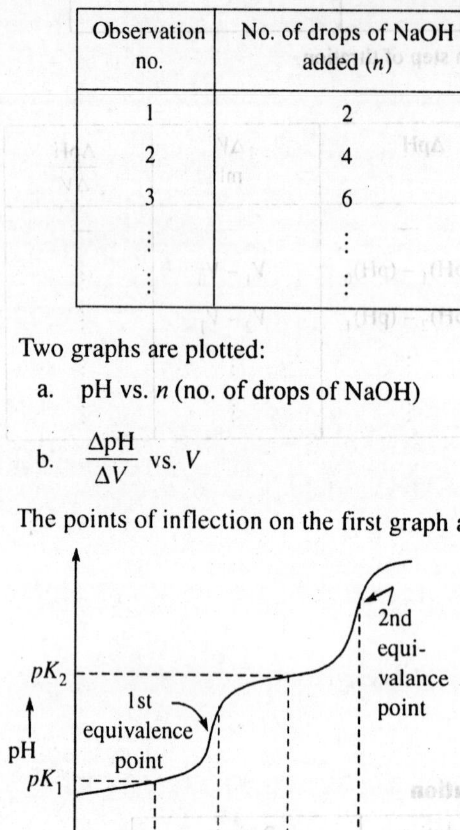

Fig. 6.34: Plot of pH vs. n (no. of drops NaOH added) for the titration of H_3PO_4 with standard NaOH solution

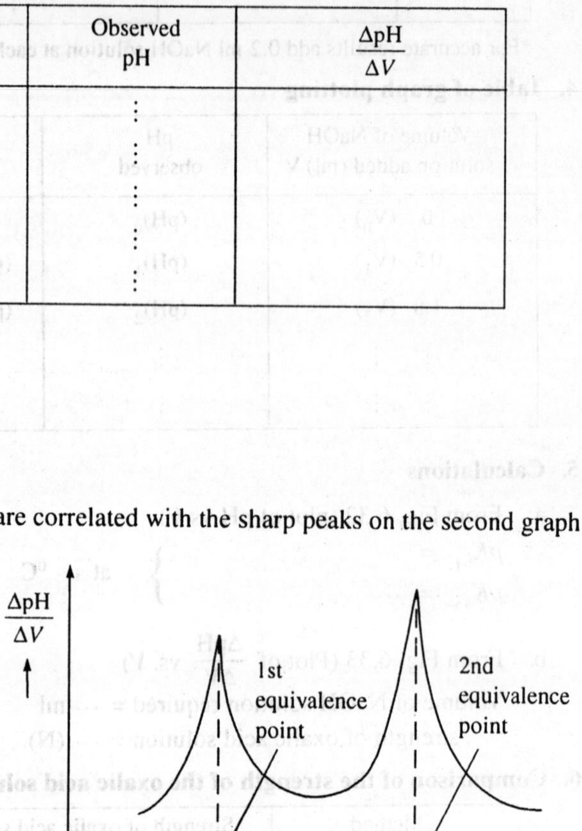

Fig. 6.35: Plot of $\frac{\Delta pH}{\Delta V}$ vs. (Vol. of NaOH solution added) for the titration of H_3PO_4 with standardised NaOH solution

pH at 1st $\frac{1}{2}$- neutralisation = … , $\therefore$ pK_1 = … (expected 2.12 at 25 °C)

pH at 2nd $\frac{1}{2}$- neutralisation = … , $\therefore$ pK_2 = … (expected 7.21 at 25 °C)

Determination of pK_3, using glass electrode, is unusual as glass electrodes become ineffective in the region of the third inflection point.

APPENDIX

Table 1: Standard reduction potentials (E^o) of some common redox systems at 25 °C and 1 atm. in volt.

System	Half-cell reaction	E° (volt)
$S_2O_8^{2-}/SO_4^{2-}$	$S_2O_8^{2-} + 2e \rightleftharpoons 2SO_4^{2-}$	+2.01
MnO_4^-/MnO_2	$MnO_4^- + 4H^+ + 3e \rightleftharpoons MnO_2 + 2H_2O$	+ 1.69
Ce^{4+}/Ce^{3+}	$Ce^{4+} + e \rightleftharpoons Ce^{3+}$	+ 1.61
$BrO_3^-/\frac{1}{2}Br_2$	$BrO_3^- + 6H^+ + 5e \rightleftharpoons \frac{1}{2}Br_2 + 3H_2O$	+ 1.52
MnO_4^-/Mn^{2+}	$MnO_4^- + 8H^+ + 5e \rightleftharpoons Mn^{2+} + 4H_2O$	+ 1.51
$Cr_2O_7^{2-}/Cr^{3+}$	$Cr_2O_7^{2-} + 14H^+ + 6e \rightleftharpoons \frac{1}{2}Cr^{3+} + 7H_2O$	+1.33
$IO_3^-/\frac{1}{2}I_2$	$IO_3^- + 6H^+ + 5e \rightleftharpoons \frac{1}{2}I_2 + 3H_2O$	+ 1.20
Cu^{2+}/CuI	$Cu^2 + I^- + e \rightleftharpoons CuI$	+ 0.86
Ag^+/Ag	$Ag^+ + e \rightleftharpoons Ag$	+ 0.799
Fe^{3+}/Fe^{2+}	$Fe^{3+} + e \rightleftharpoons Fe^{2+}$	+ 0.77
Hg_2^{+2}/Hg	$Hg_2^{2+} + 2e \rightleftharpoons 2Hg$	+ 0.79
H_3AsO_4/H_3AsO_3	$H_3AsO_4 + 2H^+ + 2e \rightleftharpoons H_3AsO_3 + H_2O$	+ 0.56
I_2/I^-	$I_2 + 2e \rightleftharpoons 2I^-$	+ 0.54
$Fe(CN)_6^{3-}/Fe(CN)_6^{4-}$	$Fe(CN)_6^{3-} + e \rightleftharpoons Fe(CN)_6^{4-}$	+ 0.36
Cu^{+2}/Cu	$Cu^{2+} + 2e \rightleftharpoons Cu$	+ 0.337
$Cl^-, AgCl(s)/Ag$	$AgCl(s) + e \rightleftharpoons Ag + Cl^-$	+ 0.222
Cu^{2+}/Cu^{+1}	$Cu^{2+} + e \rightleftharpoons Cu^+$	+ 0.153
Sn^{4+}/Sn^{2+}	$Sn^{4+} + 2e \rightleftharpoons Sn^{2+}$	+ 0.15
$S_4O_6^{2-}/S_2O_3^{2-}$	$S_4O_6^{2-} + 2e \rightleftharpoons 2S_2O_3^{2-}$	+ 0.08
$AgBr(s)/Ag$	$AgBr + e \rightleftharpoons Ag + Br^-$	+ 0.071
$H^+/\frac{1}{2}H_2$	$H^+ + e \rightleftharpoons \frac{1}{2}H_2$	0.00

Table 1 (*contd.*)

System	Half-cell reaction	E^o (volt)
Fe^{3+}/Fe	$Fe^{3+} + 3e \rightleftharpoons Fe$	– 0.04
Pb^{2}/Pb	$Pb^{2+} + 2e \rightleftharpoons Pb$	– 0.13
Sn^{2+}/Sn	$Sn^{2} + 2e \rightleftharpoons Sn$	– 0.14
AgI(s)/Ag	$AgI + e \rightleftharpoons Ag + I^-$	– 0.151
I^-, CuI(s)/Cu	$CuI + e \rightleftharpoons Cu + I^-$	– 0.185
SO_4^{2-}, $PbSO_4/Pb$	$PbSO_4 + 2e \rightleftharpoons Pb + SO_4^{2-}$	– 0.355
Cd^{2+}/Cd	$Cd^{2+} + 2e \rightleftharpoons Cd$	– 0.403
Fe^{2+}/Fe	$Fe^{2+} + 2e \rightleftharpoons Fe$	– 0.44
Zn^{2+}/Zn	$Zn^{2+} + 2e \rightleftharpoons Zn$	– 0.763
Al^{3+}/Al	$Al^{3+} + 3e \rightleftharpoons Al$	– 1.66
H^-/H_2	$\frac{1}{2}H_2 + e \rightleftharpoons H^-$	– 2.25
Mg^{2+}/Mg	$Mg^{2+} + 2e \rightleftharpoons Mg$	– 2.36
$Na^+/Na(Hg)$	$Na^+ + e \rightleftharpoons Na$	– 2.71
$Ca^{2+}/Ca(Hg)$	$Ca^{2+} + 2e \rightleftharpoons Ca$	– 2.87
$K^+/K(Hg)$	$K^+ + e \rightleftharpoons K$	– 2.92
$Li^+/Li(Hg)$	$Li^+ + e \rightleftharpoons Li$	– 3.05
$H^+(Pt)$, Q/H_2Q (Quinhydrone electrode)	$C_6H_4O_2 + 2H^+ + 2e \rightleftharpoons C_6H_4(OH)_2$ ($Q + 2H^+ + 2e \rightleftharpoons H_2Q$)	+ 0.6994

Table 2: Formal potentials ($E^{o'}$) of some redox systems in different medium at 25°C

System	Medium	($E^{o'}$) (volt)
Hg_2^{2+}/Hg	Saturated KCl solution (calomel)	+ 0.2458
E^o = +0.79V		
Fe^{3+}/Fe^{2+}	1 (M) $HClO_4$	+ 0.73
Fe^{3+}/Fe^{2+}	1(M) H_2SO_4	+ 0.68
Fe^{3+}/Fe^{2+}	1 (M) HCl	+ 0.70
Fe^{3+}/Fe^{2+}	0.5 (M) H_3PO_4 + 1 (M) H_2SO_4	+ 0.61
E^o = + 0.77 V		
Ce^{4+}/Ce^{3+}	1 (M) H_2SO_4	+ 1.44
Ce^{4+}/Ce^{3+}	1 (M) HNO_3	+ 1.61
Ce^{4+}/Ce^{3+}	1 (M) $HClO_4$	+ 1.70
E^o = +1.61 V		
$[Fe(CN)_6]^{3-}/[Fe(CN)_6]^{4-}$	Equal concentrations of both cyanoferrates in 1 (M) HCl, H_2SO_4 or $HClO_4$	+ 0.71
E^o = +0.36 V		
Hg^{2+}/Hg_2^{2+}	Thiocyanate ion (SCN^-)	+ 0.743
E^o = + 0.92 V		

Table 3: Ionic equations for use in the calculation of relative equivalent mass (equivalent weight) of some oxidising and reducing agents

The equivalent of an oxident or a reductant is the mole divided by the number of electrons which 1 mole of the substance gains or loses in the reaction, e.g.

$$MnO_4^- + 8H^+ + 5e \rightleftharpoons Mn^{2+} + 4H_2O; \text{Eq.} = \frac{MNO_4^-}{5} = \frac{KMnO_4}{5}$$

$$C_2O_4^{2-} \rightleftharpoons 2CO_2 + 2e; \text{ Eq}_1 = \frac{C_2O_4^{2-}}{2} = \frac{H_2C_2O_4}{2}$$

$$\text{Thus, Eq. mass} = \frac{\text{Molar mass}}{n} = \frac{M}{n} \quad (n = \text{no. of electrons transferred})$$

Substance	Partial ionic equation	n
Pot. permanganate (acid)	$MnO_4^- + 8H^+ + 5e \rightleftharpoons Mn^{2+} + 4H_2O$	5
Pot. Permanganate (neutral)	$MnO_4^- + 2H^+ + 3e \rightleftharpoons MnO_2 + 4OH^-$	3
Pot. Permanganate (Strong alkaline)	$MnO_4^- + e \rightleftharpoons MnO_4^{2-}$	1

Table 3 (*contd.*)

Substance	Partial ionic equation	n
Pot. dichromate	$C_2O_7^{2-} + 14H^+ + 6e \rightleftharpoons 2Cr^{3+} + 7H_2O$	6
Cerium (IV) sulphate	$Ce^{4+} + e \rightleftharpoons Ce^{3+}$	1
Iodine	$I_2 + 2e \rightleftharpoons 2I^-$	2
Pot. bromate	$BrO_3^- + 6H^+ + 6e \rightleftharpoons Br^- + 3H_2O$	6
Pot. iodate	$IO_3^- + 6H^+ + 6e \rightleftharpoons I^- + 3H_2O$	6
Mohr salt	$Fe^{2+} \rightleftharpoons Fe^{3+} + e$	1
Sodium thiosulphate	$2S_2O_3^{2-} \rightleftharpoons S_4O_6^{2-} + 2e$	1
Pot. peroxy disulphate ($K_2S_2O_8$)	$S_2O_8^{2-} + 2e \rightleftharpoons 2SO_4^{2-}$	2
Zine (II)-sulphate ($ZnSO_4 . 7H_2O$)	$Zn^{2+} + 2e \rightleftharpoons Zn$	2
Pot. ferrocyanide	$[Fe(CN)_6]^{3-} + e \rightleftharpoons [Fe(CN)_6]^{4-}$	1
Oxalic acid ($H_2C_2O_4 . 2H_2O$)	$C_2O_4^{2-} + 2e \rightleftharpoons 2CO_2$	2

Table 4: Dissociation constants of some acids in water at 25 °C
Dissociation constants are expressed as pK_a [= $-\log K_a$]

Acids	pK_a	Acids	pK_a
Formic	3.75	Fumaric	3.02 (pK_1)
Acetic	4.76		4.38 (pK_2)
Propanoic	4.88	Maleic	1.92 (pK_1)
Butanoic	4.82		6.23 (pK_2)
Pentanoic	4.84	Tartaric	3.03 (pK_1)
Choroacetic	2.86		4.37 (pK_2)
Lactic	3.86		
Oxalic	1.27 (pK_1)	Citric	3.13 (pK_1)
	4.27 (pK_2)		4.76 (pK_2)
			6.40 (pK_3)
Malonic	2.85 (pK_1)	Phosphoric	2.12 (pK_1)
	5.70 (pK_2)		7.21 (pK_2)
			12.30 (pK_3)
Succinic	4.21 (pK_1)		
	5.64 (pK_2)	Benzoic	4.20
Glutaric	4.34 (pK_1)	Phenylacetic	4.31
	5.27 (pK_2)	Sulphanilic	3.23
Adipic	4.43 (pK_1)	Phthalic	2.95 (pK_1)
			5.41 (pK_2)
	5.28 (pK_2)	Cinnamic	3.88 (cis)
Pyruvic	2.49		4.44 (trans)

Table 5: pH of some standard buffer solutions at 25 °C

Solution / Mixture		pH
0.1 (M)	Potassium tetraoxalate	1.48
0.05 (M)	Potassium hydrogen phthalate	4.00
0.05 (M)	Borax	9.18
0.01 (M)	KH_2PO_4 + 0.01 (M) Na_2HPO_4	6.85
0.025 (M)	$NaHCO_3$ + 0.025 (M) Na_2CO_3	10.00
0.01 (M)	Na_3PO_4	11.72
0.05 (M)	Potassium hydrogen tartarate	3.80
1.25 (M)	NH_4Cl in 8 (M) NH_4OH (17 gm NH_4Cl + 145 ml NH_4OH is diluted to 250 ml)	≈ 10

Table 6: pH-range of some buffer solutions

Buffer solutions	pH-range
Citric acid—sodium citrate	2.5-5.6
Accetic acid—sodium acetate	3.7-5.6
Disodium hydrogenorthophosphate—sodium dihydrogenorthophosphate	6.0-9.0
Ammonia solution—Hydrochloric acid	8.2-10.2
Sodium tetraborate—sodium hydroxide	9.2-11.0

CHAPTER 7

Polarimetry and Refractometry

I. **General Use of Polarimeter**

II. **Measurement of Refractive Index of a Liquid by Abbe-Refractometer**

III. **Experiment 1**

Determination of (a) Specific rotation of sucrose, (b) Composition of sucrose solution by polarimeter.

IV. **Experiment 2**

Determination of rate constant of acid catalysed inversion of sucrose and also determination of catalytic coefficient polarimetrically.

V. **Experiment 3**

Determination of concentration of a given solution of aqueous ethanol by refractivity measurements.

VI. **Appendix**

Table 1. Refractive indices and densities of some common liquids

CHAPTER 7

Polarimetry and Refractometry

I. General use of Polarimeter

A polarimeter is used to measure the optical rotation per cm or per mm thickness of solution such as cane sugar, invert sugar, egg albumin, nicotine in water and liquids as methanol, turpentine, citron oils, etc. to find out relation between concentration and rotation in polarised light.

These polarimeters available in two different shade systems:

1. Biquartz device polarimeter works on sodium light as well as on daylight source (ordinary bulb light).
2. Half shade device polarimeter works on sodium light source.

On ordinary light the Biquartz gives pink colour at 0° mark on both sides of demarcation line and it will be very sensitive. On other position this gives half portion red and other half green colour. The instrument is workable on both light sources. On sodium light the Biquartz gives semi-dark colour at 0° mark on both sides of demarcation line and will be also very sensitive. On other position, this gives half portion black and other half white.

A polarimeter essentially consists of two Nicol prisms, co-axially placed (Fig. 7.1).

The first Nicol Prism (*P*) called polariser, through which a monochromatic beam from a sodium lamp (*S*) is introduced, is kept fixed. The second Nicol prism (*A*), called analyser, is mounted on a rotating disc. The plane polarised light emerging from the first prism (*P*) falls on the second (*A*). If the second prism be at 90° to the polariser, a complete darkness would occur, when viewed through a telescope (*T*) placed on the other side of the analyser. Initially the instrument is so adjusted that the entire field of view looked through eye-piece (*E*) would have uniform brightness (V_1). The polarimeter is set at this position, which is taken as 'zero position' of the polarimeter, and the reading of the position

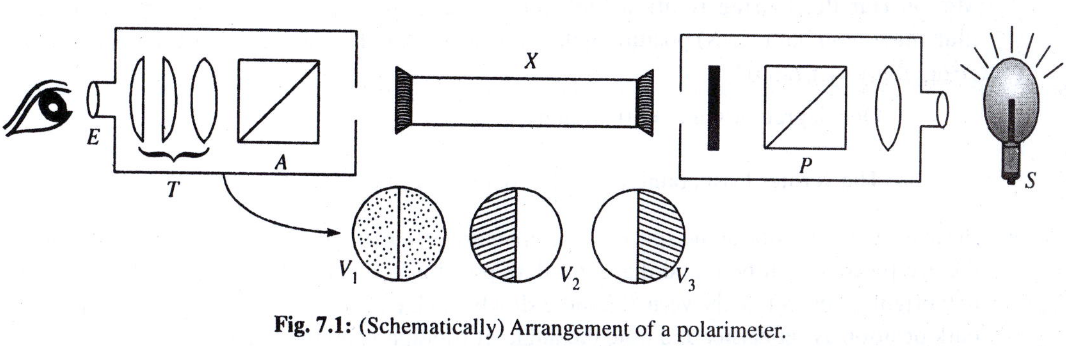

Fig. 7.1: (Schematically) Arrangement of a polarimeter.

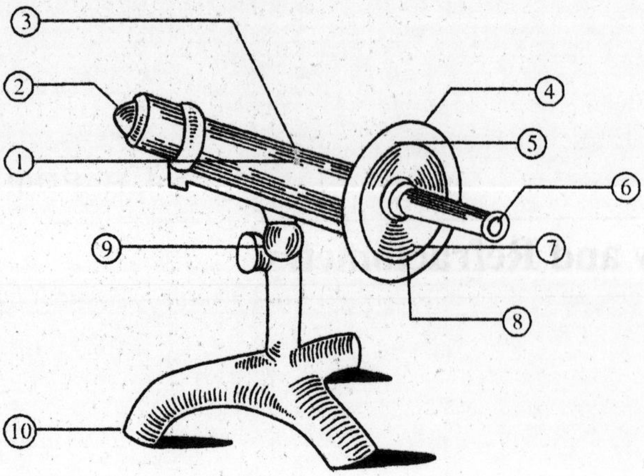

Main parts of Polarimeter

1. Space for Pol Tube
2. Input Light
3. Cover Handle
4. Scale 360° Graduated
5. Vernier 1/10° (6 minute)
6. Eyepiece
7. Fine motion Knob
8. Coarse Motion Handle with Lock System
9. Height Adjustment Knob
10. Stand

Fig. 7.2: Biquartz Polarimeter

is taken from the circular disc provided with angular scales. If the scale is moved towards right-hand or left hand from this position, the field view V_2 or V_3 would be seen through the eye-piece (E). X is the polarimeter tube, which is to be filled with the solution, of which the angle of rotation is to be measured. Generally, the polarimeter tube is 2 decimetre in length.

For the measurement of angle of rotation of an optically active substance say sucrose solution, follow the procedure mentioned below:

1. Prepare and keep ready sugar solution before use of polarimeter.
2. Filter the solution twice through filter paper.
3. Clean polarimeter tube with running water and brush and dry it. Place cover glass on one end, then place rubber washer in cap and tighten the cap.
4. Fill the tube with solution and place cover glass in the same way (mentioned above) and tighten the cap. Clean the tube from outside by a cotton cloth or tissue paper, clean both the cover glasses from outside by cloth or tissue paper and look into the tube through cover glass and check clearness.
5. For best transmission of light, solution must be filtered twice.
6. Set the polarimeter in front of light source and adjust the height according to the light gate.
7. Look into the eye-piece and focus the Biquartz device by adjusting eye-piece outside or inside.
8. Set the device illumination, on equally dark (semi-dark) position by rotating (clockwise) Coarse Motion-Handle. During rotation look into the eye-piece and note that semi-dark (equal illumination on both sides) position will come near 0° on the scale. Note also the instrumental error, if any, ± from 0°.

   One degree of scale = 60 minutes

   $$\text{The vernier least count } = \left(\frac{1}{10}\right)^{\circ} = 6 \text{ minutes.}$$

9. Place the solution tube at the space between polariser and analyser in polarimeter and look into the eye-piece. It will be found that semi-dark position is disturbed (one-half is dark and other half bright). Now rotate the vernier handle slowly in clockwise direction and again set the semi-dark position as set earlier and note the angle of rotation from the scale.

II. Measurement of refractive index of a liquid by Abbe-refractometer

Now-a-days, Abbe-refractometer is most widely used to measure the refractive index of a liquid in the general laboratory. This refractometer has some advantages over the Pulfrich refractometer: (i) refractive index of the liquid can be read directly with an accuracy of about 0.0002; (ii) Only a few drops of liquid are sufficient for examination and (iii) either monochromatic light or ordinary white daylight can be used. The following is the working procedure for measurement of refractive index of a liquid.

1. Open the prism box carefully and clean the surfaces of the prisms (lower and upper) with rectified spirit using cotton and then make it absolutely dry.
2. Place a few drops of distilled water on the ground glass surface of the lower prism and immediately lock the prism box. Semimico-pipette or dropper may be used to place the water on the ground-glass surface of the lower prism.
3. The mirror of the refractometer may be illuminated with the light from sodium lamp (if available) or ordinary bulb. The mirror is so adjusted that a circular light view is observed through the eye-piece. By moving the screw, fitted with the scale, the circular view is adjusted at half shaded position like below:

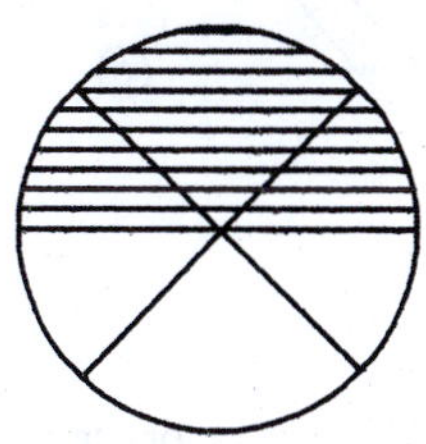

   Then note the scale. The scale should give refractive index 1.333. If the above value is not properly observed, adjust it to 1.333 by moving the scale. Then the dark shade is adjusted to exactly mid-position of the circle by moving the fine adjustment screw. This is called zero adjustment.

4. Open the prism box. Clean it with a piece of cotton wool and make it dry. Place a few drops of experimental liquid using a dropper or semimicro-pipette and close the prism box. It will be observed through the telescope that the edge of dark shade is either at lower or at upper position of intersection of the cross-hairs. The edge of the dark band is brought to the intersection of the cross-hairs. The refractive index of the liquid is then read from the scale. Generally, the scale is of the following type. Scale starts from 1.30. Thereafter shown as 31 means 1.31. Between1.30 and 1.31, there are ten small divisions. There is a diameter line on the circle. The division which coincides with the diameter line is the corresponding scale reading. For water, the diameter line coincides with the third division after 1.33. Thus refractive index of water is 1.333. This corresponds to zero of the vernier scale. So, it is called **zero setting**.

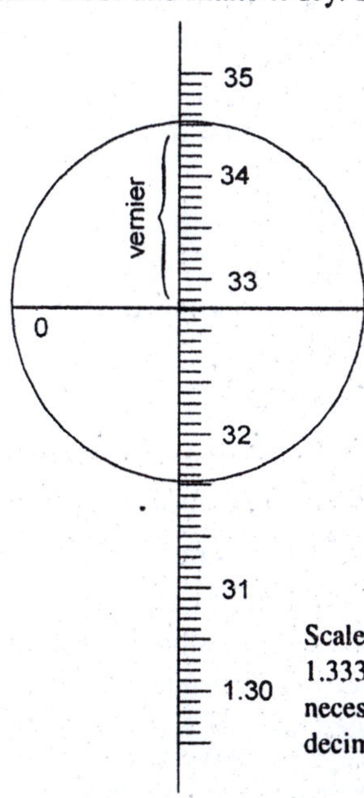

Scale reading is shown as 1.333. Vernier reading is necessary after third decimal of reading.

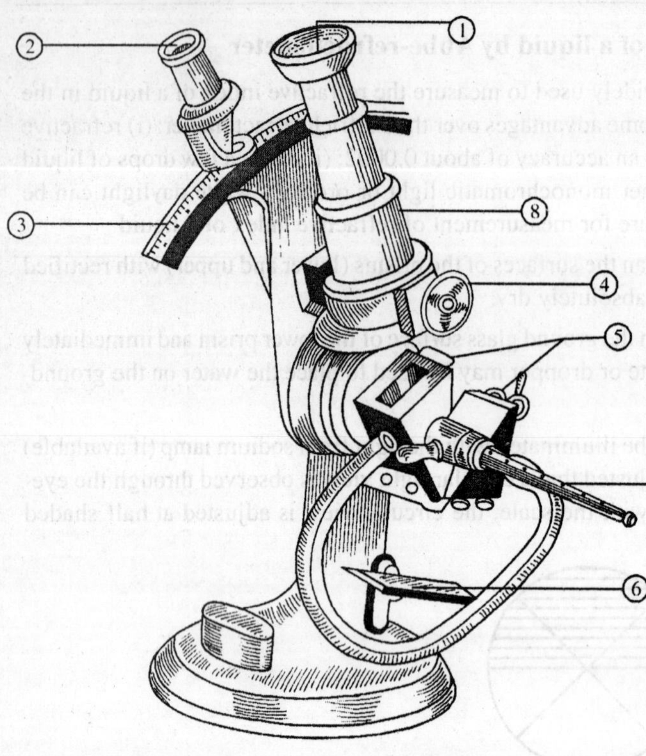

Main parts of Abbe-Refractometer

1. Telescope
2. Lens
3. Curved scale
4. Adjustable screw
5. Pair of prism
6. Mirror
7. Thermometer
8. Compensator

Fig. 7.3: Abbe-Refractometer

Experiment-1

Determination of (a) specific rotation of sucrose and (b) composition of sucrose solution by polarimeter.

Theory: The magnitude of angle of rotation of an optically active substance depends upon several factors: (i) the density or concentration of the substance, (ii) the temperature (iii) the wavelength of light and (iv) the length of the substance through which the beam has to pass, i.e., the length of the polarimeter tube.

The rotatory power of the substance is expressed in terms of specific rotation, $[\alpha]_\lambda^t$ such that

$$[\alpha]_\lambda^t = \frac{100\alpha}{l \times c}, \quad (1)$$

where α is the observed angle of rotation,

l is the length of the polarimeter tube in decimeters, and

c is the concentration of the solution in percentage $\left(\frac{W}{V}\right)$

Thus, temperature, wavelength of light and length (l) of the tube remaining constant, the angle of rotation (α) of an optically active substance is directly proportional to the concentration of that substance, as specific rotation, $[\alpha]_\lambda^t$, is a characteristic constant of an optically active substance.

If we plot the observed angles of rotation (α) against concentration (c), a straight line passing through the origin with a positive slope will be obtained and the concentration of the supplied solution of the same substance can be determined from the graph.

Again, from equation (1) $[\alpha]_\lambda^t$ can be calculated as c, l are known and α is to be measured by a polarimeter. For most of the cases l is generally 2 dm.

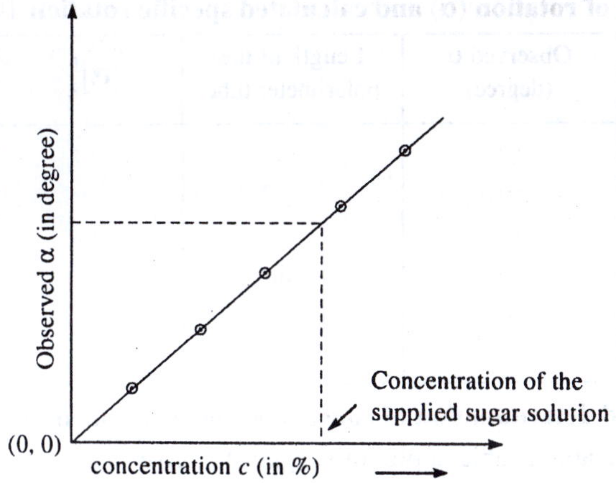

Materials required: (i) Polarimeter with sodium lamp and polarimeter tube, (ii) Sucrose, (iii) Beakers (100 ml capacity, 6 pcs), (iv) Burette or 25 ml graduated pipette, (v) Volumetric flask (250 ml capacity, 1 pc), (vi) Cotton or tissue paper, etc.

Procedure

1. Prepare 250 ml 20% sucrose solution by accurate weighing. The solution should be **colourless and homogeneous**. Filter the solution, if necessary. Then prepare 40 ml each of 15%, 10%, 7.5%, 5% and 2.5% solutions from the original 20% solution by quantitative dilution in 100 ml clean and dry beakers.
2. Measure the angle of rotation of six sugar solutions starting from lower concentration (2.5% to 20%). The detailed procedure for the measurement of angle of rotation has already been described in the general use of polarimeter of this chapter.
3. Wash the polarimeter tube with distilled water and then rinse with the supplied sugar solution of unknown concentration and measure the angle of rotation of this solution using similar procedure.
4. Plot angle of rotation (α) against the known concentrations and find the concentration of the unknown (supplied) solution from the graph.
5. Calculate the specific rotations at each concentration and make conclusion from the result.

Results and Calculations

Table 1: Recording of room temperature

	Before experiment	After experiment	Mean
Temperature →	... °C	... °C	... °C

Table 2: Preparation of sucrose solutions of different compositions

Set Nos. (100 ml beakers)	Vol. of 20% sucrose solution (ml)	Vol. of distilled water (ml)	Total Vol. (ml)	Concentration of sucrose solution
1	40	0	40	20.0%
2	30	10	40	15.0%
3	20	20	40	10.0%
4	15	25	40	7.5%
5	10	30	40	5.0%
6	5	35	40	2.5%

Table 3: Observed angle of rotation (α) and calculated specific rotation $[\alpha]_\lambda^t$ from equation (1)

Concentration of soluction (c)	Observed α (degree)	Length of the polarimeter tube	$[\alpha]_\lambda^t$	Mean $[\alpha]_\lambda^t$
2.5%	...		...	
5.0%	...		...	
7.5%	...		...	
10.0%	...	... dm	...	...
15.0%	...		...	
20.0%	...		...	

4. **Graph plotting and determination of concentration of supplied sucrose solution**
   From the data of the above table, a plot of α against c, gives a straight line passing through the origin.
   Now, the observed angle of supplied sugar solution = ··· degree.
   Hence from the graph, concentration of the supplied sucrose solution = ··· %.
5. *Conclusion:* The angle of rotation increases with increase in concentration of sucrose solution, but specific rotations $[\alpha]_\lambda^t$ are practically constant and the mean value of specific rotation of source $[\alpha]_\lambda^t$ = ...

[**N.B.** : 1. Expected value of specific rotation of sucrose $[\alpha]_D^t = +66.67 - 0.0247\ (t - 20)$ (t = 14 °C to 30 °C); D means sodium D-line.
2. *Suggestions of similar experiment*: Other optically active substances, such as tartaric acid or camphor may be studied in the same way as sucrose. Nonaqueous solvent may also be used (e.g., camphor in benzene or carbontetrachloride).

Experiment 2

Determination of rate constant of acid catalysed inversion of sucrose and also determination of catalytic co-efficient—polarimetrically.

Theory: The experiment is given in the chapter 'Reaction Kinetics'.

Experiment 3

Determination of concentration of a given solution of aqueous ethanol by refractivity measurements.

Theory: When a ray of light (preferably monochromatic) passes from one medium to another (obviously the medium should be transparent), it suffers a change of direction, which is known as refraction. According to Snell's law, the degree of refraction is such that the ratio (n) of the sine of the angle of incidence (i) to the sine of the angle of refraction (r) is constant. This ratio is equal to the ratio of the velocities of the light in the two media. Hence,

$$n = \frac{\sin i}{\sin r} = \frac{V_1}{V_2},$$

where the constant n is called the refractive index or index of refraction, V_1 is the velocity of light in vacuum and V_2 is that in the other medium. The velocity of light in a material body is less than that in vacuum (3×10^8 ms^{-1}). Therefore, it follows that the refractive indices of all substances are more than one. It is measured by Abbe-refractometer.

The aim of the present experiment is to measure refractive indices (n) of a series of aqueous ethanol solutions of known concentrations (c). Then to prepare a calibration curve by plotting n vs. c. Then to measure the n of the aqueous ethanol of unknown concentration and to determine the concentration of the supplied (unknown) aqueous ethanol from the calibration curve. The nature of graph of n vs. c plot for aqueous ethanol is shown here.

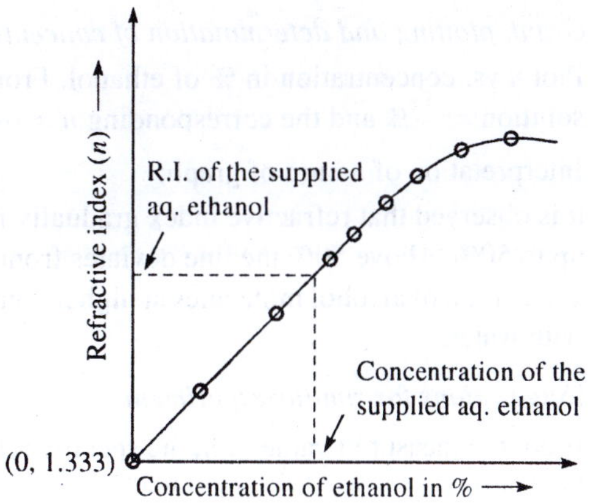

Materials required: (i) Dehydrated (100%) alcohol, (ii) Double distilled water, (iii) Weighing bottles (10-15 ml capacity) 10 nos., (iv) Graduated pipette (5 ml capacity) 1 to 2 nos., (v) Abbe-refractometer with sodium lamp, (vi) Cotton or tissue paper, (vii) Beaker, etc.

Procedure

1. Prepare 10% (V/V) to 90% (V/V) nine ethanol solutions in water with increasing order from10% (i.e., 10%, 20%, 30%, ...) with the help of a 5 ml graduated pipette in nine clean and dry weighing bottles of 10-15 ml capacity.
2. Measure the refractive index of water first and adjust the instrument at $n = 1.333$. Then measure the refractive indices of ethanol solutions starting from lowest concentration. The procedure for the measurement of n by Abbe-refractometer is described in the general discussion of this chapter.
3. Measure the refractive index of the supplied aqueous ethanol solution.
4. Plot n vs. c (in %) and determine the concentration of the supplied aqueous ethanol solution from the calibration curve.

Results and Calculations

Table 1: Recording of laboratory temperature

	Before experiment	After experiment	Mean
Temperature ⟶	... °C	... °C	... °C

Table 2: Composition of sets and observed R.I. (n)

Set Nos.	Vol. of water (ml)	Vol. of absolute alcohol (ml)	Concentration of ethanol	Observed R.I. (n)
1	5.0	0	0% (water)	1.333
2	4.5	0.5	10%	...
3	4.0	1.0	20%	...
4	3.5	1.5	30%	...
5	3.0	2.0	40%	...
6	2.5	2.5	50%	...
7	2.0	3.0	60%	...
8	1.5	3.5	70%	...
9	1.0	4.0	80%	...
10	0.5	4.5	90%	...

3. *Graph plotting and determination of concentration of the supplied aqueous ethanol solution*: Plot n vs. concentration in % of ethanol. From the graph the concentration of supplied ethanol solution = ··· % and the corresponding n = ··· .

4. Interpretation of nature of graph:

It is observed that refractive index gradually increases with increase in concentration of ethanol up to 50%. Above 50% the line deviates from linearity as shown in the theory. This is due to the association of alcohol molecules at higher concentration and formation of clathrate-like structure with water.

[*Suggestions for similar experiment*:

(i) From the measured value of n, the specific refraction (r) and molar refraction (R) of a liquid can be determined from the equations:

$$r = \frac{n^2-1}{n^2+2}\cdot\frac{1}{d};\ R = \frac{n^2-1}{n^2+2}\cdot\frac{M}{d}$$

d = density and M = molecular weight of the liquid.

(ii) Determination of R of a pair of liquids, e.g., ethanol + benzene, methanol + benzene, benzene + toluene, etc.]

APPENDIX

Table 1: Refractive indices and densities of some common liquids at 20 °C

Liquids	Refractive indices	Density (g/cc.)
Acetic acid	1.3721	1.049
Acetone	1.3588	0.792
Aniline	1.5863	1.622
Benzene	1.5011	0.879
Carbon tetrachloride	1.4630	1.595
Carbon disulphide	1.6255	1.263
Chloroform	1.4464	1.498
Ethyl alcohol	1.3611	0.785
n-Heptane	1.3876	0.684
n-Hexane	1.3749	0.660
Methanol	1.3288	0.792
Pyridine	1.5092	0.982
Toluene	1.4970	0.867

CHAPTER 8

The Reaction Kinetics

I. **Introductory Words**

- Order and Molecularity of reactions
- Examples of first, second and third order reactions
- Instruments.

II. **Experiment 1**

Studies on the kinetics of acid catalysed hydrolysis of methylacetate/ethylacetate

i. Determination of rate constants and

ii. Determination of catalytic coefficient.

III. **Experiment 2**

Determinations of rate constant of inversion of sucrose by polarimeter and verification of the effect of catalyst on the rate constant.

IV. **Experiment 3**

Kinetic study of the decomposition of hydrogen peroxide in presence of ferric chloride solution and effect of catalyst on the rate constant.

V. **Experiment 4**

Determinations of rate constant of the reaction between potassium persulphate and potassium iodide by a titration method.

[For U.G./P.G.]

VI. **Experiment 5**

Studies on the kinetics of iodination of acetone (Iodine effect).

[For P.G.]

VII. **Experiment 6**

Studies on the kinetics of iodination of acetone (Acid effect).

[For P.G.]

VIII. **Experiment 7**

Studies on the kinetics of iodination of acetone (Acetone effect).

[For P.G.]

IX. **Experiment 8**

Dilatometric study of the kinetics of alkali catalysed decomposition of diacetone alcohol. [For P.G.]

X. **Experiment 8A**

Dilatometric study of the kinetics of acid catalysed hydrolysis of acetal. [For P.G.]

XI. **Experiment 9**

Kinetics study of the reaction between hydrogen peroxide and iodide ion in acid medium using iodine clock reaction.

XII. **Experiment 10**

Studies on the kinetics of saphonification of ethyl acetate by sodium hydroxide.

i. Determination of rate constant and

ii. Studies on the influence of ionic strength on the rate constant (Salt effect) [For U.G./P.G.]

XIII. **Experiment 11**

Kinetics study of the alkaline hydrolysis of crystal violet spectrophotometrically.

i. Determination of rate constant and

ii. Determination of order of the reaction with respect to alkali. [For P.G.]

XIV. **Experiment 12**

Studies on the kinetics of the reaction between $S_2O_8^{2-}$ and I^- by colorimetric method.

i. Determinations of order of the reaction and

ii. Determination of rate constant. [For U.G./P/G.]

XV. **Appendix**

Table 1. Some reactions that can be studied kinetically with their orders.

CHAPTER 8

The Reaction Kinetics

I. Introductory Words

All chemical reactions approach to attain equilibrium. But the rate with which they proceed towards equilibrium differs from reactions to reactions. A reaction may be a single step process or may include a number of steps, all having different velocities. Of those the slowest one is called the **rate-determining step.**

Knowing the rate, various aspects regarding mechanism of the reaction is ensured. Hence, chemical kinetics provides an important tool to study the reaction mechanism. Since almost all chemical reactions are path-dependent changes, thermodynamics fails to provide its application in reaction kinetics.

The rate of change of a reaction is studied using a **suitable property of the reaction**, which may be chemical or physical. The change in concentration of the reactants or products may be studied by a simple **acid-base titration** or a **redox titration,** if the characters of the reactants or products demand such method. The change may be estimated **conductometrically, potentiometrically** or **pH-metrically** or through observation of **angle of rotation** (like inversion of cane sugar). If there be any change in colour in either of the reactants or products, the rate can well be studied **colorimetrically** or **spectrophotometrically**. Thus there are a good many physical means to study the reactant kinetics.

- **Rate of reaction as a function of concentrations**

Let us consider the reaction

$$gA + hB \rightarrow xC + yD, \tag{1}$$

where g, h, ··· are strichiometric coefficients of the respective substances. Then in a small time interval dt, the rate of concentration changes will be

$$-\frac{1}{g}\frac{d[A]}{dt} = -\frac{1}{h}\frac{d[B]}{dt} = \frac{1}{x}\frac{d[C]}{dt} = \frac{1}{y}\frac{d[D]}{dt}. \tag{2}$$

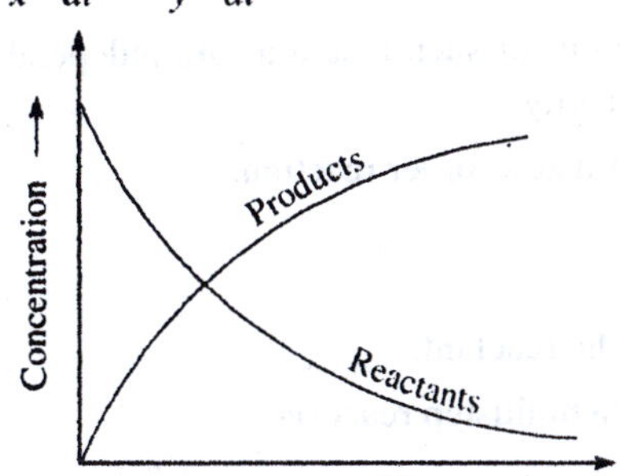

Fig. 8.1

The graphical representation of the concentration changes of the reactants and products will be as The graph shows the fall of concentrations of the reactants and rise of concentrations of the products with the progress of the reaction, i.e., with time.

• Order and molecularity of reactions

Generally, the rate of a reaction at any instant of time is proportional to the concentration of the reactants at that instant. Examples are there, where the rate is somehow proportional to the reciprocal of the concentration of the product (e.g., HBr formation from its elements). But the question is that whether the rate is directly proportional to the concentration of the reactant, i.e., to its first power or to any other exponent of it, say square, square root, etc. The problem is solved experimentally and a kinetic term 'order' is used to express the concentration-dependence of rate of a reaction.

Order of a reaction is defined as the power of the concentration of a reactant with which the rate of the reaction varies and it is ascertained experimentally.

Let it is seen that the rate of the reaction (1) is experimentally formd to be

$$r \propto [A]^{\alpha}\,[B]^{\beta} \qquad (3)$$

Then the order of the reaction is said to be α with respect to A and β with respect to B, so that the overall order of the reaction will be $(\alpha+\beta)$.

This may also be represented as

$$-\frac{d[A]}{dt} \propto [A]^{\alpha} \quad \text{and} \quad -\frac{d[B]}{dt} \propto [B]^{\beta} \qquad (4)$$

α or β may be 1, 2, 3, ··· or any fractional number and even may be negative. Accordingly the reaction will be said as first order, second order etc. However, the value of α or β will be decided by experiment.

Molecularity is defined as the minimum number of molecules of a reactant which actually takes part in the reaction. The stoichiometry of a reaction cannot tell about the molecularity. It is ascertained from some proposed mechanism of the reaction which satisfies the experimental observation.

I. Zero order reaction

For the reaction, $A \rightarrow \text{Products}$

$$\text{the rate } r = -\frac{d[A]}{dt} = k_0[A]^0$$

$$\text{or,} \quad \boxed{r = k_0} \qquad (5)$$

where $[A]$ is the concentration of A at time t. The **unit of k_0**, the zero order rate constant, is **mol.dm^{-3}s^{-1}**.

Equation (5) indicates that the rates of such reactions are independent of the concentration of the reactant and proceed at constant velocity.

The half-decomposition time for a zero-order reaction,

$$\boxed{t_{\frac{1}{2}} = \frac{a}{2k_0}} \qquad (6)$$

a being the initial concentration of the reactant.

A zero order reaction should be a multistep reaction.

Examples of zero-order reaction

1. $CH_3COCH_3 + I_2 \xrightarrow{H^+} CH_3COCH_2I + HI$
   This is zero-order with respect to iodine.

2. $2HI\ (gas) \xrightarrow{Au} H_2 + I_2$

3. $\text{Substrate} \xrightarrow{\text{Enzyme}} \text{Products}$

II. **First order reactions:**

For the reaction, $A \rightarrow \text{Product}$,

$$\text{the rate, } r = -\frac{d[A]}{dt} = k_1 [A] \qquad (7)$$

The integrated form finally becomes

$$\boxed{k_1 = \frac{2.303}{t} \log_{10} \frac{a}{a-x}} \qquad (8)$$

where k_1 = 1st order rate constant

a = initial concentration of the reactant

x = concentration of the product in time t.

Unit of k_1 is s^{-1}.

Half-decomposition period of a first order reaction.

$$\boxed{t_{\frac{1}{2}} = \frac{0.693}{k_1}} \qquad (9)$$

Hence line required to complete half of the amount of the reactant does not depend on its initial concentration. It is a constant quantity.

Examples of 1st order reactions

1. Decomposition of H_2O_2 solution:

$$H_2O_2\ (\text{solution}) \rightarrow H_2O\ (l) + \frac{1}{2} O_2 (g)$$

2. Inversion of cane sugar:

$$\underset{\text{cane sugar}}{C_{12}H_{22}O_{11}} + H_2O \xrightarrow{H^+} \underset{\text{Glucose}}{C_6H_{12}O_6} + \underset{\text{Fructose}}{C_6H_{12}O_6}$$

3. Acid hydrolysis of ester:

$$\underset{\text{Ethylacetate}}{CH_3COOC_2H_5} + H_2O \rightleftharpoons \underset{\text{Acetic acid}}{CH_3COOH} + \underset{\text{Ethyl alcohol}}{C_2H_5OH}$$

This is an example of a **pseudo first order reaction**. Though water is a reactant, the rate is independent of its concentration, which remains practically constant.

4. Decomposition ofdinitrogen pentoxide:

$$N_2O_5\ (g) \rightarrow 2NO_2\ (g) + \frac{1}{2} O_2 (g)$$

5. Decomposition of thionyl chloride:

$$SO_2Cl_2\ (g) \rightarrow SO_2\ (g) + Cl_2\ (g)$$

6. All radioactive changes.

III. **Second order reaction**

For the reaction of the type

2A → Products

the rate, $r = -\frac{d[A]}{dt} = k_2 (a-x)^2$. (10)

The integrated form becomes

$$\boxed{k_2 = \frac{x}{t\,a(a-x)}}$$ (11)

where k_2 = second order rate constant

a = initial concentration of the reactant

x = concentration of the products in time t.

Unit of k_2 is $mol^{-1}\ dm^3\ s^{-1}$.

Half decomposition-period of a first order reaction.

$$\boxed{t_{1/2} = \frac{1}{k_2 a}}$$ (12)

i.e., $t_{1/2} \propto \frac{1}{a}$ (13)

Hence larger the initial concentration of the reactant, shorter will be the time for decomposition of half-amount of the reactant.

For two different reactants with different initial concentrations of the type

A + B → Products

$r = \frac{dx}{dt} = k_2 (a-x)(b-x)$ (14)

which on integration becomes

$$\boxed{k_2 = \frac{2.303}{b-a} \log_{10} \frac{b-x}{a-x}}$$ (15)

$t_{1/2}$ cannot be calculated from this equation.

Examples of 1st order reactions

1. Alkali hydrolysis of an ester (Saponification reaction):

$CH_3COOC_2H_5 + OH^- \rightarrow CH_3COO^- + C_2H_5OH$

2. Reaction of persulphate with iodide ions

$S_2O_8^{2-} + 2I^- \rightarrow 2SO_4^{2-} + I_2$

3. Formation of urea from ammonium cyanate:

$NH_4^+ + CNO^- \rightarrow CO(NH_2)_2$

IV. **Third order reactions**

For the simple third order reactions of the type

3A → Products

$$r = -\frac{d[A]}{dt} = k_3\,(a-x)^3 \quad (16)$$

The integrated form is

$$\boxed{k_3 = \frac{1}{2t}\left[\frac{1}{(a-x)^2} - \frac{1}{a^2}\right]} \quad (17)$$

Unit of k_3 is $mol^{-2}\ dm^6\ s^{-1}$.

$$\boxed{t_{1/2} = \frac{3}{2k_3a^2}} \quad (18)$$

Examples of 1st order reactions

1. $2NO + O_2 \rightarrow 2NO_2$
2. $2NO + H_2 \rightarrow N_2O + H_2O$
3. $2NO + Cl_2 \rightarrow 2NOCl$
4. $2NO + Br_2 \rightarrow 2NOBr$
5. $2NO + D_2 \rightarrow N_2O + D_2O$

IV. ***n*-th order reactions**

$$r = -\frac{d[A]}{dt} = k\,(a-x)^n \quad (19)$$

The integrated form is

$$k_n = \frac{1}{t(n-1)}\left[\frac{1}{(a-x)^{n-1}} - \frac{1}{a^{n-1}}\right] \quad (20)$$

(except for 1st order reaction)

and $t_{1/2} = \dfrac{k'}{a^{n-1}}$. (21)

General unit of k_n: $(dm^3)^{n-1}\ mol^{1-n}\ s^{-1}$

• **Instruments**

As already mentioned, kinetic experiment find its use in a number of instruments including **mechanical titration**, i.e., pipette-burette-indicator. **Polarimeter** (angle of rotation measured), **Spectrophotometer** (for optical density measurement of both colour and colourless solutions), **Conductometer**, **pH-meter**, **Potentiometer** and even **Stalagmometer** are in use.

Different experiments in relevant chapters of this book have been discussed with the above instruments.

For better result, it is suggested to use a **thermostat**, as rate of reaction is highly influenced even by a small variation of temperature.

As the term involves, all the kinetic experiments require a **stopwatch**.

Experiment 1

Studies on kinetics of acid catalysed hydrolysis of methyl acetate / ethylacelate: (i) Determination of rate constants. (ii) Determination of catalytic co-efficient.

Theory

(i) When methyl acelate reacts with water it undergoes hydrolysis forming methanol and acetic acid, the reaction being catalysed by acid (e.g., HCl)

$$\underset{\text{Methyl acetate}}{CH_3COOCH_3} + H_3O^+ \rightleftharpoons \underset{\text{acetic acid}}{CH_3COOH} + \underset{\text{methanol}}{CH_3OH} + H^+$$

If water present in large excess, the reaction becomes pseudo first order for which the expression of rate constant K is given by

$$K = \frac{2.303}{t}\log\frac{a}{a-x}, \tag{1}$$

where a is the initial concentration of easter and $(a - x)$ is the concentration of ester remaining at time t. Since the ester hydrolysed is directly proportional to the acetic acid formed, the progress of the reaction can be studied by withdrawing aliquotes from the reaction mixture from time to time and at the end of the reaction and titrating with alkali. The volume of alkali consumed at any stage is equivalent to the sum of acetic acid produced and the catalyst acid which remains constant throughout the reaction. Let V_0, V_t and V_∞ the volumes of alkali required for a particular volume of reaction mixture initially, at time t and at the end of the reaction respectively, then

$$a \propto V_\infty - V_0 \text{ and}$$

$$(a-x) \propto V_\infty - V_t$$

So the equation (1) becomes

$$K = \frac{2.303}{t}\log\frac{V_\infty - V_0}{V_\infty - V_t}$$

$$\text{or, } \log(V_\infty - V_t) = -\frac{Kt}{2.303} + \log(V_\infty - V_0). \tag{2}$$

Hence a graph of $\log(V_\infty - V_t)$ against t should give a straight line with slope equal to $-\frac{K}{2.303}$, from which the rate constant (K) can be calculated. Some times the alternative form of equation (2) is used in which V_0 is eliminated taking the difference in time such that

$$\Delta t_n = t_n - t_1$$

$$= \frac{2.303}{K}\log\frac{V_\infty - V_1}{V_\infty - V_n}, \quad (n = 2, 3, \ldots) \tag{3}$$

where V_1 and V_n are the volumes of alkali required at time t_1 and at time t_n $(n = 2, 3, \ldots)$ from the initiation of the reaction.

From equation (3), it is evident that a plot of $\log\frac{V_\infty - V_1}{V_\infty - V_n}$ against Δt_n should give a straight line passing through the origin, from the slope of which the rate constant (K) can be determined. The nature of curves are:

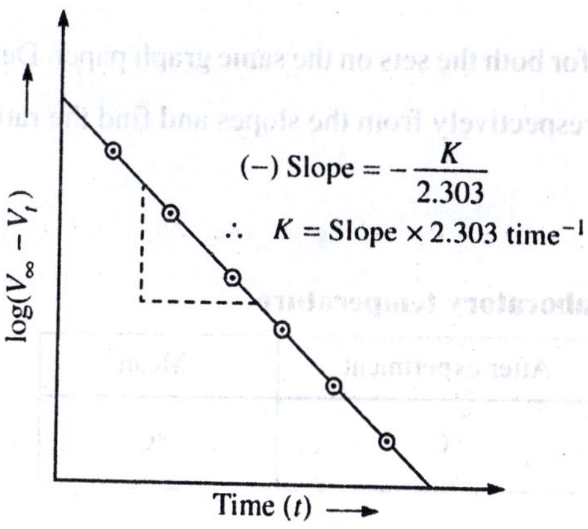

Fig. 8.2

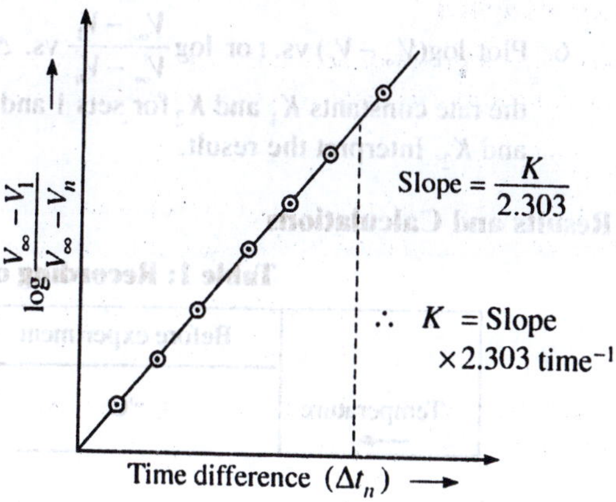

Fig. 8.3

(ii) If experiment is performed in two sets using different concentrations of the catalyst (i.e., HCl) and the rate constants be K_1 and K_2 for, say $\frac{N}{2}$ HCl and $\frac{N}{4}$ HCl respectively, then the ratio $\frac{K_1}{K_2}$ is the catalytic coefficient. For homogeneous catalyst such as HCl in the hydrolysis of methyl acetate the expected value of the ratio is 2.

Materials required: (i) $\frac{N}{2}$ and $\frac{N}{4}$ HCl solutions; (ii) $\frac{N}{20}$ NaOH solution; (iii) Methyl acetate, (iv) Ice; (v) Phenolphthalein indicator, (vi) Burettes, pipettes, conical flasks, etc.

Procedure

1. Prepare 250 ml $\approx \frac{N}{2}$ HCl (10.5 ml conc. HCl is diluted to 250 ml) and from this solution prepare 100 ml $\frac{N}{4}$ HCl solution by proper dilution (50 ml $\frac{N}{2}$ HCl + 50 ml water).
2. Prepare 500 ml $\approx \frac{N}{20}$ NaOH solution ($\approx$ 1 gm NaOH pallets).
3. Take six to seven 125 ml conical flasks each containing 25 ml of chilled water.
4. **Set-I:** Take 50 ml of $\frac{N}{2}$ HCl solution in a 250 ml clean and dry conical flask. Now pipette (clean and dry) 5ml of pure methyl acetate into HCl solution starting the stopwatch when the pipette is half-emptied. Shake the reaction mixture well. After 5 minutees withdraw 2 ml of the reaction mixture (use 2 ml pipette) and run into the chilled water contained in the 125 ml conical flask noting the time of half-discharge. Immediately titrate it against $\frac{N}{20}$ NaOH solution using phenolphthalein indicator. Note the burette reading. The time recorded is t_1 and the corresponding burette reading in V_1. Similar titrations are made after 10, 15, 20, 25, 30, 35 and 40 minutes from the initiation of the reaction and note $V_2, V_3, \cdots, V_8$. The remaining reaction mixture is boiled on a water-bath for half an hour fitting with a refluxing condenser. Cool the reaction mixture at room temperature. Pipette 2 ml of that reaction mixture and then run into 25 ml of distilled water and titrate it against $\frac{N}{20}$ NaOH solution using phenolphthalein indicator. Note the burette reading. This is the value of V_∞.
5. Repeat the experiment with $\frac{N}{4}$ HCl solution (Set 2).

6. Plot $\log(V_\infty - V_t)$ vs. t or $\log\frac{V_\infty - V_1}{V_\infty - V_n}$ vs. Δt_n for both the sets on the same graph paper. Determine the rate constants K_1 and K_2 for sets 1 and 2 respectively from the slopes and find the ratio of K_1 and K_2. Interpret the result.

Results and Calculations

Table 1: Recording of laboratory temperature

	Before experiment	After experiment	Mean
Temperature →	°C	°C	°C

Set 1: Recording of data with $\frac{N}{2}$ HCl solution

Time (t) minute	Δt_n (minute)	Volume of reaction mixture taken (ml)	Volume of alkali required V_t or V_n (ml)	Value of V_∞ (ml)	$\log(V_\infty - V_t)$ or $\log\frac{(V_\infty - V_t)}{(V_\infty - V_n)}$
t_1	—	2	V_1		—
t_2	$t_2 - t_1$	2	V_2		⋮
t_3	$t_3 - t_1$	2	V_3	…	⋮
⋮	$t_4 - t_1$	⋮	⋮		⋮
⋮	⋮	⋮	⋮		⋮

Set 2 : Recording of data with $\frac{N}{4}$ HCl solution

Time (t) minute	Δt_n (minute)	Volume of reaction mixture taken (ml)	Volume of alkali required V_n or V_t (ml)	Value of V_∞ (ml)	$\log(V_\infty - V_t)$ or $\log\frac{(V_\infty - V_t)}{(V_\infty - V_n)}$
t_1	—	2	V_1		⋮
t_2	$t_2 - t_1$	2	V_2		⋮
t_3	$t_3 - t_1$	2	V_3	…	⋮
⋮	⋮	⋮	⋮		⋮

Conclusion: From the graph of plot of $\log(V_\infty - V_t)$ vs. t or $\log\frac{V_\infty - V_t}{V_\infty - V_n}$ vs. Δt_n, it is found that

Set 1 : $K_1 = \text{slope} \times 2.303\ \text{mint}^{-1}$

Set 2 : $K_2 = \text{slope} \times 2.303\ \text{mint}^{-1}$

$\therefore$ The ratio $\frac{K_1}{K_2} = \cdots$

The expected value of the ratio is 2. since the concentration of catalyst (i.e., HCl of the Set 1 is just twice than that for the set 2. If it is, we can conclude that in this reaction HCl acts as a homogeneous catalyst.

Experiment 2

Determination of rate constant of inversion of sucrose by polarimeter and also verification of the effect of catalyst on the rate constant.

Theory: The acid catalysed hydrolysis of sucrose into an equimolar mixture of glucose and fructose, called inverted sucrose,

$$\underset{\text{Sucrose}}{C_{12}H_{22}O_{11}} + H_3O^+ \rightarrow \underset{\text{(glucose)}}{C_6H_{12}O_6} + \underset{\text{(fructose)}}{C_6H_{12}O_6} + H^+$$

is a pseudo first order reaction, as water is present in large excess and its concentration practically remains constant. The reaction can be studied with the help of a polarimeter which directly measures the angle of rotation. Sucrose is dextrorotatory (+) while fructorse is laevorotatory (–) and glucose is dextrorotary (+). Since laevorotation (–) of fructose is greater than the dextrorotation (+) of glucose, so the invert sugar (equimolar mixture of glucose and fructose) is laevorotatory, so that, with the progress of the reaction, the angle of rotation decreases and ultimately reaches to negative value.

Let θ_0 represents the initial angle of rotation, i.e., at time $t = 0$, θ_t be the angle of rotation at anytime $t = t$ and θ_∞ be the angle of rotation after complete inversion has occurred. Then the initial concentration of sucrose, *a*, is proportional to the total range of angle change, i.e., $\theta_0 - \theta_\infty$ and the concentration of sucrose remaining, $(a - x)$, at time *t* is proportional to the range of angle change at time *t*, i.e., $\theta_t - \theta_\infty$. Now, the expression for rate constant (*K*) for a first order reaction,

$$K = \frac{2.203}{t} \log \frac{a}{a-x} \quad \text{reduces to the expression}$$

$$K = \frac{2.203}{t} \log \frac{\theta_0 - \theta_\infty}{\theta_t - \theta_\infty}$$

$$\text{or,} \quad \log(\theta_t - \theta_\infty) = -\frac{Kt}{2.303} + \log(\theta_0 - \theta_\infty) \qquad (1)$$

Hence, a graph of $\log(\theta_t - \theta_\infty)$ against time *t* should give a straight line with negative slope equal to $-\frac{K}{2.303}$, from which the rate constant (*K*) can be calculated as follows:

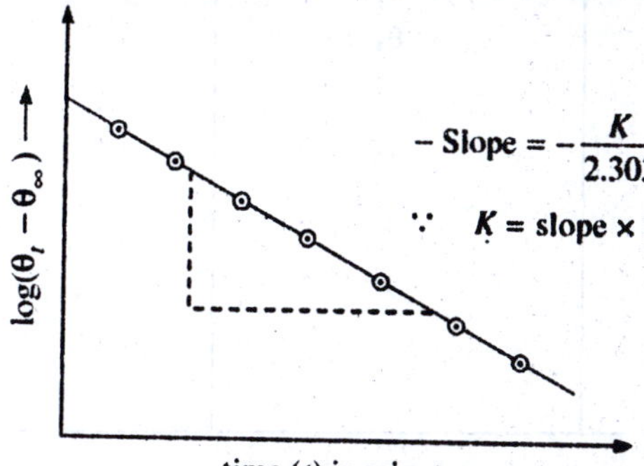

Fig. 8.4: Plot of $\log(\theta_t - \theta_\infty)$ vs. time (t)

The reaction is studied using two different concentrations of the catalyst (i.e., HCl solution) and determined the rate constants and then calculated the ratio of the two rate constants.

Materials: (i) 20% sucrose solution, (ii) Hydrochloric acid solution (~4N and 2N). (iii) Polarimeter with polarimeter tube. (iv) Conical flasks, beakers, pipettes, measuring cylinder, tissue paper, etc.

Procedure

1. Prepare 100 ml ≈ 20% sucrose solution.
2. Prepare 100 ml ≈ 4(N) HCl solution (from concentrated HCl, 12N) and prepare 100 ml of ≈ 2(N) HCl from this 4(N) HCl by proper dilution (50 ml HCl 4N + 50 ml water).
3. Fill up the polarimeter tube with distilled water and measure its angle of rotation. Generally, the instrument shows 0.0°. If any instrumental error is there, then note it. Then the actual angle of rotation of any optically active colourless solution will be $(y \pm x)°$, where $\pm x°$ is the instrumental error and y is the angle of rotation measured by that polarimeter.
4. Take 25 ml of 4(N) HCl solution in a conical flask. Pipette out 25 cc of sucrose solution and run into the HCl solution starting the stopwatch when the pipette is half-emptied. Fill the polarimeter tube with the reaction mixture. Measure the angle of rotations $\theta_1, \theta_2, \theta_3, ...$ at times $t_1, t_2, t_3, ...$ (take time interval intially 2-3 minutes and then 5 minutes, up to the negative value of θ_t). The reaction mixture is then heated for half an hour at a temerature of 60-70 °C. Cool it and measure θ_∞
5. Repeat the same experiment using ≈ 2N HCl solution.
6. Plot $\log(\theta_t - \theta_\infty)$ vs. t both sets on the same graph paper and determine K_1 and K_2. Find their ratio and then intepret the result.

Results and Calculations

Table 1: Recording of laboratory temperature

	Before experiment	After experiment	Mean
Temperature →	°C	°C	°C

Table 2: Study of the reaction with 4(N) HCl (Set 1)

No. of observation	Time in (second) or (mint)	Observed angle of rotation θ_t (in degree)	Observed value of θ_∞	$\log(\theta_t - \theta_\infty)$
1	t_1	θ_1		...
2	⋮	θ_2		...
3	⋮	⋮	...	...
4	⋮	⋮		...
5	⋮	⋮		...
6	⋮	⋮		...
7	⋮	⋮		...
8	t_8	⋮		...

3. Study of the reaction with 2(N) HCl (Set 2). The table is same as above.
4. The plot of $\log(\theta_t - \theta_\infty)$ vs. t

K_1 = slope × 2.303 time^{-1} and K_2 = slope × 2.303 time^{-1}.

5. The actual concentrations of HCl in the two sets become 2(N) and 1(N) as 25 ml acid solution is mixed with 25 ml sucrose solution.

Thus, $\dfrac{K_{2(N)}HCl}{K_{1(N)}HCl} = \cdots$. The expected value is 2. If it is, we may conclude that in this reaction HCl acts as a homogeneous catalyst.

Experiment 3

Kinetic study of the decomposition of hydrogen peroxide in presence of ferric chloride solution and effect of catalyst on the rate constant.

Theory: Hydrogen peroxide undergoes decomposition in accordance with the equation

$H_2O_2 = H_2O + \frac{1}{2}O_2$ and the rate of decomposition is markedly accelerated by the presence of various catalysts. In this experiment we will study the above decomposition reaction in presence of $FeCl_3$ solution which acts as a positive catalyst and whose concentration remains unchanged during the course of the reaction. The course of the reaction may be followed by titrating the hydrogen peroxide with potassium permanganate in acid medium, the volume of permanganate solution required at any time being proportional to the concentration of hydrogen peroxide remaining at that time.

Let V_0 be the volume of potassium permanganate solution required corresponding to the initial concentration of hydrogen peroxide a at time $t = 0$, and V_t be the volume of potassium permanganate solution required corresponding to the concentration of hydrogen peroxide remaining $(a - x)$ at time t. Since the reaction follows first order kinetics, the expression of rate constant at laboratory temperature is given by

$$K = \frac{2.303}{t}\log\frac{a}{a-x} \quad \text{or, i.e.,} \quad K = \frac{2.303}{t}\log\frac{V_0}{V_t}.$$

It is difficult to measure V_0 as the decomposition starts as soon the catalyst is mixed with hydrogen peroxide and it can be eliminated by taking the difference in time $\Delta t_n = t_n - t_1$ such that

$$K = \frac{2.303}{\Delta t_n}\log\frac{V_1}{V_n} \quad [n = 2, 3, 4, \cdots]$$

$$\text{or,} \quad \log\frac{V_1}{V_n} = \frac{K}{2.303}\Delta t_n, \tag{1}$$

where V_1 is the volume of permanganate solution required for a particular volume (say, 5 ml) of reaction mixture at time t_1 after the initiation of the reaction and V_n is that at time t_2, t_3, ··· after the initiation of the reaction. If $\log\frac{V_1}{V_n}$ is plotted against Δt_n, a straight line passing through the origin will be obtained from the slope of which K can be determined. The reaction is studied using two different concentrations of $FeCl_3$ solution, K_1 and K_2 are determined and the result is interpreted from ratio $\dfrac{K_1}{K_2}$.

Materials: (i) 2(V) hydrogen peroxide, (ii) 1.0% and 0.5% $FeCl_3$ solution, (iii) 1(N) H_2SO_4, (iv) $\frac{N}{50}$ $KMnO_4$ solution, (v) Ice, (vi) Pipettes, burettes, conical flasks, etc.

Procedure

1. In the laboratory, the hydrogen peroxide which is available is labelled as 30% (100 volumes). Prepare 250 ml 2(V) H_2O_2 from this concentrated H_2O_2 (5 ml is diluted to 250 ml with distilled water).
2. Prepare 100 ml 1% anhydrous $FeCl_3$ solution (1 gm is dissolved in 100 ml water) and then from this solution prepare 50 ml 0.5% $FeCl_3$ solution by exact dilution (25 ml $FeCl_3$ solution 1% + 25 distilled water).
3. Prepare 250 ml 1(N) H_2SO_4 from concentrated H_2SO_4 [36 (N)]. Dilute $\approx$ 7 ml concentrated H_2SO_4 to 250 ml carefully.
4. Prepare 500 ml $\frac{N}{50}$ $KMnO_4$ solution (dissolved $\approx$ 0.3 ml $KMnO_4$ in 500 ml water).
5. Take 50 ml of 2(V) H_2O_2 in a 250 ml stoppered bottle and add 10 ml of 1% $FeCl_3$ solution starting the stop-watch when the pipette is half-emptied. Shake the reaction mixture well and withdraw 5 ml portions of the reaction mixture at intervals of 3 minutes and then run it into 10 ml ice-cold (IN) H_2SO_4, noting the time of the half-discharge of the reaction mixture. Immediately titrate it against $\frac{N}{50}$ $KMnO_4$ solution. Note the burette readings when just pink colour appears in each case. Take at least seven readings.
6. Repeat the same using 0.5% $FeCl_3$ solution.

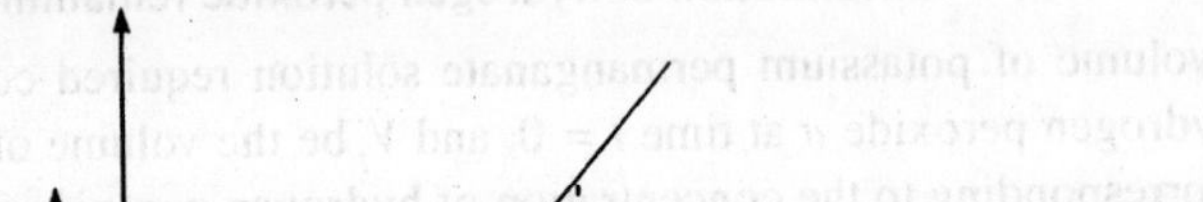

Fig. 8.5: Plot of $\log \frac{V_1}{V_n}$ vs. Δt_n

vii. Plot $\log \frac{V_1}{V_n}$ vs. Δt_n for the two sets a data on the same graph paper and determine K_1 and K_2 for two sets. The expected nature of graph will be as follows.

viii. Interpret the result.

N.B.:

a. Before starting the experiment a blank titration of H_2O_2 is necessary. For this purpose take 50 ml of 2(V) H_2O_2 (prepared) and add 10 ml of distilled water instead of $FeCl_3$ solution. Withdraw 5 ml of that H_2O_2 and then run into 10 ml of 1(N) H_2SO_4. Titrate it against $\frac{N}{50}$ thio and note the burette reading. If the burette reading lies in the range 20 ml to 30 ml, then start the experiment. Otherwise adjust the strength of H_2O_2 so that the burette reading lies approximately on the above range. The blank titration of H_2O_2 is necessary because the strength of H_2O_2, which is available in the laboratory, is not correctly known, some of it may decompose and its strength may change from 30% i.e., 100 volumes.

b. During performing the experiment there may always some bubbling at the time of pipetting. Minimize the rate of bubbling by adjusting the strength of $FeCl_3$ solution. Whatever may be the strength of $FeCl_3$ solution adjusted, the strength of one should be just half of the other.]

Results and Calculations

1. Recording of laboratory temperature

	Before experiment	After experiment	Mean
Temperature →	°C	°C	°C

2. Titration results of set 1 (using 1% $FeCl_3$ solution)

Time (sec)	Δt_n (sec)	Volume of $KMnO_4$ required (ml)	$\log \frac{V_1}{V_n}$ [n = 2, 3, 4, ⋯]
t_1	—	V_1	—
t_2	$t_2 - t_1$	V_2	...
t_3	$t_3 - t_1$	V_3	...
t_4	$t_4 - t_1$	V_4	...
t_5	$t_5 - t_1$	V_5	...
t_6	$t_6 - t_1$	V_6	...

3. Titration results for set 2 (using 0.5% $FeCl_3$ solution)

Time (sec)	Δt_n (sec)	Volume of $KMnO_4$ required (ml)	$\log \frac{V_1}{V_n}$ [n = 2, 3, 4, ...]
t_1	—	V_1	—
t_2	$t_2 - t_1$	V_2	...
t_3	$t_3 - t_1$	V_3	...
t_4	$t_4 - t_1$	V_4	...
t_5	$t_5 - t_1$	V_5	...
t_6	$t_6 - t_1$	V_6	...

4. Calculations and Conclusion

From the plot of $\log \frac{V_1}{V_n}$ vs. Δt_n .

For set 1 K_1 = slope × 2.303 sec^{-1}

For set 2 K_2 = slope × 2.303 sec^{-1}

∴ the ratio $\frac{K_1}{K_2} = \cdots$.

[It has been found that $\frac{K_1}{K_2} \neq 2$. The expected value of the ratio is nearly 1.5.

If it is, we can conclude that $FeCl_3$ does not act as a homogenous catalyst. In case of a homogeneous catalyst the expected value of the ratio is 2].

Experiment 4

Determination of rate constant of the reaction between potassium persulphate and potassium iodide by a titration method.

Theory: The second order reaction involving the oxidation of iodide by persulphate is represented as:

$$2I^- + S_2O_8^{2-} \rightarrow I_2 + 2SO_4^{2-}$$

The rate equation may be written as below:

$$\frac{dx}{dt} = K_2 (a - x)(b - 2x), \qquad (1)$$

where a and b initial molar concentrations of persulphate and iodide respectively.

If one of the reactants is present in excess such that its concentration remains sensibly constant during the course of the reaction, the reaction will follow first order kinetics.

Suppose the concentration of iodide in the above reaction is so large that it will be approximately unaltered during the course of the reaction. Equation (1) will reduce to

$$\frac{dx}{dt} = K_2 (a - x) b.$$

Integrating and applying the condition that $x = 0$ at $t = 0$, we have

$$bK_2 = \frac{2.303}{t} \log \frac{a}{a - x}$$

$$\text{or,} \quad K' = \frac{2.303}{t} \log \frac{a}{a - x}, \qquad (2)$$

where $K' = bK_2$.

Hence a plot of $\log \frac{a}{a - x}$ vs. t will be a straight line passing through the origin and K' is determined from the slope.

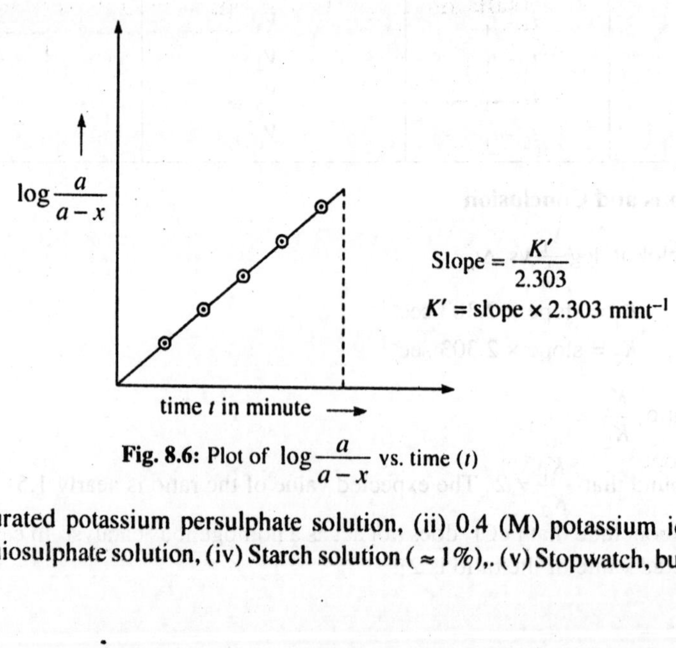

Fig. 8.6: Plot of $\log \frac{a}{a - x}$ vs. time (t)

Materials: (i) Saturated potassium persulphate solution, (ii) 0.4 (M) potassium iodide solution, (iii) 0.01, (N) sodium thiosulphate solution, (iv) Starch solution ($\approx 1\%$),. (v) Stopwatch, burettes, pipettes, conical flasks, etc.

Procedure

1. Prepare 100 ml ≈ 0.1(N) sodium thiosulphate solution (dissolve about 2.5 gm) and from this stock solution prepare 250 ml 0.01 (N) thio solution (25 ml is diluted to 250 ml).
2. Prepare 50 ml saturated solution of potassium persulphate. 20 ml of the saturated solution is diluted to 100 ml (five times diluted) with water and divide the solution into two parts (50 ml each) and take seperately in two 250 ml conical flasks (absolutely dry).
3. Prepare 100 ml 0.4 (M) KI solution (6.64 gms KI). Pour 50 ml of 0.4(M) KI solution into the potassium persulphate solution (50 ml) of the first conical flask and start the stopwatch simultaneously. Shake the reaction mixture well. It is important to place the stopper on the mouth of the conical flask containing the reactants as soon as they have mixed in order to prevent the loss of iodine by volatilization from the reaction mixture. At measured and recorded intervals of time (3, 8, 15, 20, 30, 40, 50, 60 minutes), withdraw 10 ml of the reaction mixture (use 10 ml pipette) of ice-cold water. This dilution has the effect of slowing down the reaction considerably whilst the titration is completed. Each 10 ml reaction mixture is titrated with freshly prepared 0.01 (N) sodium thiosulphate solution (x ml) using starch indicator.
4. Mix 50 ml of KI solution (0.4 ml) with the remainder of the diluted persulphate solution of the second conical flask and cover it with stopper. Heat it at 60 °C for 30 minutes. Cool the reaction mixture at room temperature and titrate 10 ml of this mixture with 0.01 (N) $Na_2S_2O_3$ solution using starch indicator (a ml).
5. Plot $\log\frac{a}{a-x}$ vs. time and determine the rate constant K' from the slope.

Results and Calculations

1. Recording of laboratory temperature

	Before experiment	After experiment	Mean
Temperature →	°C	°C	°C

2. Recording of titration results and calculation of data for graph plotting in tabular form

No. of observation	Time (t) (minute)	Titration a ml	Titration x ml	$\log\frac{a}{a-x}$
1	t_1		x_1	
2	t_2		x_2	
3	t_3		x_3	
⋮	⋮	...	⋮	
⋮	⋮		⋮	
⋮	⋮		⋮	

3. Conclusion: From the plot of $\log\frac{a}{a-x}$ vs. t, it is found that slope = ...

∴ $K' = \text{slope} \times 2.303\ \text{min.}^{-1}$

Experiment 5

Studies on the kinetics of iodination of acetone [Iodine effect]

Theory: Iodine reacts with acetone to form acetonyl iodide as follows

$$CH_3COCH_3 + I_2 = CH_3COCH_2I + HI$$

The reaction is catalysed by an acid. If acetone is in large excess, its concentration may be regarded as constant. It has been observed that the rate of the reaction is independent of concentration of iodine, i.e., it is a zero order reaction with respect to iodine, i.e.,

$$\text{rate} = \frac{dx}{dt} = K[I_2]^\circ$$

On integration, $x = Kt + \text{constant}$, (1)

where x is the concentration of iodine consumed at any time t and K is the rate constant.

The rate of the reaction is determined by withdrawing samples from the reaction mixture at known times and measuring the concentration of iodine by titration with sodium thiosulphate solution. If V_0 and V_t be the volumes of thiosulphate required corresponding to the initial concentration of iodine and the concentration of iodine at time t respectively, then obviously x, the concentration of iodine that consumed at time t is proportional to $V_0 - V_t$. Then equation (1) reduces to

$$V_0 - V_t = Kt + \text{constant}. \quad (2)$$

Measurement of V_0 is difficult as the reaction starts as soon as iodine mixed with acetone and it may be eliminated by taking the difference $\Delta t_n = t_n - t_1$ such that

$$V_1 - V_n = K\Delta t_n \quad [n = 2, 3, 4, \cdots]$$

i.e., $$\Delta V_n = K\Delta t_n, \quad (3)$$

where $\Delta V_n = V_1 - V_n$ and $\Delta t_n = t_n - t_1$. V_1 and V_n are the titre values after time t_1 and t_n respectively.

A plot of ΔV_n vs. Δt_n will give a straight line passing through the origin from the slope of which the rate constant K can be determined.

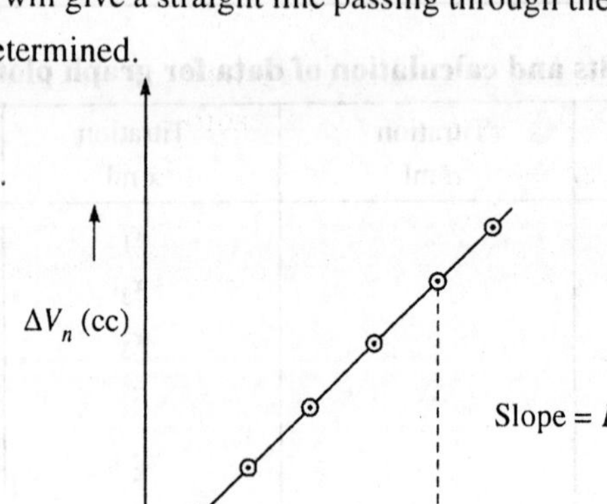

Fig. 8.7: Plot of ΔV_n vs. Δt_n

The experiment is studied in two sets with differing the concentration of I_2 to determine K_1 and K_2 for two sets and the ratio $\frac{K_1}{K_2}$ and then the result is interpreted.

• **Materials:** (i) Acetone (pure), (ii) ≈ 0.1 (N) Iodine solution in KI, (iii) ≈ 0.05 (N) Iodine solution in KI, (iv) ≈ 1.0(N) H_2SO_4, (v) $\approx \frac{N}{10}$ sodium bicarbonate solution, (vi) Starch solution, (vii) Burettes pipettes, conical flasks, volumetric flasks, etc.

Procedure

1. Iodine solution (~0.1 N) is prepared by dissolving ~12.7 gm of resublimed Iodine in a saturated solution of 16.6 gm of potassium iodide and diluting it to 1000 ml. Use proportionately less amount of iodine and potassium iodide to prepare less volume at a time. Take 50 ml of ~0.1 (N) iodine solution and dilute it to 100 ml so that the concentration of the solution thus prepared is exactly half (i.e. ≈ 0.05 N) of that for the original solution.
2. Prepare 500 ml each of $\frac{N}{50}$ sodium thiosulphate solution (≈ 2.48 gm), $\frac{N}{50}$ sodium bicarbonate solution (≈ 4.2 gm) and 100 ml ≈ 1.0 (N) H_2SO_4 (≈ 3 ml conc. H_2SO_4) and also prepare 50 ml starch solution.
3. **Composition of sets**

Set No.	Volume of distilled water (ml)	Volume of acetone (ml)	Volume of 1(N) H_2SO_4 (ml)
1	100	20	10
2	100	20	10

The above composition of sets should be made in 250 ml stoppered bottles so that no evaporation takes place.

4. Add 25 ml of 0.1 (N) iodine solution to the solution mixture of Set No. 1 starting the stopwatch when the pipette is half-emptied and mix it well.
5. Withdraw several 10 ml portions of the reaction mixture at intervals of 3 minutes and run into 10 ml 0.1 (N) $NaHCO_3$ solution contained in a 100 ml conical flask, nothing the time of half-discharge of the reaction mixture. Titrate the reaction mixture in each case with $\frac{N}{50}$ thiosulphate solution, using starch as indicator. Note the burette readings in each case when the reaction mixture just becomes colourless (take at least eight readings).
6. Add 25 ml of 0.05 (N) iodine solution in the solution mixture of Set No. 2 starting the stopwatch when the pipette is half-emptied and repeat the same as done for Set 1.
7. Plot ΔV_n against Δt_n for the two sets on the same graph paper. Determine K_1, K_2 and find the ratio $\frac{K_1}{K_2}$ and interpret your result.

Results and Calculations

1. **Recording of laboratory temperature**

	Before experiment	After experiment	Mean
Temperature →	°C	°C	°C

2. Titration Results (for Set 1)

No. of observations	Volume of reaction mixture taken (ml)	Time recorded (sec)	Δt_n (sec)	Volume of $\frac{N}{50}$ thio required (ml)	ΔV_n (ml)
1.	10	t_1	—	V_1	—
2.	10	t_2	$t_2 - t_1$	V_2	$V_1 - V_2$
3.	10	t_3	$t_3 - t_1$	V_3	$V_1 - V_3$
4.	10	⋮	$t_4 - t_1$	V_4	$V_1 - V_4$
5.	10	⋮	$t_5 - t_1$	V_5	$V_1 - V_5$
6.	10	⋮	$t_6 - t_1$	V_6	$V_1 - V_6$
7.	10	⋮	$t_7 - t_1$	V_7	$V_1 - V_7$
8.	10	t_8	$t_8 - t_1$	V_8	$V_1 - V_8$

3. **Similar table for Set 2.**

4. **Graph plotting and calculations**

From the graph of the plot of ΔV_n vs. Δt_n

$K_1 = \cdots$ ml sec^{-1}

$K_2 = \cdots$ ml sec^{-1}.

The ratio : $\frac{K_1}{K_2} = \cdots$

5. **Interpretation of results:** The expected value of the ratio $\frac{K_1}{K_2} \approx 1$. If the ratio be 1, then we can conclude that the rate of the reaction is independent of iodine concentration, i.e., the reaction is zero order with respect to iodine.

Experiment 6

Studies on kinetics of iodination of acetone [Acid Effect].

Theory: The theoretical portion is same just like previous experiment (Iodine Effect).

Since the reaction is zero order with respect to iodine, the rate determining step does not involve the interaction between the ketone and the iodine molecule. The usually accepted meachanism is as follows:

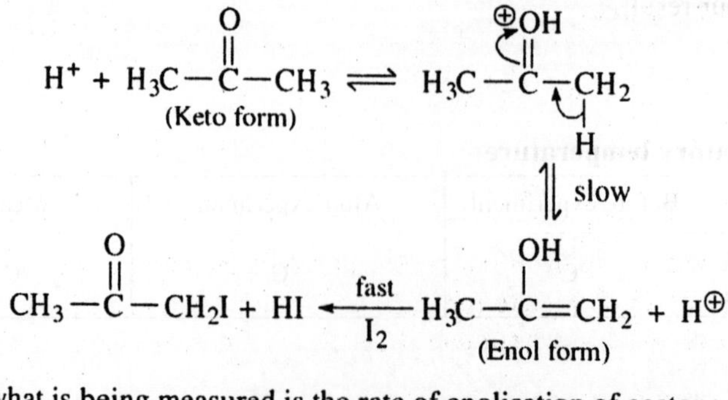

So in reality what is being measured is the rate of enolisation of acetone.

Materials: Same as previous experiment.

Procedure

1. Prepare the solutions mentioned in the procedure of the previous experiment and the following composition of sets are made—

Composition of sets

Stoppered bottle no.	Volume of distilled water (ml)	Volume of acetone (ml)	Volume of ~1(N) H_2SO_4 (ml)	Volume of 0.5(N) H_2SO_4 (ml)
1	100	20	10	×
2	100	20	×	10

2. Add 25 ml of 0.1 (N) iodine solution to the solution mixture of set 1 starting the stopwatch when the pipette is half-emptied. Shake well and study the experiment just like the previous experiment.
3. Repeat the same for set 2.

   In both the sets, the same concentration of I_2 should be used, i.e., 0.1 (N).

Results and Calculations

1. **Recording of laboratory temperature**

	Before experiment	After experiment	Mean
Temperature →	°C	°C	°C

2. Titration table, graph plotting and calculations are same as previous experiment.
3. **Interpretation:** Find the ratio $\frac{K_1}{K_2}$.

   The expected value of the ratio is 2. If it is, than we can conclude that the reaction is 1st order with respect to the acid.

Experiment 7

Studies on the kinetics of iodination of acetone [Acetone Effect].

Theory: The theoretical part, materials, procedure, etc. are same as previous experiments (acid effect or iodine effect). Only the difference is that the experiment is studied with same acid concentration and same iodine concentration but the concentration of acetone is varied. So, the composition table for the present experiment will be as under:

Bottle Nos. (250 ml stoppered)	Volume of distilled water taken (ml)	Volume of pure acetone taken (ml)	Volume of 1(N) H_2SO_4 taken (ml)	Total volume (ml)
1	100	20	10	130
2	110	10	10	130

Add 25 ml of 0.1 (N) I_2 solution to the mixture of solution of set no. 1 starting the stopwatch when the pipette is half-emplied (use 25 ml pipette for addition of iodine solution). Shake well and perform similar work to that of previous experiment (Iodine effect). Repeat the same for set 2 (bottle no. 2). Find the rate constants K_1 and K_2 for sets 1 and set 2 respectively and also calculate the ratio $\frac{K_1}{K_2}$.

Actually the overall iodination of acetone in presence of acid follows the rate expression:

$$\frac{dx}{dt} = K\,[I_2]^{\alpha}\ \ [\text{Acid}]^{\beta}\ \ [\text{Acetone}]^{\gamma}$$

The expected value of $\frac{K_1}{K_2} \approx 2$, i.e., the reaction is first order with respect to acetone. Thus $\gamma = 1$.

Experiment 8

Dilatometric study of the Kinetics of alkali-catalysed decomposition of diacetone alcohol.

Theory: The alkali catalysed cleavage of diacetone alcohol

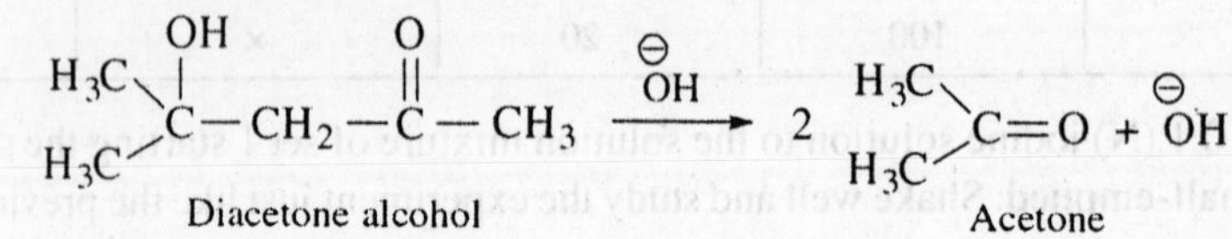

is a pseudo first order reaction being dependent on the first power of diacetone alcohol concentration and on the first power of the hydroxide ion concentration. The hydroxide ion concentration, however, remains constant during the course of the reaction.

For such reactions, if x_1, x_2, x_3, etc., correspond to a suitable physical property (e.g. ,volume) measured at time t_1, t_2, t_3, etc., the rate expression may be written as

$$\left(x_1 - x_\infty\right) = \left(x_0 - x_\infty\right)e^{-Kt_1}, \tag{1}$$

where x_0 and x_∞ are the physical properties measured at $t = 0$ and $t = \infty$ respectively. When the readings during the mid-portion of the kinetic determination are more accurate than the initial and final readings, Guggenheim's method of graphical plot becomes more convenient. According to this method, if a second series of readings are made after a constant increment of time Δt such that x_1', x_2', x_3', etc., are the measurements of the physcial property made at time $(t_1 + \Delta t), (t_2 + \Delta t)$, etc., then

$$\left(x_1' - x_\infty\right) = \left(x_0 - x_\infty\right)e^{-K(t_1 + \Delta t)} \tag{2}$$

Substracting equation (1) from equation (2), we get

$$\left(x_1' - x_1\right) = \left(x_0 - x_\infty\right)e^{-Kt_1}\left(e^{-K\Delta t} - 1\right)$$

or, $Kt_1 + \ln\left(x_1' - x_1\right) = \text{constant}$

In general terms this may be written as

$$\ln\left(x_n' - x_n\right) = -Kt_n + \text{constant}$$

i.e., $\log\left(x_n' - x_n\right) = -\frac{K}{2.303}\,t_n + \text{constant}$ (3)

Hence, a graph of $\log\left(x_n' - x_n\right)$ vs. t_n gives a straight line with negative slope, from which K, the rate constant of the decomposition reaction, can be determined as

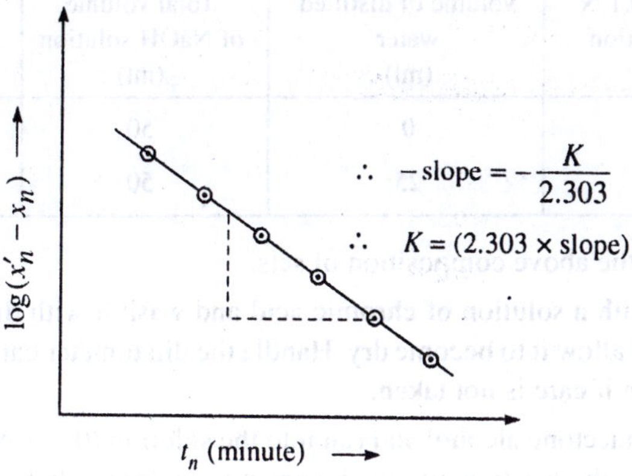

Fig. 8.8: $\log(x'_n - x_n)$ vs. t_n plot

A suitable physical property by which the reaction may be followed is the volume which is readily measured with a dilatometer. The volume of the reaction mixture is derectly proportional to the height of the capillary tube of the dilatometer. So, measured physical property is the height of the solution in the dilatometer capillary at different times. Obviously, x'_n is always greater than x_n.

The reaction is investigated at different concentrations of sodium hydroxide and K-values are compared.

Materials: (i) Caustic soda, (ii) Diacetone alcohol, (iii) Dilatometer, (iv) Volumetric flasks (250 ml, 100 ml), pipettes (10 ml), conical flasks (125 ml), measuring cylinder, etc.

[**N.B.:** Since dilatometer is a special type of appratus, a brief description of a dilatometer used in the general laboratory is given below:

The dilatometer is shown in the following Fig.8.9. A is a 50 ml capacity bulb and B is a fine capillary with mm scale and C is a stopcock. Generally, the height of capillary is taken as 1 ft.]

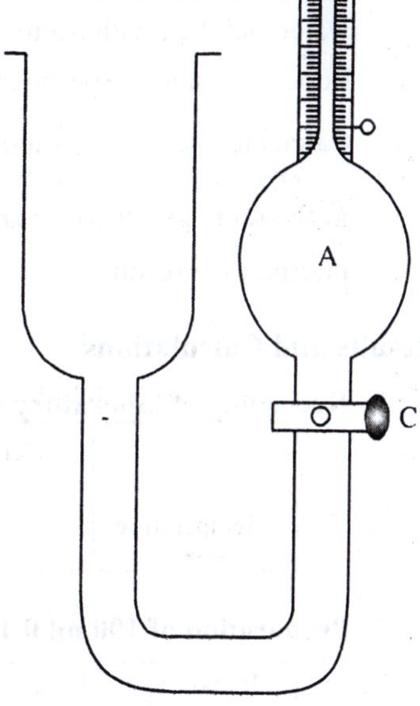

Fig. 8.9: Dilatometer

Procedure

i. Prepare 100 ml 0.1 (N) oxalic acid solution by accurate weighing.

ii. Prepare 250 ml ≈ 0.1 N (slightly higher) NaOH solution and standardise it against $\frac{N}{10}$ standard oxalic acid solution (prepared), and from this solution prepare 100 ml exact 0.1 (N) NaOH solution by quantitative dilution.

iii. Take two clean and dry conical flasks (capacity 125 ml) and make the following composition:

Set No.	Volume of 0.1 N NaOH solution (ml)	Volume of distilled water (ml)	Total volume of NaOH solution (ml)	Concentration of NaOH solution
1	50	0	50	0.1 (N)
2	25	25	50	0.05 (N)

Use burette for making the above composition of sets.

iv. **Clean the dilatometer with a solution of chromic acid and wash it with distilled water. Finally, wash it with acetone and allow it to become dry. Handle the dilatometer carefully, as it is made of glass and may be broken if care is not taken.**

v. **Pipette out 10 ml pure diacetone alcohol and run into the solution (0.1 N NaOH solution) of set no. 1. Mix it well, stirring the reaction mixture by one way motion. Fill the dilatometer bulb with the reaction mixture carefully so that no air bubble remains present there and adjust the lower meniscus of solution column in the minimum level (i.e., zero mark) of the capillary by moving the stopcock carefully. When solution (i.e., reaction mixture) level touches the minimum mark of the capillary tube the stopcock is closed and immediately start the stopwatch. Thus the volume of the reaction mixture is taken as fixed at the start of the time counting. With time it will be observed that the height of the solution column in the capillary increases gradually from minimum mark. Note the heights of the solution column in the capillary at intervals of 1 minute. Record the readings for 10 minutes, i.e., take 10 readings. According to the theory, the readings will be x_1, $x_2, x_3, \cdots, x_{10}$ mm at $t_1, t_2, t_3, \cdots, t_{10}$ minutes respectively. Then take an interval of 20 minutes. After the constant interval of 20 minute, record another set of 10 readings denoted as $x'_{31}, x'_{32}, x'_{33}, \cdots x'_{40}$ at $t'_{31}, t'_{32}, t'_{33}, \cdots t'_{40}$ minutes respectively, i.e., according to the theory $\Delta t = 20$ minute.**

vi. **Drain out the reaction mixture from the dilatometer by opening the stopcock. Wash it with distilled water and then with acetone. It is then allowed to become dry as before.**

vii. **Repeat the same experiment using NaOH solution (0.05 N) of set 2.**

viii. **Calculate $\log(x'_n - x_n)$ at times $t_1, t_2, \cdots, t_{10}$ minutes for both set 1 and set 2. Plot $(x'_n - x_n)$ vs. t_n in the same graph paper and calculate K_1 and K_2 from the slopes. Then find the ratio $\frac{K_1}{K_2}$ and interpret the result.**

Results and Calculations

1. **Recording of laboratory temperature**

	Before experiment	After experiment	Mean
Temperature →	... °C	... °C	... °C

2. **Preparation of 100 ml 0.1 (N) oxalic acid solution**

Initial mass (g)	Final mass (g)	Mass of oxalic acid transferred (g)	Mass of oxalic acid to be taken (g)	Strength of oxalic acid sol.
			0.63	... (N)

3. **Standardisation of NaOH solution: Volume of oxalic acid taken = 10 ml**

Observation number	Vol. of NaOH solution required (ml)	Mean vol. of NaOH solution required (ml)	Strength of NaOH solution
			... (N)

For preparation of exact 0.1 (N) 100 ml NaOH solution, ... ml is diluted to 100 ml.

4. **Recording of heights at different times (for Set 1)**

Time t_n (minute)	Height X_n (mm)	gap (Δt)	Time ($t_n + \Delta t$) minute	Height x'_n (mm)
1			31	
2			32	
⋮		20 minute	⋮	
⋮			⋮	
10			40	

5. Recording of heights at different times (Set 2):
Table is same for Set-1.

6. **Calculation of data for graph plotting**

	Set -1	Set -2
Time t_n (min)	$\log\ (x'_n - x_n)$	$\log(x'_n - x_n)$
1	...	...
2	...	...
3	...	...
⋮	...	...
⋮	...	...
10	...	...

7. From the graph of the plot of $\log\left(x'_n - x_n\right)$ vs. t_n, it is found that

$$K_1 = \text{slope} \times 2.303 \text{ mint}^{-1} = \ldots \ \ldots \text{ mint}^{-1}$$

and $$K_2 = \text{slope} \times 2.303 \text{ mint}^{-1} = \ldots \ \ldots \text{ mint}^{-1}$$

Thus, the ratio $\dfrac{K_1}{K_2} = \ldots$

8. Interpretation: The expected value of the ratio $\dfrac{K_1}{K_2}$ is 2. If we find that the experimental ratio be 2, then we can conclude that NaOH acts as a homogeneous catalyst in the alkaline decomposition of diacetone alcohol.

Experiment 8A

Suggestions for performing similar type of experiment.

Dilatometric study of the kinetics of acid catalysed hydrolysis of acetal.

The acid catalysed hydrolysis of acetal

$$CH_3CH(OC_2H_5)_2 + H_2O \xrightarrow{H^+} CH_3CHO + 2C_2H_5OH + H^+$$

is also a pseudo first order reaction, in which the volume of the reaction mixture increases with time. Therefore, it can be studied by a dilatometer. The principle, procedure, etc. are same as the previous experiment of kinetics study of alkali-catalysed decomposition of diacetone alcohol. Only the difference is that, in the present experiment 0.1(N) and 0.05 (N) HCl solutions are used instead of 0.1(N), 0.05 (N) NaOH solutions, and 0.4 (M) freshly distilled acetal is used instead of pure diacetone alcohol.

Experiment 9

Kinetic study of the reaction between hydrogen peroxide and iodide ion in acid medium using iodine clock reaction.

Theory: The overall reaction

$$H_2O_2 + 2HI \rightarrow 2H_2O + I_2$$

results from a sequence of steps. The slow rate determining step is thought to be the formation of HIO. A more rapid step follows to complete the reaction.

i. $H_2O_2 + HI \xrightarrow{\text{slow}} H_2O + HIO$

ii. $HIO + HI \xrightarrow{\text{rapid}} H_2O + I_2$ (or, $HIO + H^+ + I^- \rightarrow H_2O + I_2$)

The most likely ways in which the HIO is formed are:

a. $H_2O_2 + I^- \xrightarrow{K_1} OH^- + HIO$

$$\text{Rate} = K_1 [H_2O_2][I^-]$$

b. $H_2O_2 + H^+ \underset{K_3}{\overset{\text{rapid}\; K_2}{\rightleftharpoons}} H_3O_2^+$

$$K_2[H_2O_2][H^+] = K_3 [H_3O_2^+]$$

$$I^- + H_3O_2^+ \xrightarrow[\text{slow}]{K_4} H_2O + HIO$$

$$\text{Rate} = K_4 [H_3O_2^+][I^-]$$

$$= \frac{K_4K_2}{K_3}[H_2O_2][H^+][I^-] \quad \text{(substituting } [H_3O_2^+] \text{ from } b\text{)}.$$

If the HIO is formed by both (a) and (b), then

rate of overall reaction = rate of HIO formation.

Hence we have

$$\text{rate of overall reaction} = K_1 [H_2O_2][I^-] + \frac{K_4K_2}{K_3}[H_2O_2][H^+][I^-]$$

$$= \left\{K_1 + \frac{K_4K_2}{K_3}[H^+]\right\}[H_2O_2][I^-]$$

$$= K [H_2O_2][I^-], \text{ where } K \text{ is a constant and} \quad (1)$$

$$K = \left\{K_1 + \frac{K_2K_4}{K_3}[H^+]\right\}.$$

The reaction is kinetically second order. Thus, in the above equation (1) K is the second order rate constant. If during the course of the reaction $[I^-]$ remains constant, then the kinetics follow the first order reaction.

Experimentally this is achieved by continually adding small volume of sodium thiosulphate solution to remove iodine as soon as it is liberated and to regenerate iodide ion according to the reaction

$$2S_2O_3^{=} + I_2 \rightarrow S_4O_6^{=} + 2I^-$$

The rate of the reaction depends only on $[H_2O_2]$ and temperature. Since one molecule of H_2O_2 is destroyed for every molecule of I_2 liberated, the volume of thiosulphate solution required provides the values of the amount of H_2O_2 used up as a funtion of time and permits the calculation of the first order velocity constant for the rate of reaction of H_2O_2 at constant concentration of iodide. To calculate the corresponding second order rate constant (K) we assume the reaction first order with respect to iodide.

The volume of thiosulphate solution required, V_t, is proportional to the concentration of H_2O_2 $(a - x)$ remaining at time t after the initiation of the reaction. If V_0 be the volume of thiosulphate solution required at $t = 0$ for total H_2O_2, then V_0 is proportional to the initial concentration of H_2O_2, a. Then the expression for the first order rate constant is given by

$$K' = \frac{2.303}{t}\log\frac{a}{a-x} = \frac{2.303}{t}\log\frac{V_0}{V_t}$$

$$= \frac{2.303}{t_n}\log\frac{V_0}{(V_0 - V_n)} \quad (n = 1, 2, 3, \cdots), \qquad (2)$$

where $V_1, V_2, \cdots, V_n$ are volumes of thiosulphate solution consumed at times $t_1, t_2, \cdots, t_n$ respectively. If we plot $\log\left(\frac{V_0}{V_0 - V_n}\right)$ vs. t_n, a straight line passing through the origin will be obtained, from the slope slope of which the first order rate constant K' can be determined.

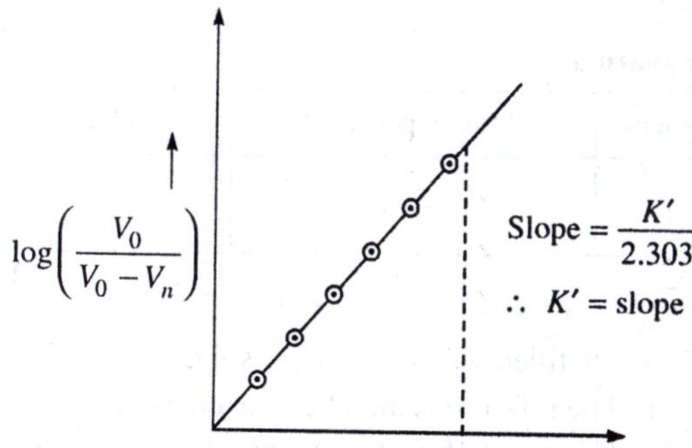

Fig. 8.10: Plot of $\log\left(\frac{V_0}{V_0 - V_n}\right)$ vs. time (t_n)

The second order rate constant K can then be determined as follows:

$$K[H_2O_2][I^-] = K'[H_2O_2]$$

i.e., $K' = K[I^-]$

$$\therefore\ K = \frac{K'}{[I^-]}. \qquad (3)$$

Thus, if iodide ion concentration is known, which remains constant during the course of the reaction, K can be calculated from equation (3).

Materials: (i) Hydrogen peroxide 30% (100 V), (ii) Potassium iodide, (iii) Sulphuric acid (conc.) (iv) Starch solution, (v) Sodium thiosulphate, (vi) Burettes, pipettes, conical flasks, etc.

Procedure

1. **Preparation of 0.4% 500 ml potassium iodide solution**

Initial mass (g)	Final mass (g)	Mass of KI transferred (g)	Mass of KI to be taken (g)	Strength of KI solution
			2.0	.... %

2. Preparation of 250 ml 2(V) H_2O_2 solution:

$30\%\ H_2O_2 \equiv 100(V)\ H_2O_2$

$250 \times 2 = 100 \times x$, $\therefore\ x = 5$ cc, i.e., **5 cc of 30% H_2O_2 is taken in a 250 ml volumetric flask and the volume is made up to the mark by adding distilled water.**

3. Preparation of 1 : 2 H_2SO_4 solution:

25 cc of concentrated H_2SO_4 is mixed carefully with 50 cc of distilled water to prepare 75 ml 1 : 2 V/V H_2SO_4 solution.

4. Preparation of 500 ml $\frac{N}{10}$ sodium thiosulphate solution:

Relative equivalent mass of $Na_2S_2O_3 \cdot 5H_2O = 248$.

$\therefore$ $\approx$ 12.4 gms of sodium thiosulphate is dissolved in distilled water and the volume of the solution is made to 500 cc.

5. **Prepare $\approx$ 50 ml starch solution by usual method and cool it before use.**

Results and Calculations

1. **Recording of laboratory temperature**

	Before experiment	After experiment	Mean
Temperature ⟶	... °C	... °C	... °C

2. **Determination of V_0**

In a 250 ml conical flask, take 50 cc distilled water + $\approx$ 2 gm solid KI + 5cc. (1 : 2) H_2SO_4 + 10 cc 2(V) H_2O_2 + 2cc starch solution. The mixture is then kept in a dark place, covering the mouth of the conical with watch glass, for 45 minutes for completion of the reaction. Then it is titrated with $\frac{N}{10}$ thio solution and note the burette reading which is V_0. Use three sets and take the mean of V_0-values.

Sets for V_0 determination	Volume of thiosulphate solution required (ml)	Mean value of V_0 (ml)
1	V_0'	$\frac{V_0' + V_0'' + V_0'''}{3} = V_0$
2	V_0''	
3	V_0'''	

3. **Titration results using 0.4% KI solution (Set 1):** In a 250 ml conical flask take (20 cc 0.4% KI solution + 5 cc 1 : 2 H_2SO_4 + 2 cc starch solution + 30 cc distilled water) + add 10 cc of 2(*V*) H_2O_2 from a pipette starting the stopwatch when the pipette is half-emptied. Shake the mixture well. Titrate the reaction mixture with $\frac{N}{10}$ sodium thiosulphate solution when the blue colour disappears. Note the burette reading which is V_1. Note the time when the blue colour just reappear, which is t_1. Continue this procedure by adding small portion of $\frac{N}{10}$ thio to disappear the blue colour and note the time of reappearence of blue colour. Note in this way $V_2, V_3 \cdots V_n$ and time $t_2, t_3, \cdots, t_n$ (take at least eight readings). In tabular form the result may be respresented as below:

Time of reappearance of blue colour (min) t_n	Volume of thio required to discharge the blue colour (ml) (V_n)	V_0 (ml)	$\log \frac{V_0}{V_0 - V_n}$
t_1	V_1		
t_2	V_2		
t_3	V_3		
t_4	V_4	...	
t_5	V_5		
t_6	V_6		
t_7	V_7		
t_8	V_8		

4. **Titration result using 0.2% KI solution (Set 2)**
In a 250 ml conical flask take (10 cc 0.4% KI solution + 5 cc 1 : 2 H_2SO_4 + 2 cc starch solution + 40 cc distilled water) + add 10 cc 2(V) H_2O_2 from a pipette, starting the stopwatch when the pipette is half-emptied. Shake the mixture well and proceed similarly as before (using 0.4% KI solution, Set 1) and $V_1, V_2, \cdots, V_n$ at $t_1, t_2, \cdots, t_n$ respectively. These may be represented in the similar tabular form:

Time of reappearance of blue colour t_n (min)	Volume of thio required to appear the blue colour (V_n) (ml)	V_0 (ml)	$\log \frac{V_0}{V_0 - V_n}$
t_1	V_1		
t_2	V_2		
⋮	V_3		
⋮	⋮	...	
⋮	⋮		
t_8	V_8		

5. **Graph plotting:** From the plot of $\log \frac{V_0}{V_0 - V_n}$ vs. t_n it is found that

$$K' \text{ (Set 1)} = 2.303 \times \text{slope min}^{-1}$$

and $$K'' \text{ (Set 2)} = 2.303 \times \text{slope min}^{-1}.$$

Thus, the ratio $\frac{K'}{K''} = 2$ (expected).

6. **Calculation of second order rate constants for the two sets**

For Set 1

The concentration of KI solution used

$$= 0.4\% = 4 \text{ gm/lit} = \frac{4}{166}(M) = 2.41 \times 10^{-2} (M)$$

Total volume of set 1 is 67 cc.

∴ concentration of KI solution in set 1

$$= \frac{20 \times 2.41 \times 10^{-2}}{67} = 7.2 \times 10^{-3} \text{ (M)}.$$

∴ Second order rate constant $K = \frac{K'}{7.2 \times 10^{-3}}$ lit mol^{-1}. mint^{-1} [From equation (3)]

For Set 2

Similarly, concentration of KI solution $= \frac{7.2 \times 10^{-3}}{2} = 3.6 \times 10^{-3}$ (M).

∴ second order rate constant $K = \frac{K''}{3.6 \times 10^{-3}}$ lit mol^{-1} min^{-1}. [From equation (3)]

The ratio $\frac{K_{\text{1st set}}}{K_{\text{2nd set}}} = 1$ (expected).

7. **Interpretation of results:** (To be given by the students).

Experiment 10

Studies on the kinetics of saponification of ethyl acetate by sodium hydroxide (i) Determination of rate constant, (ii) Studies on the influence of ionic strength on the rate constant (Salt effect).

Theory

i. **Determination of rate constant:** The saponification of ethyl acetate,

$$CH_3COOC_2H_5 + OH^- = CH_3COO^- + C_2H_5OH$$

proceeds by a second order reaction. Hence, if the initial concentration of both the reactants are same and equal to *a* moles/litre and if *x* is the number of moles/litre which have reacted after a time *t*, then

$$\frac{dx}{dt} = (a - x)^2.$$

Integration of this equation gives

$$\frac{1}{a-x} = Kt + \frac{1}{a}, \tag{1}$$

where $(a - x)$ is the concentration (in moles/litre) of reactant remaining at time t and K is second order rate constant which is expressed in lit mol^{-1} sec^{-1}.

If we plot $\frac{1}{a-x}$ vs. t, a straight line with positive slope will be obtained and its slope will be the value of K (Fig. 8.11).

ii. **Studies on the influence of ionic strength on the rate constant (Salt effect)**

According to Bronsted-Bjerrum equation, for a dilute solution,

$$\log K = \log K_0 + 1.018\, Z_A Z_B \sqrt{\mu}$$

which is later slightly modified by Günterberg as

$$\log K = \log K_0 + 1.018 Z_A Z_B \frac{\sqrt{\mu}}{1+\sqrt{\mu}}, \tag{1}$$

where K is the rate constant at ionic strength μ; K_0 is equal to the rate constant at $\mu = 0$; Z_A, Z_B are the charge of the ions in the reaction; μ is called ionic strength and it is given as

$\mu = \frac{1}{2}\Sigma\left(C_i Z_i^2\right)$, C_i is the concentration in moles/lit and Z_i is the valency of the ion. The sum is over all the different kinds of ions in solution. In the present reaction, the reactant $CH_3COOC_2H_5$ is a neutral molecule, and hence is not taken into account. The only reactant ion is OH^- and its charge is -1. Hence, the above equation (1) may be written as

$$\log K = \log K_0 - 1.018 \frac{\sqrt{\mu}}{1+\sqrt{\mu}}. \tag{2}$$

If the reaction is studied in a medium of potassium chloride solutions of different concentrations, then μ will vary. By determining the rate constants at different ionic strengths a graph of log K against $\frac{\sqrt{\mu}}{1+\sqrt{\mu}}$ can be plotted. This should be a straight line with negative slope, the intercept of which gives log K_0, from which K_0 can be determined (Fig. 8.12).

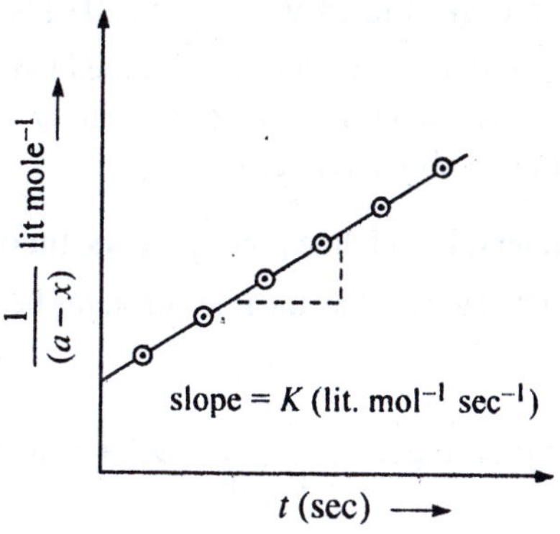

Fig. 8.11: Plot of $\frac{1}{a-x}$ vs. t

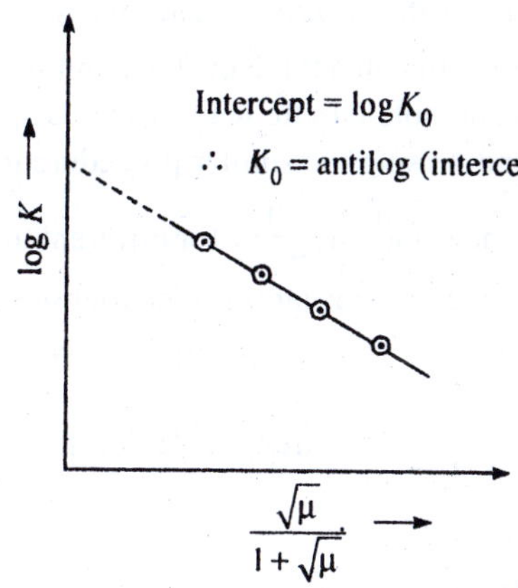

Fig. 8.12: Plot of log K vs. $\frac{\sqrt{\mu}}{1+\sqrt{\mu}}$

Materials: (i) Hydrochloric acid, (ii) Caustic soda, (iii) Potassium chloride, (iv) Ethyl acetate, (v) Burettes, pipettes, stoppered bottles, conical flasks, etc. (vi) Oxalic acid.

Short Procedure

1. Prepare 100 ml 0.1 (N) oxalic acid solution by accurate weighing.
2. Prepare ≈ 0.1 (M) HCl solution and ≈ 0.1(M) NaOH solutions (500 ml each).
3. Standardise NaOH solution against standard oxalic acid solution and then HCl solutions against standardised NaOH solution.
4. Prepare 1 lit each of $\frac{M}{30}$ HCl and $\frac{M}{30}$ NaOH solution by quantitative dilution.

5. Prepare 250 ml 1 (M) KCl solution by accurate weighing and 250 ml $\frac{M}{6}$ ethyl acetate from its molecular weight and density data.
6. The following reaction mixtures are to be prepared in four clean and dry stoppered bottles of capacity 500 ml.

Bottle No.	Volume of $\frac{M}{30}$ NaOH Solution (ml)	Volume of water (ml)	Volume of 1(M) KCl solution (ml)	Volume of $\frac{M}{6}$ ethyl acetate (ml)	Total Volume (ml)
1	125	100	0	25	250
2	125	50	50	25	250
3	125	25	75	25	250
4	125	0	100	25	250

Add the first three components (alkali, water and KCl solution) of the four compositions first and mix them well. Take the solution of bottle no. 1, 25 ml of the $\frac{M}{6}$ ester is pipetted out and run into it, starting the stopwatch when the pipette is half-emptied, mix it well. 25 ml of the reaction mixture is withdrawn and run into a solution of 25 ml $\frac{M}{30}$ HCl solution in a 100 ml conical flask noting the time of half-discharge of the HCl from the pipette (time should be 3-4 minutes from the starting of the reaction). The excess HCl is back titrated with $\frac{M}{30}$ NaOH solution using phenolphthalein indicator. Note the burette reading. Repeat the same at the time interval of about 4 minutes and note burette readings in each case. Take at least six readings. Study the same for sets 2, 3 and 4 using the similar procedure mentioned above for set 1.

7. Calculate the values of $\frac{1}{a-x}$ for different time intervals and then plot $\frac{1}{a-x}$ vs. time (t) for each set. Determine rate constants for the four sets from the graph. Say the rate constants be K_1, K_2, K_3 and K_4 for sets 1, 2, 3 and 4 respectively.
8. Calculate μ, the ionic strengths for each set and plot $\log K$ vs. $\frac{\sqrt{\mu}}{1+\sqrt{\mu}}$. Determine K_0 from the intercept.

Results and Calculations

1. **Recording of laboratory temperature**

	Before experiment	After experiment	Mean
Temperature →	°C	°C	°C

2. **Preparation of 100 ml 0.1 (N) oxalic acid solutions**

Initial mass (g)	Final mass (g)	Mass of oxalic acid transferred (g)	Mass of oxalic acid to be taken (g)	Strength of oxalic acid solution
			0.63	... (N)

3. **Standardisation of NaOH solution by oxalic acid solution**

No. of observations	Volume of oxalic acid solution taken (ml)	Volume of NaOH solutions required (ml)	Mean volume of NaOH solution (ml)	Strength of NaOH solution
1	10	...		
2	10	...	...	... (M)
3	10	...		

4. **Standardisation of HCl solution against standardised NaOH solution**

Observation No.	Volume of HCl solution taken (ml)	Volume of NaOH solution required (ml)	Mean volume of NaOH solution (ml)	Strength of HCl solution
1	10	...		
2	10	...	...	...(M)
3	10	...		

5. 1 lit. $\frac{M}{30}$ NaOH solution and 1 lit. $\frac{M}{30}$ HCl solution are prepared by quantitative dilution. ... ml HCl is diluted to 1 lit and ... ml NaOH is diluted to 1 lit.

6. **Recording of data for set no. 1**

The strengths of NaOH and HCl solutions initially are made equal (i.e., $\frac{M}{30}$).

Therefore 25 ml$\left(\frac{M}{30}\right)$ HCl solution ≡ 25 ml$\left(\frac{M}{30}\right)$ NaOH solution.

No. of observations	Time (sec)	Titre value of $\frac{M}{30}$ NaOH solution x_i (ml)	Volume of $\frac{M}{30}$ unreacted NaOH in the reaction mixture $(25-x_i)$ (ml)	$(a-x) = \frac{(25-x_i)}{25} \times \frac{1}{30}$ (moles/lit)	$\frac{1}{a-x}$ (lit/mol)
1	t_1 ...	$x_1 =$	$25 - x_1$	...	...
2	t_2 ...	$x_2 =$	$25 - x_2$	...	...
3	t_3 ...	$x_3 =$	$25 - x_3$	...	...
4	t_4 ...	$x_4 =$	$25 - x_4$	...	...
5	t_5 ...	$x_5 =$	$25 - x_5$	...	...
6	t_6 ...	$x_6 =$	$25 - x_6$	...	...
7	t_7 ...	$x_7 =$	$25 - x_7$	...	...

Present similar tables for sets 2, 3 and 4.

7. **Table for graph plotting (for rate constants determination)**

Set 1			Set 2			Set 3			Set 4		
Time (sec)	$\frac{1}{a-x}$ lit mol^{-1}	K_1 lit mol^{-1} sec^{-1}	Time (sec)	$\frac{1}{a-x}$ lit mol^{-1}	K_2 lit mol^{-1} sec^{-1}	Time (sec)	$\frac{1}{a-x}$ lit mol^{-1}	K_3 lit mol^{-1} sec^{-1}	Time (sec)	$\frac{1}{a-x}$ lit mol^{-1}	K_4 lit mol^{-1} sec^{-1}

8. **Graph plotting for K_0 determination**

Set No.	Concentration of KCl solution (M)	Concentration of NaOH solution (M)	$\frac{\sqrt{\mu}}{1+\sqrt{\mu}}$	K values (lit mol^{-1} sec^{-1})
1	0	$\frac{1}{60}$		
2	$\frac{1}{5}$	$\frac{1}{60}$		
3	$\frac{3}{10}$	$\frac{1}{60}$		
4	$\frac{2}{5}$	$\frac{1}{60}$		

*** If KCl solution be made exactly 1(M) then the data in column 2 may be directly used. If the concentration of KCl be made *f* 1(M) [*f* is a factor], then the above data will be multiplied by factor *f*.**

9. **Conclusion**

From the plot of $\log K$ vs. $\frac{\sqrt{\mu}}{1+\sqrt{\mu}}$ it is found that,

$\log K_0$ = intercept, $\therefore$ K_0 = antilog of intercept = ··· lit mol^{-1} sec^{-1} at ··· °C.

Experiment 11

Kinetics study of the alkaline hydrolysis of crystal violet spectrophotometrically. (i) Determination of rate constant and (ii) Determination of the order of the reaction with respect to alkali.

Theory

i. The reaction is

Crystal Violet (CV) + NaOH $\rightarrow$ Products.

The rate of the reaction may be represented as

$$-\frac{d[\mathrm{CV}]}{dt} = K\,[\mathrm{CV}]^m\,[\mathrm{OH}^-]^n$$

where, [CV] = molar concentration of CV at any time t

$[OH^-]$ = molar concentration of OH^- ions at time t.

m = order of the reaction with respect to CV

n = order of the reaction with respect to OH^-

If $[OH^-]$ be in excess compared to the dye concentration, then

$$-\frac{d[CV]}{dt} = K'[CV]^m, \text{ where } K' = K[OH^-]^n$$

If $m = 1$, then the above equation becomes

$$-\frac{d[CV]}{dt} = K'[CV] \quad \text{or,} \quad -\frac{d[CV]}{[CV]} = K'\,dt$$

On integration, $\log\frac{[CV]_0}{[CV]_t} = \frac{K'}{2.303}t$

where, $[CV]_0$ = molar concentration of CV at time $t = 0$

$[CV]_t$ = molar concentration of CV at time $t = t$.

Now, by considering molar concentration of CV directly proportional to the optical density D of the reaction mixture the above equation becomes,

$$\log\frac{D_0}{D_t} = \frac{K'}{2.303}t$$

$$\text{or,} \quad \log D_t = \log D_0 - \frac{K}{2.303}t \qquad (3)$$

Where D_0 = initial optical density (O.D.) of the dye solution and D_t = O.D. of the reaction mixture at time t.

By measuring the O.D. of the solution at a regular interval of time as the hydrolysis proceeds, and by plotting log D_t vs. t, the value of K' can be determined from the slope. The nature of graph will be as below:

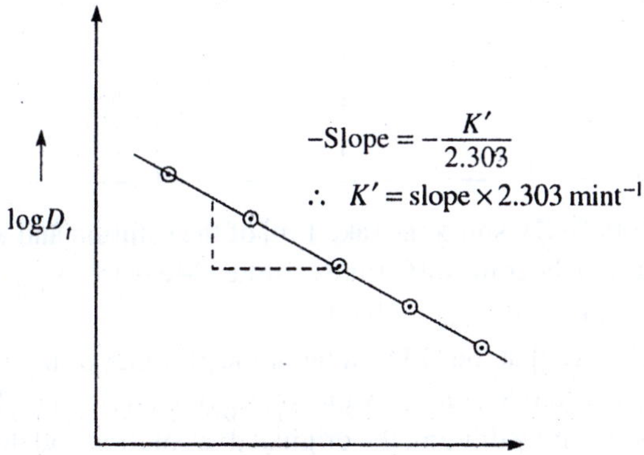

Fig. 8.13: Plot of logD_t vs. time (t)

ii. Again, $K' = K[OH^-]^n$

$$\text{or,} \quad \log K' = \log K + n\log[OH^-] \qquad (2)$$

Thus, by plotting $\log K'$ vs. $\log[OH^-]$, we can obtain the order (n) of the reaction from the slope and rate constant K from the intercept as follows:

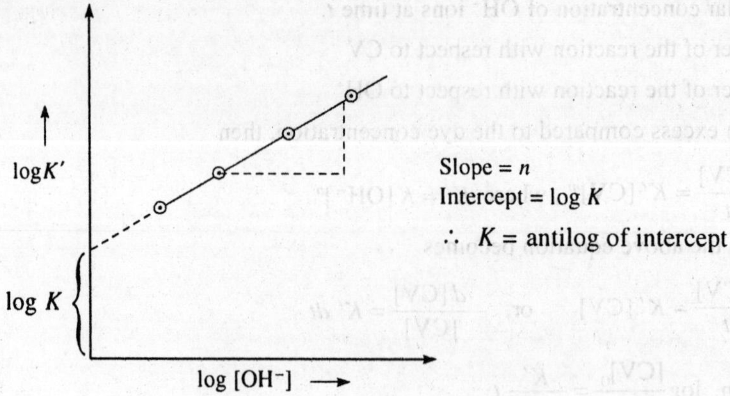

Fig. 8.14: Plot of logK' vs. log $[OH^-]$

Materials: (i) Oxalic acid, (ii) Caustic soda, (iii) Crystal violet dye, (iv) Pipettes & burettes, (v) Test tubes (5 pcs) of 20 ml capacity, (vi) Tissue paper, (vii) Beaker, (viii) Spectrophotometer (visible), etc.

Procedure

1. Prepare 100 ml 0.1 (N) oxalic acid solution by accurate weighing.
2. Prepare 100 ml $\approx$ 0.05 (N) NaOH solution and standardise it against standard oxalic acid solution. Then prepare exact 50 ml 0.05 (N) NaOH solution by quantitative dilution.
3. Prepare five solutions of NaOH of different concentrations in five test tubes of 20 ml capacity according to the following composition of sets:

Test tube nos.	Volume of 0.05 (M) NaOH solution (ml)	Volume of distilled water (ml)
1	1	8
2	2	7
3	3	6
4	4	5
5	5	4

4. Prepare 50 ml $\approx$ 0.002% CV solution. Take 1 ml of the solution and add 9 ml of distilled water to it so that dye solution becomes 10 times dilute. Measure λ_{max} taking this solution using spectrophotometer (expected λ_{max} = 605 nm).
5. It is important to observe that the O.D. value of the 10 times diluted dye solution at λ_{max} should be within the range 0.5 to 0.8. It is the value of D_0 of equation (1). If the D_0 value does not fall within this range, adjust it by diluting the original dye solution and then diluting it 10 times using procedure (4). Otherwise the kinetics study will not be successful.
6. Adjust λ_{max} in the spectrophotometer. Take set no. 1 and add 1 ml of the original dye solution (use 1 ml pipette) to it starting the stopwatch when the pipette is half-emptied. Mix it well and immediately place it in the spectrophotometer. Note O.D. value of the reaction mixture just after 1 minute. This is the value of D_t, i.e., D_1. Similarly record D_2, D_3, ... at a constant interval of 1 minute (take at least ten readings). Repeat the same for other four sets. Remember that for a set $D_1 > D_2 > D_3 \cdots$
7. Plot log D_t vs. t for each set and determine K_1', K_2', K_3', K_4' and K_5'.

8. Calculate the concentration of OH^- ions in each set and then plot $\log K_n'$ vs. $\log[OH^-]$. Determine the order (n) and the value of K, from the slope and intercept respectively.

Results and Calculations

1. **Recording of laboratory temperature**

	Before experiment	After experiment	Mean
Temperature →	°C	°C	°C

2. **Preparation of 100 ml 0.1 (N) oxalic acid solution**

Initial mass (g)	Final mass (g)	Mass of oxalic acid transferred (g)	Mass of oxalic acid to be taken (g)	Strength of oxalic acid solution
...	...	...	0.63	.... (N)

3. **Standardisation of NaOH solution**

Observation nos.	Volume of oxalic acid solution taken (ml)	Volume of NaOH solution required (ml)	Mean volume of NaOH solution (ml)	Strength of NaOH solution
1	10	...		
2	10	...	...	...(N)
3	10	...		

50 ml exact 0.05 (N) NaOH solution is prepared by diluting ··· ml to 50 ml.

4. **Determination of λ_{max} of the dye solution**

Observation No.s.	Wave length (nm)	Optical density (O.D.) observed
1	550	...
2	560	...
3	⋮	...
4	⋮	...
⋮	⋮	...
⋮	650	...

From the plot of O.D. vs. λ, it is found that λ_{max} = ··· nm.

5. **Recording of D_t values of different sets at λ_{max}**

Obs. No.	Time (t) (minutes)	D_t-values				
		Set 1	Set 2	Set 3	Set 4	Set 5
1	1					
2	2					
3	3					
4	⋮					
5	⋮					
6	⋮					
7	⋮					
8	⋮					
9	⋮					
10	⋮					

6. **From the graphs of the plot of log D_t vs. t, it is found that $K_1' =$ slope × 2.303 mint^{-1}, $K_2' =$ slope × 2.303 mint^{-1}, $K_3' =$ slope × 2.303 mint^{-1}, $K_4' =$ slope × 2.303 mint^{-1} and $K_5' =$ slope × 2.303 mint^{-1}**

7. **Table for graph plotting for order determination**

Set Nos.	$[OH^-]$ moles/lit	log $[OH^-]$	K_n' (mint^{-1})	log K_n' (n = 1 to 5)
1	0.005			
2	0.010			
3	0.015			
4	0.020			
5	0.025			

From the plot of $\log K_n'$ vs. $\log[OH^-]$, it is found that slope = n = ...

Expected value of n = 1, i.e., the reaction is first order with respect to alkali.

Experiment 12

Studies on the kinetics of the reaction between $S_2O_8^{2-}$ and I^- by colorimetric method: (i) Determination of order of the reaction and (ii) Determination of rate constant.

Theory

i. The overall reaction between $S_2O_8^{2-}$ and I^- is

$$S_2O_8^{2-} + 2I^- \rightarrow 2SO_4^{2-} + I_2$$

If 'a' equivalent/litre of both reactants, $S_2O_8^{2-}$ and I^- are mixed, and if 'n' be the overall order of the reaction, then the time (t) required for a definite fraction of the reactants to react will be inversely proportional to a^{n-1}, i.e.,

$$t \propto \frac{1}{a^{n-1}}, \qquad (1)$$

Thus, if a_1 and a_2 be the initial concentrations (in gm- equivalent/litre) of the two reactants and t_1 and t_2 be the times required for a definite fraction of the reactants to react, then according to equation (1),

$$\frac{t_1}{t_2} = \left(\frac{a_2}{a_1}\right)^{n-1}$$

or, $$\log\left(\frac{t_1}{t_2}\right) = (n-1)\log\left(\frac{a_2}{a_1}\right).$$

$$\therefore \quad n = 1 + \frac{\log\left(t_1/t_2\right)}{\log\left(a_2/a_1\right)}. \tag{2}$$

To find out the order of the reaction, optical densities, D_t, are plotted against t for two sets of experiments with two different initial concentrations. The order n can then be determined from equation (2).

ii. Experimentally the value of n is found to be 2, i.e., the reaction is second order, being first order with respect to $S_2O_8^{2-}$ and first order with respect to I^- and the rate law may be expressed according to,

$$\text{rate} = -\frac{d[S_2O_8^{2-}]}{dt} = K[S_2O_8^{2-}][I^-],$$

where K is second order rate constant.

If x (in gm equivalent/litre) be the decrease in concentration of each of the reactants (and also the concentration of iodine formed) after time t, then the above rate expression can be written as

$$\frac{dx}{dt} = K(a-x)^2.$$

On integration and taking the range $x = 0$ at $t = 0$ and $x = x$ and $t = t$, the above equation reduces to

$$Kt = \frac{x}{a(a-x)}. \tag{3}$$

Since one of the product, I_2, is coloured $(\lambda_{max} \approx 525 \text{ nm})$, its optical density D_t at any instant of time t is proportional to its concentration x, provided Beer's Law is obeyed. Then equation (3) becomes

$$Kt = \frac{1}{a}\cdot\frac{D_t}{D_\infty - D_t}, \tag{4}$$

where D_t and D_∞ are the optical densities at time t and after the completion of the reaction respectively.

Equation (4) can be re-written as

$$\frac{1}{Kat} = \frac{D_\infty}{D_t} - 1$$

Dividing both sides by D_∞, we get

$$\frac{1}{Kat}\cdot\frac{1}{D_\infty} = \frac{1}{D_t} - \frac{1}{D_\infty}$$

or, $$\frac{1}{D_t} = \frac{1}{KaD_\infty}\cdot\frac{1}{t} + \frac{1}{D_\infty}. \tag{5}$$

Thus, a plot of $\left(\frac{1}{D_t}\right)$ against $\left(\frac{1}{t}\right)$ will give a straight line with positive slope equal to $\left(\frac{1}{Ka\,D_\infty}\right)$ and intercept equal to $\frac{1}{D_\infty}$.

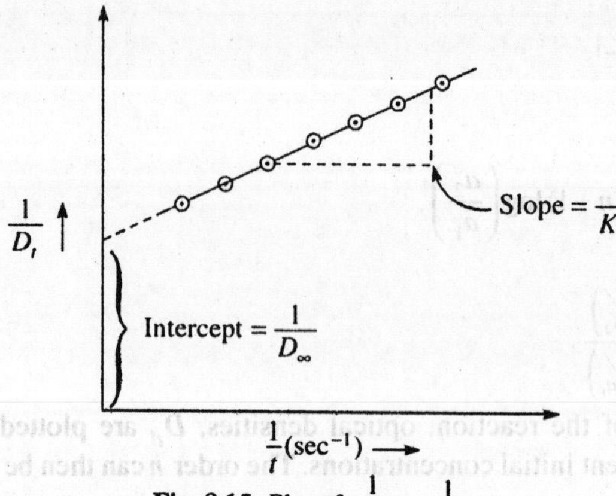

Fig. 8.15: Plot of $\frac{1}{D_t}$ vs. $\frac{1}{t}$

Hence the rate constant (K) may be evaluated using the relation:

$$K = \frac{1}{a} \cdot \frac{\text{Intercept}}{\text{Slope}} \quad (\text{lit-eqv}^{-1}\ \text{sec}^{-1}). \tag{6}$$

Materials: (i) Potassium persulphate, (ii) Potassium iodide, (iii) Potassium dichromate, (iv) Sodium thiosulphate, (v) Starch solution (~1%), (vi) Spectrophotometer, (vii) Burette, pipette, volumetric flask, conical flask, beaker, etc.

Procedure

1. Prepare:
   a. 100 ml 0.1 (N) $K_2Cr_2O_7$ solution by accurate weighing.
   b. 100 ml 0.1 (N) KI solution by accurate weighing. First prepare slightly higher than 0.1 (N) and then make 50 ml exact 0.1 (N) KI solution by dilution.
   c. 100 ml 0.1 (N) $K_2S_2O_8$ solution (slightly higher than 0.1 N so that 0.1 (N) can be prepared by dilution after standardisation with thiosulphate solution iodometrically). Molar mass of $K_2S_2O_8$ = 270 i.e., eqv. mass = 135.
   d. 250 ml 0.1 (N) $Na_2S_2O_3$ solution using a rough balance (take ~ 6.2 gm of $Na_2S_2O_3 \cdot 5H_2O$ crystals).
2. Standardise the thiosulphate solution against $\frac{N}{10}$ standard $K_2Cr_2O_7$ solution using the following procedure:

   Place 100 ml of distilled water in a 250 ml conical flask preferably with glass stoppered. Add ≈ 3gm of iodate-free potassium iodide and 2 gm of pure sodium bicarbonate and shake until the solid dissolves. Add 6 ml of conc. HCl slowly and gently rotae the flask in order to mix the liquids. Run 25 ml of standard $K_2Cr_2O_7$ solution (0.1 N) into the mixture (use pipette) shake well and allow to stand for 5 minutes in dark in order to complete the reaction, covering the mouth of the flask with stopper or watch glass. Titrate the liberated iodine with $Na_2S_2O_3$ solution $\left(\approx \frac{N}{10}\right)$; add 2 ml of 1% starch solution when straw-yellow colour appears. Continue the addition of thio solution dropwise until one drop changes the colour from greenish blue to light green.

3. The strength of the prepared $K_2S_2O_8$ solution is determined as follows:

   Pipette out 10 ml of $K_2S_2O_8$ soloution in a 250 ml stoppered conical flask, add 2 gm of solid iodate free KI and 2 ml of glacial acetic acid. Cover the conical flask with stopper or watch glass. Shake the mixture well and keep it in a dark place for 25-30 minutes for complete liberation of iodine. Add 90 ml distilled water and then titrate the liberated iodine with standard $\left(\sim\frac{N}{10}\right)$ thiosulphate solution using starch as indicator. Calculate the strength of $K_2S_2O_8$ solution.

   Prepare exact $\frac{N}{10}$ 50 ml $K_2S_2O_8$ solution by proper dilution.

4. *Composition of sets*:

   Mix the reactants in the following two sets and note the time of half-discharge of any one of the reactants in each case.

Set No.	Volume of $\frac{N}{10}$ KI solution (ml)	Volume of distilled water (ml)	Volume of $\frac{N}{10}$ $K_2S_2O_8$ solution (ml)
1	10	0	10
2	5	10	5

5. Record the optical densities (D_t) of these two sets of solutions at 525 nm wavelength at an interval of about 2 minutes for both the sets.
6. Plot D_t against t for each of the two sets. Select any D_t for set 1 and find the corresponding time t_1 from the graph. Similarly, from the graph of set 2 find the time t_2 for the value of D_t that is just half of the D_t chosen for set 1. Calculate the order of the reaction using equation (2).
7. Plot $\frac{1}{D_t}$ vs. $\frac{1}{t}$ for both the sets and find the values of K from the slopes and intercepts of the resulting straight lines using equation (6).

Results and Calculations

1. **Recording of laboratory temperature**

	Before experiment	After experiment	Mean
Temperature →	°C	°C	°C

2. **Preparation of 0.1 (N) 100 ml $K_2Cr_2O_7$ solution**

Initial mass (g)	Final mass (g)	Mass of $K_2Cr_2O_7$ transferred (g)	Mass of $K_2Cr_2O_7$ to be taken (g)	Strength of $K_2Cr_2O_7$ solution
...	...	...	0.49	.... (N)

3. **Preparation of 100 ml 0.1 (N) KI solution**

Initial mass (g)	Final mass (g)	Mass of KI transferred (g)	Mass of KI to be taken (gm)	Strength of KI solution
...	...	...	1.66	.... (N)

50 ml exact 0.1 (N) KI solution is prepared by diluting ··· ml to 50 ml.

4. **Standardisation of $Na_2S_2O_3$ solutions**

No. of observations	Volume of $K_2Cr_2O_7$ taken (ml)	Volume of thio solution required (ml)	Mean volume of thio required (ml)	Strength of thio solution
1	25			
2	25	...	...	... (N)
3	25	...		

5. **Standardisation of $K_2S_2O_8$ solution**

No. of observations	Volume of $K_2Cr_2O_7$ taken	Volume of thio solution required (ml)	Mean volume of thio required (ml)	Strength of $K_2S_2O_8$ solution
1	10	...		
2	10	...	...	... (N)
3	10	...		

50 ml exact 0.1 (N) $K_2S_2O_8$ solution is prepared by diluting ⋯ ml to 50 ml.

6. **Recording of O.D. of Set 1**

No. of observation	time (mint), t	Observed transmittance or O.D. (D_t)
1 2 3 ⋮	⋮	

7. **Recording of O.D. of Set 2**

No. of observation	time (mint), t	Observed transmittance or O.D. (D_t)
1 2 3 ⋮	⋮	⋮

8. **Determination of order (n): From the plot of D_t vs. t**

For set 1, t_1 = ⋯ mint., corresponding D_t = ⋯

For set 2, t_2 = ⋯ mint., corresponding to $\frac{D_t}{2}$ of set 1.

Hence n = ⋯ [equation (2)].

9. **Determination of rate constant (K):** From the plot of $\frac{1}{D_t}$ vs. $\frac{1}{t}$,

For set 1, Slope = ⋯, intercept = ⋯ , a = 0.05 (N)

∴ K_1 = ⋯ lit eqv.$^{-1}$ mint^{-1} [equation (6)]

For set 2, slope = ⋯, intercept = ⋯, a = 0.025 (N). [equation (6)]

∴ K_1 = ⋯ (lit eqv.$^{-1}$ mint^{-1})

[N.B. : $S_2O_8^{2-} + 2e \rightleftharpoons 2SO_4^{2-}$. Thus equivalent mass of $K_2S_2O_8 = \frac{\text{Molecular weight}}{2}$]

APPENDIX

Table 1: Some reactions that can be studied kinetically with their order

Reaction type	Order
1. Acid hydrolysis of methyl/ethyl acetate	1 (Pseudo)
2. Inversion of sucrose by acid catalyst	1 (Pseudo)
3. Decomposition of hydrogen peroxide in presence of $FeCl_3$ catalyst	1
4. Reaction between potassium persulphate and potassium iodide	2 1 (Pseudo)
5. Iodination of acetone in acid medium	
i. with respect to iodine	0
ii. with respect to acid	1
iii. with respect to acetone	1
6. Alkali catalysed decomposition of diacetone alcohol	1
7. Reaction between hydrogen peroxide and iodide ion in acid medium	2 1 (Pseudo)
8. Acid hydrolysis of acetal	1
9. Saponification of ethyl/methyl acetate by NaOH solution	2
10. Alkaline hydrolysis of crystal violet	1
11. Reaction between potassium bromate and potassium iodide	2

CHAPTER 9

Distribution Law, Phase Rule and Chemical Equilibrium

I. **General Discussion**

II. **Experiment 1**

i. Determination of partition coefficient of I_2 between CCl_4 and water and
ii. equilibrium constant of $KI + I_2 \rightleftharpoons KI_3$ by partition method.

III. **Experiment 2**

To show that benzoic acid dimerises in benzene.

IV. **Experiment 3**

Determination of dimerisation constant of benzoic acid in benzene medium. [For P.G.]

V. **Experiment 4**

Determination of coordination number of Cu^{2+} ion in cuprammonium complex in aqueous medium by partition method. [For P.G.]

VI. **Experiment 5**

Determination of critical solution temperature of phenol-water system.

VII. **Experiment 6**

A study of phase diagram of three-component liquid system at room temperature (benzene-acetic acid-water) [For P.G.]

VIII. **Experiment 7**

Determination of composition of the azeotropic mixture of benzene-methanol/benzene-ethanol system (binary mixture) by boiling point method [For P.G.]

IX. **Appendix**

Table 1. Boiling point of some pure liquids.

Table 2. Some azeotropic mixtures.

Table 3. C.S.T. of some liquid pairs.

Table 4. Partition coefficient of some solutes in two immiscible liquid pairs.

CHAPTER 9

Distribution Law, Phase Rule and Chemical Equilibrium

I. General Discussion

The Distribution law:

It is known that the chemical potential of a component present in different phases of a system at equilibrium is same in every phase. If a substance be present in two liquid phases, denoted by suffix, I and II, at equilibrium, then

$$\mu_{I} = \mu_{II}$$

$$\text{or,}\quad \mu_{I}^{0} + RT\ln a_{I} = \mu_{II}^{0} + RT\ln a_{II}$$

Then at constant temperature.

$$\frac{a_{I}}{a_{II}} = K\ (\text{constant})\,.$$

where a_{I} and a_{II} are the activities of the components in two liquid phases.

Thus, if the activity of the components in one phase be fixed, its activity in another phase at equilibrium is also fixed. In ideal solutions or dilute solutions, the activities may be replaced by concentrations C_{I} and C_{II}.

$$\text{Then,}\ \frac{C_{I}}{C_{II}} = K_{D}, \tag{1}$$

where K_D is called the 'Distribution Coefficient' or 'Partition Coefficient' of the solute between the two immiscible solvents at the given temperature. Thus, the concentrations of a solute distributed between two phases at equilibrium and at constant temperature bear a fixed ratio. This is **Nernst Distribution Law**.

The equation (1) is valid only when the molecular species is same in two phases. If the solute molecule undergoes dissociation or association in any of the phases, the distribution law as such will not hold good. Here, we will discuss, with proper example, the three cases:

i. Distribution of a solute which remains as a single molecule in two immiscible solvents (phases):

   When iodine is distributed between carbontetrachloride and water, or benzene and water; or ammonia distributed between chloroform and water, there is no change in molecular species of the solutes in either of the solvents.

Hence, $K_D = \frac{C_I}{C_{II}}$, which is confirmed from experiment (see experiments in this chapter)

ii. Distribution of a solute which dissociates in one of the phases:

Such a situation is found in the distribution of oxalic acid between the solvents ether and water. In ether layer, molecules of oxalic acid remain unchanged and suppose its concentration is C_I. In aqueous layer, oxalic acid undergoes dissociation as

$$\begin{array}{ccccc} H_2(OX) & \rightleftharpoons & H^+ & + & H(OX)^- \\ C_{II}(1-\alpha) & & \alpha C_{II} & & \alpha C_{II} \end{array}$$

where C_{II} is the total concentration of oxalic acid in aqueous layer and α is the degree of dissociation, which can be evaluated from the dissociation constant of the acid, K_a.

In such a case the ratio, $\frac{C_{II}}{C_I}$, of total concentrations would not be constant but the ratio of concentrations of undissociated species in the two layers should be constant, i.e.,

$$\frac{C_{II}(1-\alpha)}{C_I} = \text{constant.}$$

iii. Distribution of a solute when it associates in one of the phases:

When benzoic acid is added to a mixture of benzene and water, it suffers dimerisation in benzene layer but in aqueous layer it remains as a single molecule (very small ionisation being ignored). In the benzene layer, there is an equilibrium between dimers and the monomers as below:

$$\begin{array}{ccc} A_2 & \rightleftharpoons & 2A \\ C_{II}(1-\alpha) & & 2\alpha C_{II} \end{array} \qquad [A = \text{benzoic acid}]$$

where C_{II} is the total concentration of benzoic acid in benzene layer and α is the degree of dissociation of dimers into monomers. The equilibrium constant of the above dissociation is given by

$$K = \frac{4\alpha^2 C_{II}^2}{C_{II}(1-\alpha)} \approx 4\alpha^2 C_{II}^2 \quad [\because \ \alpha << 1]$$

$$\therefore \ \alpha = \frac{\sqrt{K}}{2\sqrt{C_{II}}}.$$

Therefore, the concentration of monomer in benzene layer

$$= 2\alpha C_{II} = 2 \cdot \frac{\sqrt{K}}{2\sqrt{C_{II}}} \cdot C_{II}$$

$$= \sqrt{K} \cdot \sqrt{C_{II}} = C_B \text{ (say)}.$$

Thus, the distribution coefficient (K_D) of benzoic acid between benzene and water is

$$K_D = \frac{\text{Concentration of monomer in aqueous layer}}{\text{Concentration of monomer in benzene layer}} = \frac{C_w}{\sqrt{K} \cdot \sqrt{C_B}}.$$

$$\therefore \quad K_D^2 \cdot K = \frac{C_w^2}{C_B}.$$

$$\therefore \quad \frac{C_w}{\sqrt{C_B}} = \text{constant.}$$ [where C_w is the concentration of benzoic acid in aqueous layer]

To consider a general case where '*n*' molecules of solute associate in one phase, i.e., $nA \rightleftharpoons A_n$, if C_{II} be the concentration of the solute in solvent II where the solute is practically polymarised, and C_I be the concentration of solute in solvent I, where solute is present as a single molecule it can be shown in the same way that,

$$\frac{C_I}{\sqrt[n]{C_{II}}} = \text{constant}$$

Application of the 'Nernst Distribution Law':

i. To determine of distribution co-efficient or partition co-efficient of a solute between two immiscible liquids (solvents).

ii. To determine the dimerisation constant of benzoic acid/salicylic acid in benzene.

iii. To determine the equilibrium constant of a reaction, e.g.,

$$KI(aq.) + I_2\,(aq.) \rightleftharpoons KI_3\,(aq.)$$

iv. To determine the co-ordination number of metal ions, e.g., co-ordination number of Cu^{2+}, etc.

• The Phase rule and heterogenous equilibria

The condition for equilibrium in heterogeneous systems is given by the phase rule:

$$F = C - P + 2$$

where F is the degree of freedom, C, the number of components, P, the number of phases present and additional '2' for temperature and pressure of the system. For details about F, C and P, consult any standard text book of physical chemistry.

When $P = C + 2$, the system has no degrees of freedom; the temperature and pressure under which it can exist are perfectly definite. If one of the variables, temperature or pressure, is fixed, then equilibrium will be definitely fixed when $P = C + 1$. Thus, if the pressure is fixed (say, atmospheric), the equilibrium between a solid and a liquid phase of a single substance will be defined by a definite temperature, namely the melting point.

Owing to the importance of heterogeneous equilibria a number of different cases (experiments) are given in this chapter.

Experiment 1

Determination of (i) the partition co-efficient (or distribution coefficient) of iodine between Carbon tetrachloride and water, and (ii) the equilibrium constant for the reaction: $KI\,(aq.) + I_2\,(aq.) \rightleftharpoons KI_3(aq.)$ *by partition method.*

Theory

i. The simple 'Nernst Distribution Law' is obeyed quite well for the distribution of iodine between carbontetrachloride (in which iodine is highly soluble) and water (in which iodine is comparatively low soluble). At a constant temperature, the partition (or distribution) co-efficient is given by

$$K_D = \frac{C_{org.}}{C_{aq.}} \qquad (1)$$

where $C_{org.}$ and $C_{aq.}$ represent the concentrations (molar) of iodine in carbon tetrachloride and water respectively (determined by titration).

ii. The reversible reaction

$$KI + I_2 \rightleftharpoons KI_3$$

occurs in aqueous solution and according to the law of mass action the equilibrium constant K_c is given by:

$$K_c = \frac{[KI_3]}{[KI]\,[I_2]} \qquad (2)$$

The equilibrium may be investigated by studing the distribution of iodine between an organic solvent (e.g. CCl_4) and water, followed by a similar study for the distribution of iodine between the same organic solvent and an aqueous solution of potassium iodide.

Since the distribution law only applies to the species common to both layers, the concentration of free iodine in aqueous potassium iodide layer can be determined from relationship (1).

Let the concentration of iodine in organic layer = C_1 moles/lit, concentration of iodine in aqueous KI layer = C_2 moles/lit.

$\therefore$ concentration of free iodine in aqueous layer = $\frac{C_1}{K_D}$ moles/lit and

the concentration of KI_3 = total iodine concentration in aqueous layer – free iodine concentration

$$= \left(C_2 - \frac{C_1}{K_D}\right) \text{ moles/lit.}$$

Let the initial concentration of KI = C_3 moles/lit.

Since one mole of KI_3 is formed out of one mole of KI, therefore, concentration of free KI at equilibrium

$$= \left[C_3 - \left(C_2 - \frac{C_1}{K_D}\right)\right] \text{ moles/lit.}$$

Hence equation (2) becomes:

$$K_c = \frac{\left(C_2 - \frac{C_1}{K_D}\right)}{\left[C_3 - \left(C_2 - \frac{C_1}{K_D}\right)\right]\left(\frac{C_1}{K_D}\right)} \text{ moles}^{-1}\text{lit.} \qquad (3)$$

Thus we can determine K_D from equation (1) and K_c from equation (3).

Materials

i. Saturated solution of iodine in carbon tetrachloride. ($\approx$ 100 ml).

ii. Potassium dichromate solution $\frac{N}{10}$ (100 ml).

iii. Sodium thiosulphate solution 500 ml $\left(\approx \frac{N}{10}\right)$ and 500 ml $\left(\approx \frac{N}{100}\right)$

iv. 250 ml $\frac{N}{10}$ KI solution.

v. Approx. 10% KI solution (250 ml)

vi. 1% starch solution.

vii. Stoppered bottles (250 ml capacity, 4 Nos.)

viii. Conical flasks (125 ml, 4 Nos.), burette, pipettes (25 ml, 10 ml, 5 ml), measuring cylinder (100 ml), volumetric flasks (100 ml, 250 ml, 500 ml).

ix. Separating funnel.

Procedure

1. **Determination of the partition coefficient (K_D) of iodine between carbon tetrachloride and water**

a. Prepare 100 ml saturated solution of iodine in carbon tetrachloride (shake $\approx$ 5 gm resublimed iodine in about 100 ml carbon tetrachloride for an hour so that after shaking solid iodine is left at the bottom of the stoppered bottle).

b. Make the following composition of sets:

Set Nos. (250 ml stoppered bottles)	Volume of saturated I_2 in CCl_4 (ml)	Volume of pure CCl_4 (ml)	Volume of distilled water (ml)
1	25	5	120
2	20	10	120

Shake the above bottles well for an hour and then allow to settle them for 15 minutes so that the two layers become distinct.

c. Standardise sodium thiosulphate solution $\left(\approx \frac{N}{10}\right)$ against $\frac{N}{10}$ standard dichromate iodometrically as follows:

Pipette out 10 ml of $K_2Cr_2O_7$ solution in a 250 ml conical flask — add 5 ml conc. HCl — add 100 ml distilled water. Then add $\approx$ 2 gm of solid KI. Keep in dark place for 10 minutes sealed with stoppered. Then titrate it against $\frac{N}{10}$ thio solution adding 1 ml 1% starch solution when straw-yellow colour appears. Continue the titration and end point is noted when just light green colour appears.

Calculate the strength of this solution and prepare 10 times diluted $\frac{N}{100}$ thio solution (250 ml) from it.

d. Separate the layers using a separating funnel. Pipetted 5 ml of organic layer (the pipette must be dry) in a 125 ml conical flask — add 10 ml 10% KI solution and titrate it against $\frac{N}{10}$ thio solution using starch as indicator. The end point is noted when the solution becomes calourless from blue. Repeat the same titration for bottle 2 (each titration should be repeated at least twice).

Pipette out 25 ml of aqueous layer of bottle 1 and titrate it against $\frac{N}{100}$ thio solution. Repeat the same for bottle 2.

e. Calculate the strengths of I_2 of organic and aqueous layer for two bottles. Calculate K_D from equation (1) and take mean of the two sets.

Results and Calculations

1. **Recording of laboratory temperature**

	Before experiment	After experiment	Mean
Temperature →	°C	°C	°C

2. Preparation of 100 ml 0.1 (N) potassium dichromate solution

Initial mass (g)	Final mass (g)	Mass of $K_2Cr_2O_7$ transferred (g)	Strength of $K_2Cr_2O_7$ solution
w_1	w_2	$(w_1 - w_2)$	$\frac{(w_1 - w_2)}{0.49}\left(\frac{N}{10}\right)$

3. Standardisation of $Na_2S_2O_3$ solution against standard $K_2Cr_2O_7$ solution

Observation No.	Volume of $K_2Cr_2O_7$ solution (ml)	Volume of thio $\frac{N}{10}$ required	Mean volume of thio (ml)	Strength of thio solution
1	10	V_1	$\frac{V_1 + V_2}{2}$	... (N)
2	10	V_2		

4. Titration result and determination of K_D

Set No.	Volume of each layer taken (ml) Organic	Aqueous	Volume of $\frac{N}{10}$ thio for org. layer V_{org} (ml)	Volume of $\frac{N}{100}$ thio for aq. layer V_{aq} (ml)	C_{I_2} in organic layer (M/L)	C_{I_2} in aq. layer (M/L)	K_D	Mean K_D
1	5	25	...	...	...	...	...	...
2	5	25	...	...	...	...	...	

5. Calculations

[N.B.: (1) Expected value of $K_D = 85 \pm 5$ at 25 °C-30 °C. (2) Benzene, chloroform, hexane may be used as organic solvent instead of CCl_4].

2. Determination of equilibrium constant of the reaction: $KI + I_2 \rightleftharpoons KI_3$.

Procedure

a. Prepare 250 ml $\frac{N}{10}$ KI solution by accurate weighing and make the following sets of composition.

Composition of sets

Set Nos. 250 ml stoppered (bottles)	Volume of saturated I_2 in CCl_4 (ml)	Volume of pure CCl_4 (ml)	Volume of 0.1 (N) KI solution (ml)	Volume of water (ml)
1	25	5	120	0
2	20	10	60	60

Burette must be used for the preparation of aqueous layer composition, as the exact strength of KI is required for calculation.

Shake the sets for an hour and allow to settle for 15 minutes so that distinct layers are observed.

b. Pipette out 5 ml of organic layer from bottle 1 (after separating it using separating funnel) and run into 10 ml 10% KI solution. Titrate it against $\frac{N}{10}$ thio solution using starch as indicator.

- Repeat it twice. Similarly pipette out 10 ml of aqueous layer and titrate it against $\frac{N}{10}$ thio solution. Repeat it twice or thrice.

  Repeat the same titration for bottle 2.

c. Calculate C_1, C_2 and C_3 all in moles/lit and calculate K_c from equation (3). Take mean of the two sets. [Use K_D from the previous experiment].

Results and Calculations

a. **Preparation of 250 ml 0.1 (M) KI solution:**

Initial mass (g)	Final mass (g)	Mass of KI transferred (g)	Strength of KI solution
w_1	w_2	$(w_1 - w_2)$	$\frac{w_1 - w_2}{4.15}\left(\frac{M}{10}\right)$

b. For standardisation of thio solution, follow the previous experiment.

c. **Titration results**

Set No.	Volume of each layer taken (ml)		Volume of $\frac{N}{10}$ thio required for org. layer (ml)	Volume of $\frac{N}{10}$ thio * required for aq. layer (ml)	Strength of thio solution (Previously standardised)
	Organic	Aqueous			
1	5	10	...	...	... (N)
2	5	10	...	...	

*If the titre value is low, dilute the solution five times.

d. **Determination of K_c**

Set Nos.	Mean K_D	C_1 moles/lit	C_2 moles/lit	C_3 moles/lit	$C_2 - \frac{C_1}{K_D}$ moles/L	$C_3 - \left(C_2 - \frac{C_1}{K_D}\right)$ moles/L	K_c (mol^{-1} lit)
1	...	...	...	...	...	...	...
2		...	...	...	...	...	...

Hence mean value of equilibrium constant (K_c) = ... at t °C.

[N.B.: Normal concentration of iodine must be converted to molar concentration and should be divided by 2].

Expected value of $K_c = 7.14 \times 10^2$ at 25 °C and 6.2×10^2 at 30 °C.

Experiment 2

To show that benzoic acid dimerises in benzene.

Theory: The activities of a solute distributed between two phases at equilibrium at a constant temperature bear a fixed ratio. This is 'Nernst Distribution Law.'

This law is valid only when the molecular species is the same in the two phases. If the solute undergoes dissociation or association in any one of the phases, the distribution law as such will not hold.

According to Nernst Distribution Law when a_1 and a_2 are two activities of the solute in phases 1 and 2 respectively at equilibrium at a given temperature, then

$$a_1/a_2 = K_D \quad (K_D \text{ is unitless})$$

This K_D is called distribution coefficient or partition coefficient of the solute between phases 1 and 2 at experimental temperature.

For dilute solution, $a \approx c$ (numerically, c = molar concentration).

Then, $\frac{C_1}{C_2} = K_D$.

When solid benzoic acid is shaken with two practically immiscible liquids, benzene and water, then benzoic acid remains as monomer and almost undissociated in aqueous phase (very small ionisation being neglected) but it dimerises in benzene phase.

If benzoic acid dimerises in benzene, then it can be shown that (derivation is given in general discussion of this chapter)

$\frac{C_w}{C_B}$ is not constant but $\frac{C_w}{\sqrt{C_B}}$ remains constant at a given temperature. Here, C_w is the concentration (molar) of benzoic acid in aqueous layer and C_B is the molar concentration of benzoic acid in benzene.

Thus the aim of the present experiment is to show $\frac{C_w}{C_B}$ for different sets of composition are not constant but $\frac{C_w}{\sqrt{C_B}}$ remains constant.

Materials

i. Nearly saturated solution of benzoic acid in benzene (200 ml).
ii. Pure benzene.
iii. Stoppered bottles (250 ml capacity, 4 nos.)
iv. Caustic soda solution $\approx$ 0.1(N) 500 ml.
v. Standard oxalic acid solution 0.1 (N), 100 ml.
vi. Burettes, pipettes (25 ml, 10 ml, 5 ml), separating funnel.
vii. Phenolphthalein indicator solution..

Procedure

i. Dissolve $\approx$ 10-12 gms of solid benzoic acid in pure benzene (200 ml) to prepare nearly saturated solution of benzoic acid in benzene.
ii. Make the following composition of sets taking the solution of benzoic acid in benzene:

Composition of Sets

Bottle nos. (stoppered 250 ml)	Volume of benzoic acid in benzene (ml)	Volume of pure benzene (ml)	Volume of distilled water (ml)	Total volume (ml)
1	50	0	100	150
2	40	10	100	150
3	30	20	100	150
4	20	30	100	150

[Measuring cylinder may be used to prepare the sets.]

iii. Shake the bottles (sets) for an hour and then allow them to settle for at least 15 minutes.

iv. Prepare 500 ml $\approx \frac{N}{10}$ NaOH solution (dissolve $\approx$ 2 gm of NaOH pallets in 500 ml water) and standardise it against standard $\frac{N}{10}$ oxalic solution which is prepared by accurate weighing. Dilute this standardised NaOH solution 10 times (25 ml is diluted to 250 ml) to prepare $\frac{N}{100}$ NaOH solution.

v. Take set no. 1. Separate the two layers using a separating funnel and collect the two liquids in two separate conical flasks and cover them with watch glass. Pipette out 25 ml of aqueous layer in a 125 ml conical flask and titrate it against $\frac{N}{100}$ NaOH solution using phenolphthalein indicator. Repeat it for second and third times. Perform the same for bottles 2, 3 and 4.
Pipette out 5 ml of benzene layer and pour into 25 ml distilled water in a 125 ml conical flask and titrate it against $\frac{N}{10}$ NaOH solution using phenolphthalein indicator (shake well during titration). Repeat and perform the same for bottles 2, 3 and 4. Calculate C_W and C_B for each set.

Results and Calculations

i. **Recording of laboratory temperature**

	Before experiment	After experiment	Mean
Temperature →	°C	°C	°C

ii. **Preparation of 100 ml 0.1 (N) oxalic acid solution**

Initial mass (gm)	Final mass (gm)	Mass of oxalic acid transferred (g)	Strength of oxalic acid solution
w_1	w_2	$(w_1 - w_2)$	$\frac{(w_1 - w_2)}{0.63}\left(\frac{N}{10}\right)$

iii. **Standardisation of NaOH solution against standard oxalic acid solution**

Strength of oxalic acid	Volume of oxalic acid solution taken (ml)	Volume of NaOH solution required (ml)	Mean volume of NaOH solution (ml)	Stength of NaOH solution
	a) 10	...		
... (N)	b) 10	...	...	...(N)
	c) 10	...		

[25 ml of this NaOH solution is properly diluted to 250 ml to prepare $\frac{1}{10}$th strength of the NaOH solution, which is standardised.]

iv. Titration of aqueous layer

Set Nos.	Volume of aq. layer (ml)	Volume of NaOH solution (0.01N) required (ml)	Mean volume of NaOH solution (ml)	C_W moles/lit
1	a) 25 b) 25 c) 25	a) b) c)	...	
2	a) 25 b) 25 c) 25	a) b) c)	...	
3	a) 25 b) 25 c) 25	a) b) c)	...	
4	a) 25 b) 25 c) 25	a) b) c)	...	

v. Titration of benzene layer

Set Nos.	Volume of benzene layer (ml)	Volume of NaOH solution (0.1N) (ml)	Mean volume of NaOH solution (ml)	C_B moles/lit
1	a) 5 b) 5 c) 5	a) ... b) ... c) ...	...	...
2	a) 5 b) 5 c) 5	a) ... b) ... c) ...	...	...
3	a) 5 b) 5 c) 5	a) ... b) ... c) ...	...	...
4	a) 5 b) 5 c) 5	a) ... b) ... c) ...	...	...

vi. Table for calculated $\frac{C_w}{C_B}$ and $\frac{C_w}{\sqrt{C_B}}$

Set Nos.	$\frac{C_w}{C_B}$	$\frac{C_w}{\sqrt{C_B}}$ (moles lit^{-1})$^{1/2}$
1	...	...
2	...	...
3	...	...
4	...	...

vii. **Calculations:** Calculations of C_w, C_B, $\frac{C_w}{C_B}$ and $\frac{C_w}{\sqrt{C_B}}$

viii. **Interpretation of results:**

Since $\frac{C_w}{C_B}$ are not constants but $\frac{C_w}{\sqrt{C_B}}$ remain almost constant, so it may be concluded that benzoic acid dimerises in benzene, which may be represented as

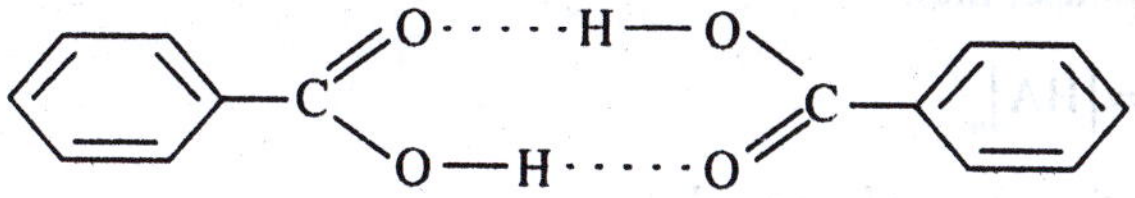

The expected value of $\frac{C_w}{\sqrt{C_B}} \approx 0.03$ at 25 °C.

Experiment 3

Determination of dimerisation constant of benzoic acid in benzene medium.

Theory: The activities of a solute distributed between two phases at equilibrium at a constant temperature bear a fixed ratio. This is 'Nernst Distribution Law'.

This law is valid only when the molecular species is same in the two phases. If the solute undergoes dissociation or association in any of the phases, the distribution law as such will not hold.

According to 'Nernst Distribution Law' when a_1 and a_2 are two activities of the solute in phases 1 and 2 respectively at equilibrium at a given temperature, then

$$\frac{a_1}{a_2} = K_D$$

This K_D is called the distribution coefficient or partition coefficient of the solute between phases 1 and 2 at experimental temperature. For dilute solution when $a \approx c$ (numerically, c = molar concentration), then

$$\frac{C_1}{C_2} = K_D$$

When solid benzoic acid (HA) is shaken with two practically immiscible solvents, benzene and water, then benzoic acid remains as monomer and almost undissociated in aqueous phase but it dimerises in benzene phase.

So, there are two equlibria,

i. HA (water) $\rightleftharpoons$ HA (benzene)

$$K_D = \frac{[\text{HA}]_B}{[\text{HA}]_W} \qquad (1)$$

and ii. 2HA (benzene) $\rightleftharpoons$ $(\text{HA})_2$ (benzene)

$$K = \frac{[(\text{HA})_2]_B}{[\text{HA}]_B{}^2}, \qquad (2)$$

where

K_D = distribution coefficient of benzoic acid between benzene and water.

K = dimerisation constant of benzoic acid in benzene medium.

$[HA]_B$ = molar concentration of monomeric HA in benzene phase.

$[HA]_W$ = molar concentration of monomeric HA in water phase.

$[(HA)_2]_B$ = molar concentration of dimeric HA in benzene phase.

Let C_w be the total molar concentration of of HA in water phase and C_B be the total molar concentration of HA in benzene phase, then

$$C_w = [HA]_w \tag{3}$$

and $$C_B = [HA]_B + 2[(HA)_2]_B. \tag{4}$$

Dividing equation (4) by equation (3), we get

$$\frac{C_B}{C_w} = \frac{[HA]_B}{[HA]_w} + 2\frac{[(HA)_2]_B}{[HA]_w}$$

$$= \frac{[HA]_B}{[HA]_w} + 2\frac{[(HA)_2]_B}{[HA]_B^2} \times \frac{[HA]_B^2}{[HA]_w^2} \times [HA]_w .$$

$$\therefore \quad \frac{C_B}{C_w} = K_D + 2K\,K_D^2\,C_w \quad \text{[from equations (1) and (2)].}$$

Thus, $$\frac{C_B}{C_w} = K_D + 2K\,K_D^2 \cdot C_w . \tag{5}$$

If we plot $\left(\frac{C_B}{C_w}\right)$ against C_W, we will get a straight line with positive slope and intercept. From the slope $(= 2K\,K_D^2)$ and intercept $(= K_D)$, we can calculate the values of K_D and K individually at experimental temperature.

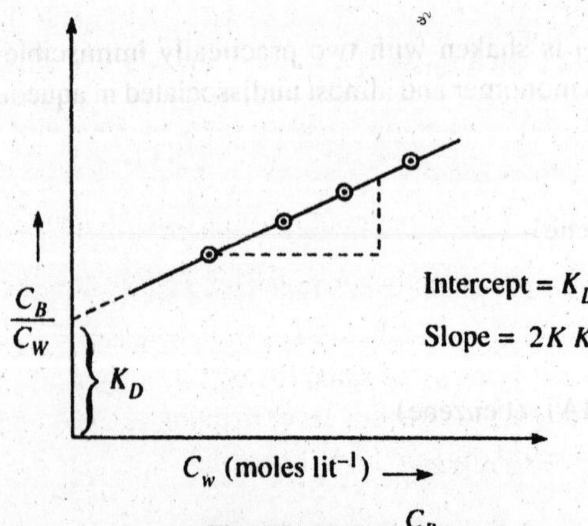

Fig. 9.1: Plot of $\frac{C_B}{C_w}$ vs. C_w

Materials

i. Saturated solution of benzoic (nearly saturated) acid in benzene ($\approx$ 200 ml).
ii. Pure benzene.
iii. Caustic soda solution 500 ml $\approx \frac{N}{10}$
iv. Oxalic acid 100 ml $\frac{N}{10}$.
v. Stoppered bottles (250 ml capacity) – 6 Nos.
vi. Conical flasks (125 ml capacity) 3-4 nos.
vii. Pipettes (10 ml, 5 ml, 25 ml).
viii. Burette, (ix) Phenolphthalein solution.
ix. Measuring cylinder (100 ml) 2 nos.

Procedure

i. Prepare $\approx$ 200 ml saturated solution of benzoic acid in benzene (dissolve 15gm of benzoic acid in 200 ml benzene) and make the following composition of sets:

Set Nos. (250 ml bottles)	Volume of benzoic acid in benzene (ml)	Volume of pure benzene (ml)	Volume of distilled water (ml)
1	50	0	100
2	40	10	100
3	30	20	100
4	20	30	100
5	10	40	100

[Measuring cylinder may be used to make the above composition of sets]
Shake the sets for an hour and settle them for at least 15 minutes.

ii. Prepare 500 ml $\frac{N}{10}$ NaOH solution ($\approx$ 2 gm NaOH should be dissolved in 500 ml) and 100 ml $\frac{N}{10}$ oxalic acid solution. Standardise NaOH solution against standard oxalic acid solution and then prepare 250 ml 10 times diluted NaOH solution (25 ml is diluted to 250 ml) so that strength of diluted NaOH solution becomes $\approx$ 0.01 (N).

iii. Separate the two layers of set 1 by a separating funnel. Pipette out 25 ml of aqueous layer and titrate it against $\frac{N}{100}$ NaOH solution using phenolphthalein indicator. Do it twice or thrice. Repeat the same for bottles 2, 3, 4 and 5.
Take 5 ml of organic layer(using a dry pipette) of bottle 1, add 25 ml distilled water and titrate it against $\frac{N}{10}$ NaOH solution using phenolphthalein indicator. During titration, the solution of conical flask should be shaken well as benzoic acid distributes from benzene layer to water and neutralises by NaOH. Do it twice or thrice. Repeat the same for sets 2, 3, 4 and 5. You may use two same burettes for organic layer titration and aqueous layer titration as the strengths of NaOH solution are different, used for aqueous and organic layer titration.

iv. Calculate C_W and C_B values in moles/lit for each set—plot $\left(\frac{C_B}{C_w}\right)$ vs. C_W and then determine K and K_D from the slope and intercept.

Results and Calculations

i. **Recording of laboratory temperature**

	Before experiment	After experiment	Mean
Temperature →	°C	°C	°C

ii. **Preparation of 100 ml 0.1 (N) oxalic acid solution**

Initial mass (gm)	Final mass (gm)	Mass of oxalic acid transferred (gm)	Strength of oxalic acid solution
w_1	w_2	$(w_1 - w_2)$	$\frac{(w_1 - w_2)}{0.63}\left(\frac{N}{10}\right)$

iii. **Standardisation of NaOH solution**

No. of observations	Volume of oxalic acid solution taken (ml)	Strength of oxalic acid solution	Volume of NaOH solution required (ml)	Mean volume of NaOH solution (ml)	Strength of NaOH solution
1	10		…		
2	10	… (N)	…	…	… (N)
3	10		…		

iv. **Titration of aqueous layer**

Set Nos.	Volume of aq. layer taken (ml)	Volume of NaOH solution required $\left(\frac{N}{100}\right)$ (ml)	C_W (moles/lit)
1	25	…	…
2	25	…	…
3	25	…	…
4	25	…	…
5	25	…	…

v. **Titration of organic (benzene) layer**

Set Nos.	Volume of org. layer taken (ml)	Volume of NaOH solution required $\left(\frac{N}{10}\right)$ (ml)	C_B (moles/lit)
1	5	…	…
2	5	…	…
3	5	…	…
4	5	…	…
5	5	…	…

vi. **Graph plotting and calculation of K and K_D:**

From the plot of $\left(\frac{C_B}{C_w}\right)$ vs. C_w

Slope = ...; Intercept = ...

$\therefore$ K_D = ... and K = ... (mol^{-1} lit)

[**N.B.**: (a) Expected value of slope = 850 ± 50 and intercept = 1.8.

(b) $K = 150 \pm 10$ mol^{-1} lit at 25 °C – 30 °C.

(c) Similar experiment may be performed using salicylic acid instead of benzoic acid].

vii. Interpretation of results (to be given by the students).

Experiment 4

Determination of coordination number of Cu^{++} *ion in cuprammonium complex in aqueous medium by partition method.*

Theory: A metal ion may combine with neutral molecules or anions (ligands) to give a new reasonably identifiable entity, called complex.

$$M^{n+} + xL \rightleftharpoons [ML_x]^{n+}$$

The number of monodentate ligands bound around a metal ion in the complex ion is called the co-ordination number of metal ion. In the above reaction a metal iron M^{n+} reacts with x moles of a neutral monodentate ligand L to form a complex $[ML_x]^{n+}$. Here x represents the coordination number of the metal ion M^{n+}.

Cu^{2+} reacts with NH_3, present in excess, according to the equation

$$Cu^{2+} + nNH_3 \rightleftharpoons \left[Cu(NH_3)_n\right]^{2+}$$

with the formation of deep blue colour. As the complex is fairly stable, we can find the value of n by partition method.

Ammonia is distributed between immiscible liquid pair (water and chloroform) and its concentration ratio between water and chloroform is fixed at equilibrium at experimental temperature (ef. Nernst distribution law), i.e.,

$$\frac{C_1}{C_2} = K_D, \tag{1}$$

where C_1 and C_2 are the equilibrium molar concentrations of NH_3 in aqueous and chloroform (organic) layers respectively. K_D is called partition or distribution coefficient of ammonia between water and chloroform. If we add $CuSO_4$ solution (aq.) in aqueous layer at equilirbium with $CHCl_3$, then concentration of complexed NH_3 is

$$C_3 = C_1' - C_2' \cdot K_D. \tag{2}$$

In equation (2) C_1' and C_2' are the molar concentrations of NH_3 in aqueous and organic layers respectively when $CuSO_4$ is added in aqueous layer.

Let the concentration of Cu^{2+} ion is C_4 moles/lit in aqueous layer, then coordination number (n) of Cu^{2+} ion is given by

$$n = \frac{C_3}{C_4}. \tag{3}$$

Materials

i. Standard oxalic acid solution 0.5 (N) (100 ml).
ii. Caustic soda solution ≈ 0.5 (N) (100 ml).
iii. Standard potassium dichromate solution 0.2 (N) (100 ml).
iv. Sodium thiosulphate solution ≈ 0.2 (N) (250 ml).
v. Ammonia solution ≈ 1 (N) 250 ml.
vi. Aqueous copper sulphate solution ≈ 1(N) 50 ml.
vii. Sulphuric acid ≈ 0.5 (N) 500 ml.
viii. Sulphuric acid ≈ 0.05 (N) 250 ml.
ix. Starch solution (~1%).
x. Phenolphthalein and methyl orange indicator.
xi. Stoppered bottles (125 ml capacity), 4 nos.
xii. Conical flasks (100 ml capacity), 3-4 pcs.
xiii. Volumetric flasks (100 ml, 250 ml and 500 ml).
xiv. Burettes and pipettes (5 ml, 10 ml and 25 ml).

Procedure

i. The above solutions are prepared and glass goods are collected.

ii. First standardise NaOH solution against standard oxalic acid solution and then H_2SO_4 solution (≈ 0.5N) against standardised NaOH solution, both using phenolphthalein as indicator. Again standardise the prepared NH_3 solution (≈ 1N) against standardised H_2SO_4 solution (≈ 0.5N) using methyl orange indicator to verify the strength of NH_3 solution (≈ 1 N). Generally, concentrated ammonium hydroxide has base normality ≈ 14 (N). Sometimes the concentrated ammonia supplied in the laboratory is of normality much less than ~14 (N). In such case, more concentrated ammonia is to be added to prepare ≈ 1 (N) than required considering it as ~14 (N). Any way, ≈ 1(N) ammonia solution is to be prepared.

iii. Standardise thiosulphate solution against standard potassium dichromate solution and then $CuSO_4$ solution against standardised thiosulphate solution, both iodometrically. This can be done as follows:

Take 10 ml $K_2Cr_2O_7$ solution (use pipette) in a 250 ml conical flask—add ≈ 2 gm solid potassium iodide, ≈ 5ml concentrated HCl and 100 ml distilled water. Shake well and allow it to keep in dark place covering with a watch glass for 10 minutes. Titrate it against ≈ 0.2 (N) thiosulphate solution adding 1 ml of 1% starch solution when straw yellow colour is appeared. The end point is detected when just light green colour is appeared.

Pipette 2 ml of $CuSO_4$ solution in a 125 ml conical flask— add 1 cc glacial acetic acid, ≈ 2 gm of solid KI and 50 ml of distilled water. Shake well and titrate it against that standardised thiosulphate solution using starch as indicator. The end point is noted when just milky white precipitate is appeared, verify whether $CuSO_4$ solution is ≈ 1 (N), or not.

Taking NH_3 solution and $CuSO_4$ solution, thus prepared, the following composition of sets are prepared. The first two sets are prepared for determination of K_D and other two sets are for determination of *n*.

iv. **Composition of sets**

Set no. (125 ml stoppered bottles)	Volume of aq. NH_3 solution ($\approx$ 1N) (ml)	Volume of pure chloroform (ml)	Volume of aq. $CuSO_4$ solution (~1N) (ml)
1	40	20	—
2	35	25	—
3	35	20	5
4	30	20	10

v. Shake the bottles for an hour and allow to settle for 20 minutes.

vi. Pipette 5 ml of aqueous layer from set 1 and titrate it against 0.5 (N) H_2SO_4 using methyl orange indicator. Repeat the same for set nos. 2, 3 and 4. Pipette 5 ml of organic layer and titrate it against 0.05 (N) H_2SO_4 using methyl orange as indicator. Repeat the same of sets 2, 3 and 4. (each titration should be done thrice and take mean of the three burette readings).

vii. Calculate the molar concentrations of NH_3 in aqueous and organic layers. Then calculate K_D and n from equations (1 to 3).

[**N.B.:** Composition 1 and 2 are used for calculation of K_D and composition 3 and 4 are used for calculations of coordination number (n).

Results and Calculations

i. **Recording of laboratory temperature**

	Before experiment	After experiment	Mean
Temperature →	°C	°C	°C

ii. **Preparation of 100 ml 0.5 (N) oxalic acid solution**

Initial mass (gm)	Final mass (gm)	Mass of oxalic acid transferred (gm)	Strength of oxalic acid solution
w_1	w_2	$(w_1 - w_2)$	$\frac{(w_1 - w_2)}{3.15}\left(\frac{N}{2}\right)$

iii. **Preparation of 100 ml 0.5 (N) $K_2Cr_2O_7$ solution**

Initial mass (gm)	Final mass (gm)	Mass of $K_2Cr_2O_7$ transferred (gm)	Strength of $K_2Cr_2O_7$ solution
w_1	w_2	$(w_1 - w_2)$	$\frac{(w_1 - w_2)}{2.45}\left(\frac{N}{2}\right)$

iv. **Standardisation of NaOH solution against standard oxalic acid solution**

Strength of oxalic acid solution	Volume of oxalic acid solution taken (ml)	Volume of NaOH solution required (ml)	Mean volume of NaOH solution (ml)	Stength of NaOH solution
··· (N)	a) 10 b) 10 c) 10	a) ··· b) ··· c) ···	···	···(N)

v. **Standardisation of sodium thiosulphate solution against standard potassium dichromate solution (iodometrically)**

Strength of $K_2Cr_2O_7$ solution	Volume of $K_2Cr_2O_7$ solution taken (ml)	Volume of thio solution required (ml)	Mean volume of thio solution required (ml)	Stength of thio solution
... (N)	a) 10 b) 10 c) 10	a) ... b) ... c) ...	...	... (N)

vi. **Standardisation of $CuSO_4$ solution against standardised thio solution (iodometrically)**

Strength of thio solution	Volume of $CuSO_4$ solution taken (ml)	Volume of thio solution required (ml)	Mean volume of thio solution required (ml)	Strength of $CuSO_4$ solution
... (N)	a) 2 b) 2	a) ... b) ...	...	... (M)

vii. **Standardisation of H_2SO_4 solution against standardised NaOH solution**

Strength of NaOH solution	Volume of H_2SO_4 solution taken (ml)	Volume of NaOH solution required (ml)	Mean volume of NaOH solution required (ml)	Strength of H_2SO_4 solution
... (N)	a) 10 b) 10	a) ... b) ...		... (N)

This standardised H_2SO_4 solution (nearly 0.5 N) is diluted 10 times to prepare nearly 0.05 (N) H_2SO_4 solution which is necessary for the titration of NH_3 in organic layers.

viii. **Determination of K_D**

Set No.	Volume of each layer taken (ml)		Volume of 0.5/0.05 (N) H_2SO_4 required (ml)	Strength of NH_3 in aq. layer moles/lit.	Strength of NH_3 in org. layer moles/lit.	$K_D = \frac{S_1}{S_2}$	Mean K_D
	Aqueous	Organic					
1	5 5	5 5	...	S_1	S_1	...	...
2	5 5	5 5	...	S_2	S_2	...	

ix. **Determination of coordination number (n)**

Set No.	Volume of each layer taken (ml)		Vol. of 0.5/0.05 (N) H_2SO_4 required (ml)	Mean vol. of H_2SO_4 0.5/0.05 (N) (ml)	C_1' moles/lit	C_2' moles/lit	C_3 (moles/lit) $= C_1' - C_2' \cdot K_D$	C_4 (moles/lit)	$n = \frac{C_3}{C_4}$
	Aqueous	Organic							
3	5 5	5 5	... ...	...	...	...	...	...	...
4	5 5	5 5	... ...	...	...	...	...	...	...

x. Calculation

xi. Average coordination number (n) = Average of 3 & 4 sets

xii. Interpretation of results (to be given by the students)

Experiment 5

Determination of critical solution temperature (C.S.T.) of phenol-water system.

Theory: At ordinary temperature phenol and water are partially miscible. If a small quantity of phenol is added to water, the substance will first dissolve completely but on further addition a stage is attained when no further dissolution takes place and two liquid layers separate out. Each layer is a saturated solution of one in the other. These two layers in equilibrium are known as conjugate solutions. The upper layer is a saturated solution of phenol in water and the lower one that of water in phenol.

The mutual solubility of phenol and water increases with the rise of temperature which causes the concentration of phenol as well as that of water in phenol. At a certain temperature two conjugate solutions change into one homogeneous solution. This temperature are known as the critical solution temperature (C.S.T.). This temperature is the characteristic of a particular system and is influenced by the presence of impurities. This temperature for phenol-water system has been found to be 66.4 °C. The variation of mutual solubility of phenol and water with temperature is shown in the following figure.

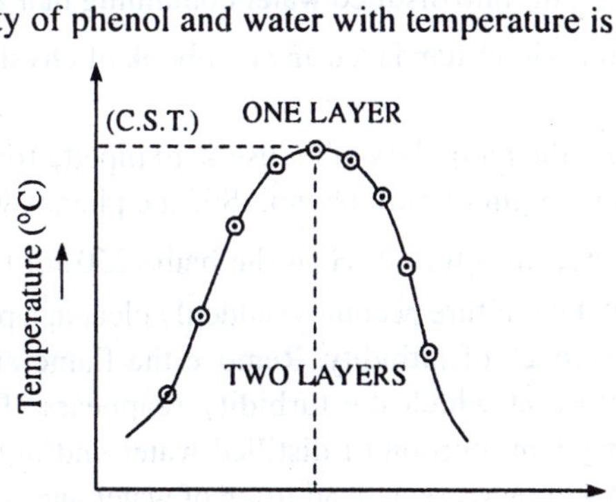

Fig. 9.2: Plot of temperature vs. wt. % of phenol

At any temperature above C.S.T., phenol-water are miscible in all proportions. Outside the curve, the system is completely homogeneous, i.e., only one layer. Below the curve always two layers will separate out and the curve gives the composition of the two conjugate solutions constituting the two layers.

In studying these solubilities, the pressure is always kept constant. There is thus a restriction, i.e., one variable is fixed. Hence,

$$F = C - P + 1.$$

For the homogenous phase of the two component liquids,

$$F \equiv 2 - 1 + 1 = 2 \cdot$$

i.e., it would be bivariant system. So, for homogeneous solutions both composition and temperature have to be defined.

For the heterogeneous equilibrium with two conjugate solutions,

$$F = 2 - 2 + 1 = 1,$$

i.e., the system would behave as a univarint one. So, if either the temperature or the composition be fixed, the other would be automatically known from the diagram.

At critical solution temperature, there is an additional restriction that the composition of the two conjugate solutions must be the same.

Hence, $F = 2 - 2 + 1 - 1 = 0$.

Thus the point, the critical solution temperature, is a non-variant point.

Materials

i. Hard glass tube fitted with thermometer and a stirrer (an 8-inch hard glass test tube)
ii. Tipod stand with clamp to hold the tube.
iii. Water bath (a 250 ml beaker half-filled with distilled water).
iv. Pure redistilled phenol, distilled water, etc.

Procedure

i. Weigh about 3 gm of pure phenol (use weighing bottle) accurately and transfer it to the hard glass test tube.
ii. Fit the hard glass test tube with a two-hole stopper holding, and attach it to a burette stand. The theromometer (100 °C) and a stirred should be fitted into the test tube through a cork. Immerse the bottom of the test tube ($\frac{1}{4}$th) into distilled water containing in a 250 ml beaker which is kept on an asbestos board over the tripod stand (see any text book of physical chemistry for setting up apparatus).
iii. Add 2 ml distilled water into the hard glass tube (use 2 ml pipette for measuring distilled water) which already contains about 3 gm of pure phenol. Stir the phenol-water solution well with the stirrer so that the mixture becomes turbid. Heat the bath (250 ml beaker containing half of it distilled water) slowly until the mixture becomes suddenly clear at a particular temperature. Note the temperature of disappearence of turbidity. Remove the flame and allow the liquid to cool slowly. Note the temperature at which the turbidity reappears. Take the mean of both the temperatures. Go on adding 1 ml portion of distilled water and note the mean temperature in each step. The temperature will increase with addition of water and will reach a maximum value. After that the solution temperature will fall. Take 5-6 more readings after maximum temperature.
iv. Taking density of water as ≈ 1 gm/ml, calculate the weight percentage of phenol for each step. Plot mean temperature vs. wt.% of phenol. Draw a smooth curve and show the C.S.T. and the corresponding wt.% of phenol in the graph.

Results and Calculations

i. **Recording of laboratory temperature**

	Before experiment	After experiment	Mean
Temperature ⟶	°C	°C	°C

ii. **Weighing of Phenol**

Initial mass phenol + weighing bottle (g)	Final mass (g)	Mass of phenol transferred (g)
w_1	w_2	$(w_1 - w_2)$

iii. **Recording of miscibility temperatures**

No. of obs.	Volume of water added (ml)	Miscibility temperatures (^{0}C)		
		Turbidity disappearence	Turbidity reappearance	Mean
1	2	...	...	...
2	3 (2 + 1)	...	...	...
3	4 (3 + 1)	...	...	...
4	5 (4 + 1)	...	...	...
⋮	⋮	...	...	...
⋮	⋮	...	...	...
⋮	⋮	...	...	...

iv. **Table for graph plotting**

No. of observations	Percentage of phenol (by weight)	Miscibility temperature °C (Mean)
1	⋮	⋮
2	⋮	⋮
3	⋮	⋮
4	⋮	⋮
⋮	⋮	⋮
⋮	⋮	⋮

v. **Conclusion**

From the plot of miscibility temperatures vs. wt% of phenol,

Critical solution temperature (C.S.T.) = ··· °C

Critical composition = ··· % of phenol by weight.

Experiment 6

A study of phase diagram of three-component liquid (ternary) system at room temperature: (Benzene – accetic acid-water system).

Theory: For a three-component system at constant temperature and pressure the phase rule reduces to

$$F = 3 - P.$$

The composition of such a system may be expressed in terms of the coordinates of an equilateral triangle, each apex corrosponding to a pure component.

Thus the point X in the following figure corresponds to a composition of 25% A, 50% B and 25% C.

The components benzene and water are essentially immiscible, hence two layers will be obtained on mixing. The third component, viz., acetic acid, is soluble in each layer and on addition to a mixutre of benzene and water distributes itself between these layers. The composition of the layers changes as more acetic acid is added. Under these conditions, P equals 2, and the system is completely defined by the composition. Thus, if the points corresponding to these compositions are plotted, a smooth curve may be drawn indicating the limits of miscibility of water-benzene mixtures in acetic acid.

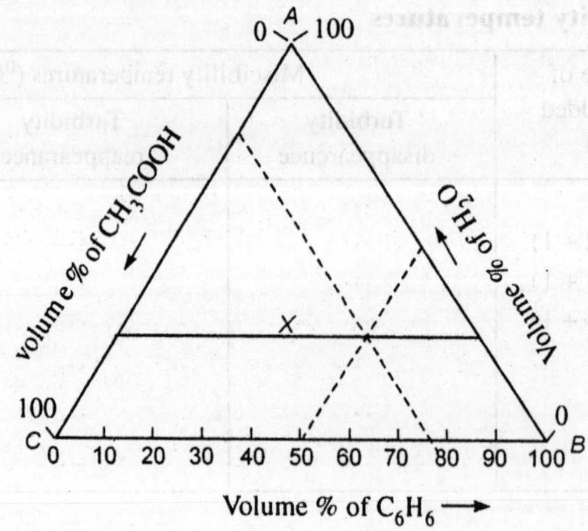

Fig. 9.3: Representation of composition in a triangular diagram

It will be noted that as more and more acetic acid is added, the water and benzene solutions become more and more miscible until complete miscibility is achieved.

This solubility curve is known as binodal curve. Any mixture within the area enclosed by this binodal curve and the base of the triangle will resolve itself into two-liquid layers and any mixture outside the area will form only one-liquid layer. The position of this curve changes with temperature.

Materials

(i) Graduated pipette, (ii) Burettes, (iii) 125 ml conical flask (~ 3 Nos.) 250 ml 1 No., (iv) Glacial acetic acid, (v) Pure benzene, (vi) Distilled water, (vii) Dry air oven, etc.

Procedure

i. 5 ml of benzene is placed in a 125 ml clean and dry conical flask (preferably stoppered). 1 ml of water (preferably doubly distilled) is added from a burette and glacial acetic acid is slowly added from a second burette until a clear solution is just obtained on vigorously shaking the mixture. The volume of acetic acid which has been added is then recorded. Successive volumes of 5 ml benzene are added to the mixture. After each addition of benzene, acetic acid is run in and the mixture is shaken vigorously until the system becomes homogeneous. The results are recorded as indicated in table 2 (series A).

ii. In series *B*, 5 ml of benzene is placed in a 125 ml dry conical flask and treated successively with 3, 3, 4, 5, 6 and 7 ml portions of water and titrated with acetic acid after each addition.

iii. In the series *C*, 2 ml of benzene is placed in a dry 250 ml conical flask and is treated successively with six 25 ml portions of water and titrated with acetic acid after each addition.

iv. Calculate the volume percentage (V/V) of benzene, water and acetic acid for each clear solution, i.e., homogeneous solution (i.e., for just completely miscible solution) and plot on a triangular graph paper.

v. In another graph paper plot "volume of benzene per 100 ml of acetic acid" against "volume of water per 100 ml of acetic acid" (two-dimensional graph) and indicate the homogeneous and heterogeneous phases on the graph paper.

Results and Calculations

i. **Table 1: Recording of laboratory temperature**

	Before experiment	After experiment	Mean
Temperature →	°C	°C	°C

ii. **Table 2: Composition of different series of experiment**

Series	Volume of water (ml)	Volumeof bezene (ml)	Minimum volume of acetic acid required (ml)
	1.00	5.00	...
	1.00	10.00	...
A	1.00	15.00	...
	1.00	20.00	...
	1.00	25.00	...
	1.00	30.00	...
	3.00	5.00	...
	6.00	5.00	...
B	10.00	5.00	...
	15.00	5.00	...
	21.00	5.00	...
	28.00	5.00	...
	25.00	2.00	...
	50.00	2.00	...
C	75.00	2.00	...
	100.00	2.00	...
	125.00	2.00	...
	150.00	2.00	...

iii. **Table 3: Determination of percentage of each component for each homogeneous (single phase) solutions**

Series A			Series B			Series C		
Vol. % of benzene	Vol. % of water	Vol. % of HAc	Vol. % of benzene	Vol. % of water	Vol. % of HAc	Vol. % of benzene	Vol. % of water	Vol. % of HAc

[**N.B.**: If the time is limited, alternate titrations in series *B* and *C* of table 1 may be carried out and the remainder is omitted.]

iv. **Table 4: Table for two dimensional graph plotting**

Series A		Series B		Series C	
Volume of benzen per 100 ml of HAc	Volume of water per 100 ml of HAc	Volume of benzene per 100 ml of HAc	Volume of water per 100 ml of HAc	Volume of benzene per 100 ml of HAc	Volume of water per 100 ml of HAc

v. **Calculation: Let it be considered in Series *A*.**

Volume of water = 1 ml

Volume of benzene = 5 ml

Volume of acetic acid = *x* ml (say) required

$$\therefore \quad \text{volume \% of water} = \frac{1\times 100}{6+x}\%$$

$$\text{Volume \% of benzene} = \frac{5\times 100}{6+x}\%$$

$$\text{Volume \% of acetic acid} = \frac{x}{6+x}\times 100\%$$

Similarly for others:

$$\text{Volume of benzene per 100 ml of HAc} = \frac{5}{x}\times 100 \text{ ml}$$

$$\text{Volume of water per 100 ml of HAc} = \frac{1\times 100}{x} \text{ ml.}$$

Similarly for others.

The nature of graph for two dimensional graph is as under

vi. **Interpretation of results**

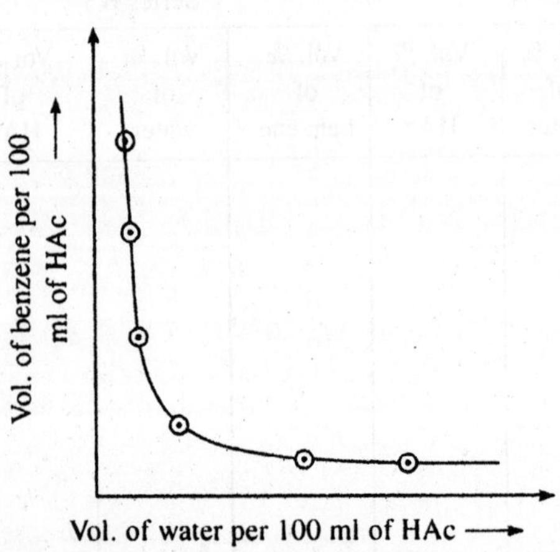

Fig. 9.4: Two dimensional graph plot

Experiment 7

Determination of composition of the azeotropic mixture of benzene-methanol or benzene-ethanol system (binary mixture) by boiling point method.

Theory: Binary liquid mixtures showing a large positive deviation from Raoult's law exhibit a maximum in the vapour pressure vs. composition curve (Fig. 9.5).

For this reason, the mixture, the composition of which corresponds to the maximum vapour pressure, will have the minimum boiling point of any mixture of the two components. Furthermore, it will be seen from Fig. 9.6 that the liquidus and vaporus curves coincide at the temperature of the minimum boiling point.

The number of degrees of freedom F, of a system consisting of C components and P phases, is given by:

$$F = C - P + 2.$$

For a binary system at constant pressure this reduces to

$$F = 3 - P.$$

On application of the phase rule to the point M (minimum point) an identity restriction must be introduced by substracting one degree of freedom since the compositions of the liquid and the vapour are the same at this point.

Thus, at constant pressure, M is an invariant point, i.e., $F = 0$.

A mixture which distils without change of composition is said to be azeotropic. The mixture of composition corresponding to the point M is thus an azeotrope.

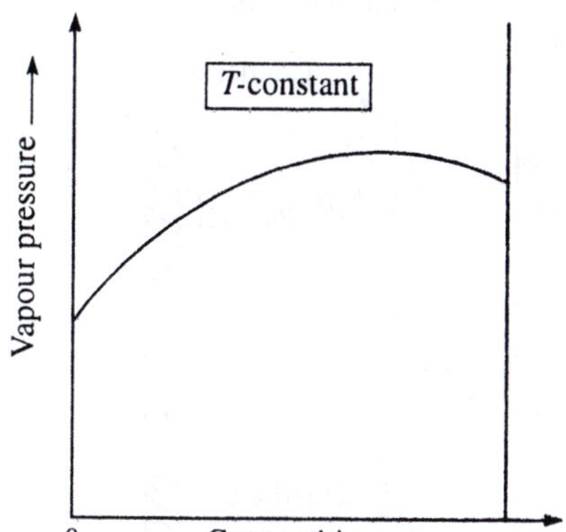

Fig. 9.5 : Vapour pressure vs. composition curve

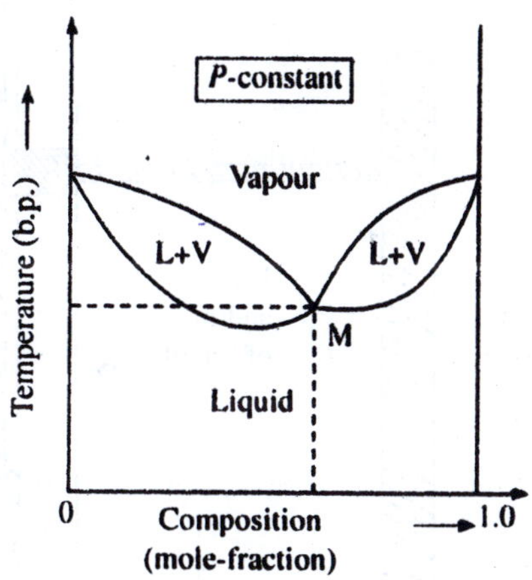

Fig. 9.6: Temperature vs. composition curve

Apparatus and Chemicals

(i) Abbe-refractometer, (ii) Thermometer (40-100 °C, graduated in 0.1 °C), (iii) Boiling point apparatus (Cottrell's apparatus), (iv) Stoppered weighing bottles (10 nos., 10 ml capacity), (v) Pipette (25 ml) and graduated pipette (5 ml), (vi) Pure benzene and methanol/ethanol, (vii) Cotton, (viii) Spirit lamp, etc.

Procedure

i. A series of mixtures of known composition (mole percent benzene in alcohol) are prepared and their refractive indices are measured. The refractive indices of pure methanol/ethanol and pure benzene are also measured using Abbe-refractometer. A plot of refractive index against mole-fraction of benzene in benzene-methanol or benzene-ethanol mixture is prepared which is called "Calibration Curve" Fig. 9.8.

Composition of sets for refractive-index measurement

[Use 5 ml graduated pipette for preparing the above composition of sets].

Set No. (10 ml weighing bottle)	Volume of methanol (ml)	Volume of benzene (ml)
1	4.5	0.5
2	4.0	1.0
3	3.5	1.5
4	3.0	2.0
5	2.5	2.5
6	2.0	3.0
7	1.5	3.5
8	1.0	4.0
9	0.5	4.5

ii. This data can be conveniently determined and correlated while distillation is carried out. Alternatively, time will be saved if a previously prepared graph is obtainable in the laboratory.

iii. A sketch of appratus for boiling point determination and distillation is shown in Fig. 9.7.

A known quantity of pure benzene (25 ml) is placed in the hard glass tube of the boiling point apparatus and its boiling point is determined. The thermometer is read with the aid of a lens. A

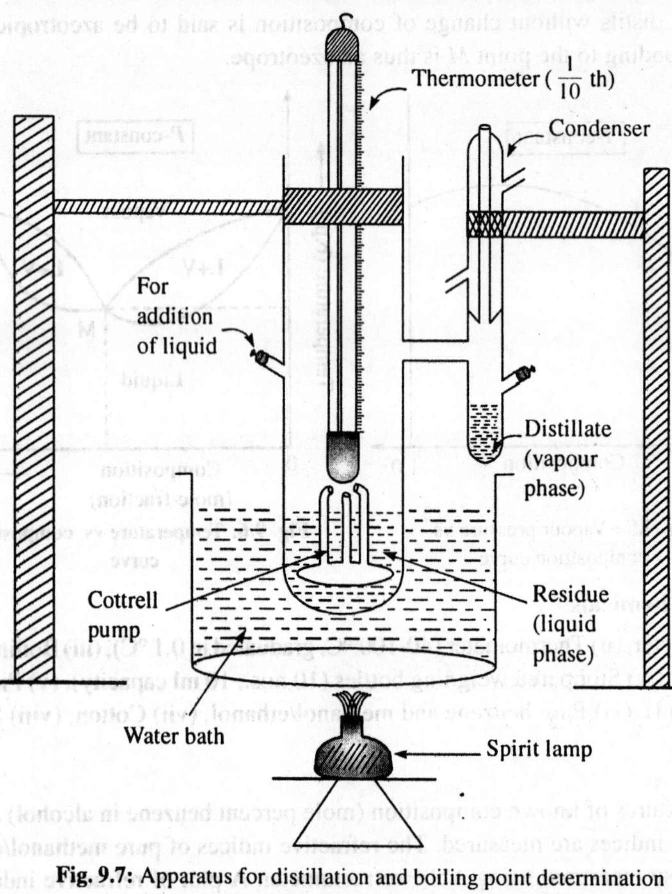

Fig. 9.7: Apparatus for distillation and boiling point determination

smaller quantity (say, 1 ml) of methanol or ethanol is added to the benzene and the mixture is refluxed for several minutes until a constant boiling point of the mixture is reached. The boiling point is noted. The spirit lamp is removed and when the boiling ceases a pipette is used to take up a few drops of the distillate from the small cup as shown in the apparatus and its refractive index is measured. A second pipette is used to take a few drops of residue from the tube and its refractive index is measured. This procedure is repeated after adding successively 1, 1, 1, 1, 2, 2, 2, 5, 5, and 5 ml of methanol/ethanol to the benzene in the hard glass tube. From the refractive indices, the composition of benzene and methanol/ethanol in both vapour phase and liquid phase are determined from the 'calibration curve'.

The hard glass tube is emptied, rinsed with methanol/ethanol and 25 ml of methanol/ethanol is taken in the tube and the process is just reversed by successive addition of benzene to the alcohol (instead of addition of alcohol to benzene).

iv. The mole percentage composition of liquid and vapour phases of the binary mixture at each boiling temperature is plotted in the temperature-composition diagram. The azeotropic boiling point and composition at this point are recorded from the graph.

Results and Calculations

i. **Recording of laboratory temperature**

	Before experiment	After experiment	Mean
Temperature ⟶	°C	°C	°C

ii. **Measurement of refractive indices of different liquids of known composition of the binary mixtures**

Set Nos.	Volume of methanol (ml)	Volume of benzene (ml)	*Mole % of benzene in the mixture	Observed refractive index
1	5.0	0.0	0	...
2	4.5	0.5	04.78	...
3	4.0	1.0	10.15	...
4	3.5	1.5	16.22	...
5	3.0	2.0	...	...
6	2.5	2.5	....	...
7	2.0	3.0	...	...
8	1.5	3.5	...	...
9	1.0	4.0	...	...
10	0.5	4.5	...	...
11	0.0	5.0	100	...

iii. ***Calculation of mole % of benzene and alcohol**

Mole fraction (mole percentage) of benzene in the binary mixture of benzene-methanol

$$x_B = \frac{\dfrac{W_B}{M_B}}{\dfrac{W_B}{M_B} + \dfrac{W_A}{M_A}} = \frac{\dfrac{V_B d_B}{M_B}}{\dfrac{V_B d_B}{M_B} + \dfrac{V_A d_A}{M_A}} \times 100\% ,$$

where V, d and M are the volumes (in ml), densities (in g ml^{-1}) and molar mass respectively. A and B represent alcohol and benzene respectively.

Now the densities and molar mass of benzene and alcohol are

$$\left.\begin{aligned} d_B &= 0.872 \text{ g ml}^{-1} \\ d_A &= 0.7917 \text{ g ml}^{-1} \end{aligned}\right\} \text{ at } 25\ ^\circ\text{C}$$

$$M_B = 78 \text{ g mol}^{-1}$$

$$M_A = 32 \text{ g mol}^{-1} \quad (\text{A = methanol})$$

∴ **For set 2:**

$$x_B = \frac{\frac{0.5\times0.872}{78}}{\frac{0.5\times0.872}{78}+\frac{4.5\times0.7917}{32}}\times100\% = 0.0478\times100\% = 4.78\%.$$

$$\therefore\quad x_A = 95.22\%.$$

∴ **For set 3:**

$$x_B = \frac{\frac{1.0\times0.872}{78}}{\frac{1.0\times0.872}{78}+\frac{4.0\times0.7917}{32}}\times100\% = 10.15\%.$$

$$\therefore\quad x_A = 89.85\%.$$

∴ **For set 4:**

$$x_B = \frac{\frac{1.5\times0.872}{78}}{\frac{1.5\times0.872}{78}+\frac{3.5\times0.7917}{32}}\times100\% = 16.22\%.$$

$$\therefore\quad x_A = 83.78\%$$

and so on.

iv. **Measurement of boiling temperatures of the binary mixtures (methanol is added to 25 ml benzene) and hence determination of mole-percentage composition of benzene in vapour and liquid phases of the mixture**

Volume of methanol added to 25 ml benzene (ml)	Boiling temperature (°C)	Refractive index		Mol.-percentage composition of benzene	
		Residue (i.e., liquid phase)	Distillate (i.e., vapour phase)	Residue (i.e., liquid phase)	Distillate (i.e., vapour phase)
0					
1 (0 + 1)					
2 (1 + 1)					
3 (2 + 1)					
4 (3 + 1)					
6 (4 + 2)					
8 (6 + 2)					
10 (8 + 2)					
15 (10 + 5)					
20 (15 + 5)					
25 (20 + 5)					

v. **Measurement of boiling temperatures of the binary mixtures (benzene is added to 25 ml methanol) and hence determination of mol-percentage composition of benzene in vapour and liquid phases of the mixtures**

Volume of benzene added to 25 ml methanol (ml)	Boiling temperatures (°C)	Refractive index		Mol.-percentage composition of benzene	
		Residue (i.e., liquid phase)	Distillate (i.e., vapour phase)	Residue (i.e., liquid phase)	Distillate (i.e., vapour phase)
0					
1 (0 + 1)					
2 (1 + 1)					
3 (2 + 1)					
4 (3 + 1)					
6 (4 + 2)					
8 (6 + 2)					
10 (8 + 2)					
15 (10 + 5)					
20 (15 + 5)					
25 (20 + 5)					

vi. **Conclusion**

From graph (plot of B.P. vs. mole % of benzene), the azeotropic mixture with minimum B.P. is obtained at ... °C with ... mole % of methanol.

vii. [**N.B.:** 1. The mole % of benzene or methanol in vapour and liquid phases at different boiling temperatures can be obtained from the corresponding refractive index data as follows:

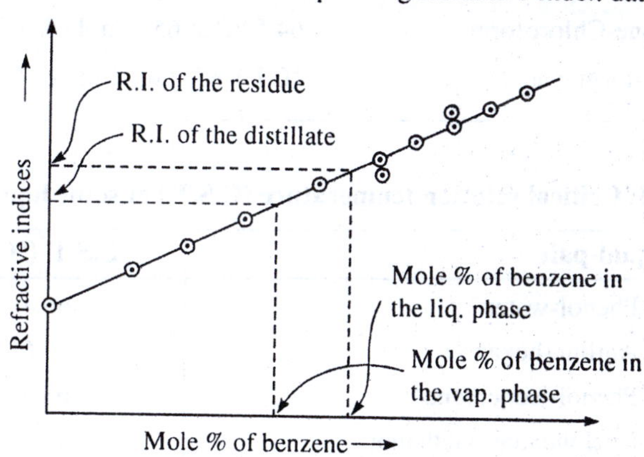

Fig. 9.8: Calibration curve

2. Since the boiling point apparatus is made of glass, it should be handled with care.
3. The chemicals (benzene, alcohol) are inflammable. So, they should be kept at the safe distances from the flame to avoid danger.
4. For measurement of refractive indices follow the procedure given in the chapter Polarimetry and Refractometry.

APPENDIX

Table 1: Boiling point (temperatures) of some pure liquids

Liquids	Boiling point (°C)
Benzene	80.2
Toluene	110.6
Methanol	64.7
Ethanol	78.0
Isopropanol	82.5
Chloroform	61.2
Carbon tetrachloride	76.8
Acetone	56.4
Formic acid	99.9
Water	100.0

Table 2: Some azeotropic mixtures

Mixtures	Azeotropic point with composition
i. Benzene-isopropanol	71.9 °C at 39.3 mole % of alcohol
ii. Benzene-methanol	58.3 °C at 61.4 mole % of alcohol
iii. Carbontetrachloride-methanol	55.7 °C at 44.5 mole % of CCl_4
iv. Acetone-Chloroform	64.5 °C at 65.5 mole % of $CHCl_3$
v. Water-formic acid	10.7.1 °C at 43.3 mole % of water.

Table 3: Critical solution temperature (C.S.T.) of some liquid pairs

Liquid-pair	C.S.T. (°C)
i. Phenol-water	66.4
ii. Aniline-hexane	59.6
iii. Phenol-isopentane	63.5
iv. Cyclohexane-methanol	49.0
v. Carbondisulphide-methanol	40.5
vi. Aniline-water	167.0

Table 4: Partition coefficient (K_D) of some solutes in two immiscible liquid pairs at 25-30 °C

Mixtures	K_D
i. Iodine between carbon tetrachloride and water	85 ± 5
ii. Iodine between benzene and water	400 ± 20
iii. Iodine between chloroform and water	120 ± 10
iv. Iodine between hexane and water	30 ± 2
v. Iodine between carbon disulphide and water	400 ± 20
vi. Oxalic acid between water and ether	10 ± 0.5
vii. Ammonia between water and chloroform	20 ± 2
viii. Benzoic acid between benzene and water	0.03 ± 0.001 $\left(C_W / \sqrt{C_B}\right)$

List of Reference Books

1. Findlay's Practical Physical Chemistry—*B.P. Levitt*
2. Practical Physical Chemistry—*S.R. Palit & S. K. De*
3. Selected Experiments in Physical Chemistry—*N. G. Mukherjee*
4. Vogel's Text Book of Quantitative Chemical Analysis
   —*G.H. Jeffery, J. Bassett, J. Mendham & R. C. Denney*
5. Experiments in Physical Chemistry
   —*J.M. Wilson, R.J. Newcombe, A.R. Denaro & R.M.W. Rickett*
6. Practical Physical Chemistry—*A. Findlay*
7. Experimental Physical Chemistry—*V.D. Athawale, P. Mathur*
8. Instrumental Analysis Analytical Chemistry—*Skoog-Holler-Crouch*
9. An Introduction to Electrochemistry—*Samuel Glasstone*
10. A Text Book of Physical Chemistry—*K.L. Kapoor*
11. Physical Chemistry—*P.C. Rakshit*
12. Physical Chemistry—*G.M. Castellan*

Appendix I

Table I: S.I. Derived Units of some Physical quantities

Physical quantity	Name of Unit	Symbol	S.I. Units
Force	newton	N	$kg\ m\ s^{-2}$ or $J\ m^{-1}$
Pressure	pascal	Pa	$kg\ m^{-1}s^{-2}$ or $N\ m^{-2}$
Energy	joule	J	$kg\ m^2\ s^{-2}$ or N m
Electric charge	coulomb	C	As (A = ampere)
Electric potential	volt	V	$kg\ m^2\ s^{-3}\ A^{-1}$ or $W A^{-1}$ or $J\ s^{-1}A^{-1}$
Electric Resistance	ohm	Ω	$kg\ m^2\ s^{-3}\ A^{-2}$ or $V\ A^{-1}$
Electric conductivity	siemens	$S = \Omega^{-1}$	$kg^{-1}m^{-2}\ s^3\ A^2$ or AV^{-1}
Frequency	hertz	Hz	s^{-1}
Volume	cubic metre	–	m^3
Capacitance	farad	F	$kg^{-1}\ m^{-2}\ s^4\ A^2$ or $C\ V^{-1}$
Magnetic flux	weber	Wb	$kg\ m^2\ s^{-2}\ A^{-1}$ or Vs
Magnetic flux density	tesla	T	$kg\ s^{-2}\ A^{-1}$ or $Wb\ m^{-2}$ or $Vs\ m^{-2}$

Table II: Conversion factors for S.I. units from C.G.S. units

C.G.S. units	S.I. units
1 erg	10^{-7} J
1 cal	4.184 J
1 dyne	10^{-5} N
1 atm = 760 torr = 1.013 bar	$1.013 \times 10^5\ Nm^{-2}$
1 eV	1.602×10^{-19} J
1 lit-atm	101.27 J
1Å	10^{-10} m = 0.1 nm = 100 pm
1 lite = 1 dm^3	$10^{-3}\ m^3$
1 poise	0.1 Pa s = 0.1 $kg\ m^{-1}\ s^{-1}$ = 0.1 $N\ m^{-2}$ s

Table III: Strengths of some concentrated Laboratory Reagents

Calculate the Molarity or Normality of the reagent from the specific gravity and weight percentage written on the label of the bottle.

If s = sp.gr. and W = wt. %, then

$$\text{Normality} = \frac{10 \times s \times W}{\text{Eqv. wt}}, \quad \text{Molarity} = \frac{10 \times s \times W}{\text{Mol. wt.}}$$

Name of the reagent	Specific Gravity	Weight Percentage	Approx. Normality (N)	Approx. Molarity (M)
Hydrochloric Acid	1.19	38	12	12*
Sulphuric Acid	1.84	96	36	18
Nitric Acid	1.42	70	15.8	15.8
Phosphoric Acid (Syrupy)	1.69	85	44	14.7
Perchloric Acid	1.54	60	9	9
Glacial Acetic Acid	1.05	99.5	17	17
Liquor Ammonia	0.88	28	14.5	14.5

* 11.3 M may be taken during preparation of dilute solution.

Table IV: Molecular weights and equivalent weights of some compounds for preparation of primary or secondary standard solutions

Compounds	Formula	Mol. wt.	Eqv. wt.
Benzoic acid	$C_6H_5 \cdot CO_2H$	122.13	122.13
Oxalic acid	$H_2C_2O_4$, $2H_2O$	126.06	63.03
Succinic acid	$H_6C_4O_4$	118.09	59.045
Salicylic acid	$H_6C_7O_3$	138.12	138.12
Zinc acetate	$(CH_3COO)_2Zn$, $2H_2O$	219.4 (Generally solutions are prepared in molarity)	109.7
EDTA (Disodium salt of ethylenediamine tetraacetic acid)	$Na_2H_2C_{10}H_{12}O_8N_2$, $2H_2O$	372.24 (Generally solutions are prepared in molarity)	186.12 (During complexation with M^{2+} ions)
Ferrous ammonium sulphate (Mohr's salt)	$FeSO_4$, $(NH_4)_2SO_4$, $6H_2O$	392.16	392.16
Potassium dichromate	$K_2Cr_2O_7$	294.2	49.03
Potassium permanganate	$KMnO_4$	158.03	(a) 31.6 (in acid medium) (b) 52.66 (in alkaline or neutral medium)
Sodium thiosulphate	$Na_2S_2O_3$, $5H_2O$	248.2	248.2
Potassium chloride	KCl	74.56	74.56
Caustic soda	NaOH	40	40
Caustic Potash	KOH	56.1	56.1

Table V: Preparation of some common solutions/reagents

1. **1% starch solution**

   Take 1 g of soluble starch and make a paste of it in a little amount of water. Pour the paste in 100 ml boiling water in a beaker and stir constantly with a glass rod for about a minute. Add a few drops of chloroform after removing from the flame. Use in a day.

2. (a) **0.4% Eriochrome Black T solution:** Dissolve 0.4 g EBT in 100 ml methanol.

   (b) **Solid EBT reagent:** Grind EBT and KNO_3 in 1 : 100 weight ratio. Use ~50 mg of the mixture for each titration.

3. **0.5% Phenolphthalein indicator solutions**

   Dissolve ~ 0.5 g of the solid indicator in a liquid mixture of 50 ml absolute alcohol + 50 ml distilled water.

4. **0.1% Bromocresol green indicator solution**

   Dissolve ~ 0.1 g of the solid indicator in 100 ml rectified spirit.

5. **0.05% Methyl Orange indicator solution**

   Dissolve 0.05 g of the dye in 100 ml distilled water.

6. **0.02% Methyl red solution**

   Dissolve 0.02 g of the dye in 60 ml ethanol and dilute to 100 ml with water.

Table VI: Relative atomic mass of elements (on C = 12.00 scale)

Atomic number with chemical symbol	Name of the element	Relative atomic mass	Atomic number with chemical symbol	Name of the element	Relative atomic mass
1 (H)	HYDROGEN	1.008	20 (Ca)	CALCIUM	40.08
2 (He)	HELIUM	4.003	21 (Sc)	SCANDIUM	45.10
3 (Li)	LITHIUM	6.340	22 (Ti)	TITANIUM	47.90
4 (Be)	BERYLLIUM	9.020	23 (V)	VANADIUM	50.95
5 (B)	BORON	10.820	24 (Cr)	CHROMIUM	52.01
6 (C)	CARBON	12.000	25 (Mn)	MANGANESE	54.93
7 (N)	NITROGEN	14.010	26 (Fe)	IRON	55.85
8 (O)	OXYGEN	16.010	27 (Co)	COBALT	58.94
9 (F)	FLUORINE	19.000	28 (Ni)	NICKEL	58.69
10 (Ne)	NEON	20.10	29 (Cu)	COPPER	63.54
11 (Na)	SODIUM	23.00	30 (Zn)	ZINC	65.38
12 (Mg)	MAGNESIUM	24.37	31 (Ga)	GALLIUM	69.72
13 (Al)	ALUMINIUM	26.97	32 (Ge)	GERMANIUM	72.60
14 (Si)	SILICON	28.06	33 (As)	ARSENIC	74.81
15 (P)	PHOSPHORUS	30.98	34 (Se)	SELENIUM	78.86
16 (S)	SULPHUR	32.07	35 (Br)	BROMINE	79.92
17 (Cl)	CHLORINE	35.46	36 (Kr)	KRYPTON	83.70
18 (Ar)	ARGON	39.94	37 (Rb)	RUBIDIUM	85.48
19 (K)	POTASSIUM	39.10	38 (Sr)	STRONTIUM	87.63

(Table VI contd.)

Atomic number with chemical symbol	Name of the element	Relative atomic mass	Atomic number with chemical symbol	Name of the element	Relative atomic mass
39 (Y)	YTTRIUM	88.92	66 (Dy)	DYSPROSIUM	162.5
40 (Zr)	ZIRCONIUM	91.22	67 (Ho)	HOLMIUM	164.9
41 (Nb)	NIOBIUM	92.01	68 (Er)	ERBIUM	167.2
42 (Mo)	MOLYBDENUM	95.95	69 (Tm)	THULIUM	169.4
43 (Tc)	TECHNETIUM	99.00	70 (Yb)	YTTERBIUM	173.0
44 (Ru)	RUTHENIUM	101.7	71 (Lu)	LUTETIUM	175.0
45 (Rh)	RHODIUM	102.90	72 (Hf)	HAFNIUM	178.6
46 (Pd)	PALLADIUM	106.70	73 (Ta)	TANTALUM	180.9
47 (Ag)	SILVER	107.90	74 (W)	TUNGSTEN	183.92
48 (Cd)	CADMIUM	112.40	75 (Re)	RHENIUM	188.3
49 (In)	INDIUM	114.80	76 (Os)	OSMIUM	190.2
50 (Sn)	TIN	118.70	77 (Ir)	IRIDIUM	193.1
51 (Sb)	ANTIMONY	121.80	78 (Pt)	PLATINUM	195.2
52 (Te)	TELLURIUM	127.60	79 (Au)	GOLD	197.23
53 (I)	IODINE	126.92	80 (Hg)	MERCURY	200.61
54 (Xe)	XENON	131.30	81 (Tl)	THALLIUM	204.39
55 (Cs)	CAESIUM	132.91	82 (Pb)	LEAD	207.21
56 (Ba)	BARIUM	137.36	83 (Bi)	BISMUTH	209.00
57 (La)	LANTHANUM	138.92	84 (Po)	POLONIUM	210.00
58 (Ce)	CERIUM	140.13	85 (At)	ASTATINE	210.00
59 (Pr)	PRASEODYMIUM	140.92	86 (Rn)	RADON	222.00
60 (Nd)	NEODYMIUM	144.27	87 (Fr)	FRANCIUM	223.02
61 (Pm)	PROMETHIUM	147	88 (Ra)	RADIUM	226.05
62 (Sm)	SAMARIUM	150.43	89 (Ac)	ACTINIUM	227.03
63 (Eu)	EUROPIUM	152.00	90 (Th)	THORIUM	232.12
64 (Gd)	GADOLINIUM	156.90	91 (Pa)	PROTOACTINIUM	231.00
65 (Tb)	TERBIUM	159.90	92 (U)	URANIUM	238.07

APPENDIX II : Logarithms and Antilogarithms

FOUR-FIGURE LOGARITHMIC TABLES

LOGARITHMS

	0	1	2	3	4	5	6	7	8	9	1	2	3	4	5	6	7	8	9
10	0000	0043	0086	0128	0170						5	9	13	17	21	26	30	34	38
						0212	0253	0294	0334	0374	4	8	12	16	20	24	28	32	36
11	0414	0453	0492	0531	0569						4	8	12	16	20	23	27	31	35
						0607	0645	0682	0719	0755	4	7	11	15	18	22	26	29	33
12	0792	0828	0864	0899	0934						3	7	11	14	18	21	25	28	32
						0969	1004	1038	1072	1106	3	7	10	14	17	20	24	27	31
13	1139	1173	1206	1239	1271						3	6	10	13	16	19	23	26	29
						1303	1335	1367	1399	1430	3	6	10	13	16	19	22	25	29
14	1461	1492	1523	1553	1584						3	6	9	12	15	19	22	25	28
						1614	1644	1673	1703	1732	3	6	9	12	14	17	20	23	26
15	1761	1790	1818	1847	1875						3	6	9	11	14	17	20	23	26
						1903	1931	1959	1987	2014	3	6	8	11	14	17	19	22	25
16	2041	2068	2095	2122	2148						3	6	8	11	14	16	19	22	24
						2175	2201	2227	2253	2279	3	5	8	10	13	16	18	21	23
17	2304	2330	2355	2380	2405						3	5	8	10	13	15	18	20	23
						2430	2455	2480	2504	2529	3	5	8	10	12	15	17	20	22
18	2553	2577	2601	2625	2648						2	5	7	9	12	14	17	19	21
						2672	2695	2718	2742	2765	2	4	7	9	11	14	16	18	21
19	2788	2810	2833	2856	2878						2	4	7	9	11	13	16	18	20
						2900	2923	2945	2967	2989	2	4	6	8	11	13	15	17	19
20	3010	3032	3054	3075	3096	3118	3139	3160	3181	3201	2	4	6	8	11	13	15	17	19
21	3222	3243	3263	3284	3304	3324	3345	3365	3385	3404	2	4	6	8	10	12	14	16	18
22	3424	3444	3464	3483	3502	3522	3541	3560	3579	3598	2	4	6	8	10	12	14	15	17
23	3617	3636	3655	3674	3692	3711	3729	3747	3766	3784	2	4	6	7	9	11	13	15	17
24	3802	3820	3838	3856	3874	3892	3909	3927	3945	3962	2	4	5	7	9	11	12	14	16
25	3979	3997	4014	4031	4048	4065	4082	4099	4116	4133	2	3	5	7	9	10	12	14	15
26	4150	4165	4183	4200	4216	4232	4249	4265	4281	4298	2	3	5	7	8	10	11	13	15
27	4314	4330	4346	4362	4378	4393	4409	4425	4440	4456	2	3	5	6	8	9	11	13	14
28	4472	4487	4502	4518	4533	4548	4564	4579	4594	4609	2	3	5	6	8	9	11	12	14
29	4624	4639	4654	4669	4683	4698	4713	4728	4742	4757	1	3	4	6	7	9	10	12	13
30	4771	4786	4800	4814	4829	4843	4857	4871	4886	4900	1	3	4	6	7	9	10	11	13
31	4914	4928	4942	4955	4969	4983	4997	5011	5024	5038	1	3	4	6	7	8	10	11	12
32	5051	5065	5079	5092	5105	5119	5132	5145	5159	5172	1	3	4	5	7	8	9	11	12
33	5185	5198	5211	5224	5237	5250	5263	5276	5289	5302	1	3	4	5	6	8	9	10	12
34	5315	5328	5340	5353	5366	5378	5391	5403	5416	5428	1	3	4	5	6	8	9	10	11
35	5441	5453	5465	5478	5490	5502	5514	5527	5539	5551	1	2	4	5	6	7	9	10	11
36	5563	5575	5587	5599	5611	5623	5635	5647	5658	5670	1	2	4	5	6	7	8	10	11
37	5682	5694	5705	5717	5729	5740	5752	5763	5775	5786	1	2	3	5	6	7	8	9	10
38	5798	5809	5821	5832	5843	5855	5866	5877	5888	5899	1	2	3	5	6	7	8	9	10
39	5911	5922	5933	5944	5955	5966	5977	5988	5999	6010	1	2	3	4	5	7	8	9	10
40	6021	6031	6042	6053	6064	6075	6085	6096	6107	6117	1	2	3	4	5	6	8	9	10
41	6128	6138	6149	6160	6170	6180	6191	6201	6212	6222	1	2	3	4	5	6	7	8	9
42	6232	6243	6253	6263	6274	6284	6294	6304	6314	6325	1	2	3	4	5	6	7	8	9
43	6335	6345	6355	6365	6375	6385	6395	6405	6415	6425	1	2	3	4	5	6	7	8	9
44	6435	6444	6454	6464	6474	6484	6493	6503	6513	6522	1	2	3	4	5	6	7	8	9
45	6532	6542	6551	6561	6571	6580	6590	6599	6609	6618	1	2	3	4	5	6	7	8	9
46	6628	6637	6646	6656	6665	6675	6684	6693	6702	6712	1	2	3	4	5	6	7	7	8
47	6721	6730	6739	6749	6758	6767	6776	6785	6794	6803	1	2	3	4	5	5	6	7	8
48	6812	6821	6830	6839	6848	6857	6866	6875	6884	6893	1	2	3	4	4	5	6	7	8
49	6902	6911	6920	6928	6937	6946	6955	6964	6972	6981	1	2	3	4	4	5	6	7	8

LOGARITHMS

	0	1	2	3	4	5	6	7	8	9	1	2	3	4	5	6	7	8	9
50	6990	6998	7007	7016	7024	7033	7042	7050	7059	7067	1	2	3	3	4	5	6	7	8
51	7076	7084	7093	7101	7110	7118	7126	7135	7143	7152	1	2	3	3	4	5	6	7	8
52	7160	7168	7177	7185	7193	7202	7210	7218	7226	7235	1	2	2	3	4	5	6	7	7
53	7243	7251	7259	7267	7275	7284	7292	7300	7308	7316	1	2	2	3	4	5	6	6	7
54	7324	7332	7340	7348	7356	7364	7372	7380	7388	7396	1	2	2	3	4	5	6	6	7
55	7404	7412	7419	7427	7435	7443	7451	7459	7466	7474	1	2	2	3	4	5	5	6	7
56	7482	7490	7497	7505	7513	7520	7528	7536	7543	7551	1	2	2	3	4	5	5	6	7
57	7559	7566	7574	7582	7589	7597	7604	7612	7619	7627	1	2	2	3	4	5	5	6	7
58	7634	7642	7649	7657	7664	7672	7679	7686	7694	7701	1	1	2	3	4	4	5	6	7
59	7709	7716	7733	7731	7738	7745	7752	7760	7767	7774	1	1	2	3	4	4	5	6	7
60	7782	7789	7796	7803	7810	7818	7825	7832	7839	7846	1	1	2	3	4	4	5	6	6
61	7853	7860	7868	7875	7882	7889	7896	7903	7910	7917	1	1	2	3	4	4	5	6	6
62	7924	7931	7938	7945	7952	7959	7966	7973	7980	7987	1	1	2	3	3	4	5	6	6
63	7993	8000	8007	8014	8021	8028	8035	8041	8048	8055	1	1	2	3	3	4	5	5	6
64	8062	8069	8075	8082	8089	8096	8102	8109	8116	8122	1	1	2	3	3	4	5	5	6
65	8129	8136	8142	8149	8156	8162	8169	8176	8182	8189	1	1	2	3	3	4	5	5	6
66	8195	8202	8209	8215	8222	8228	8235	8241	8248	8254	1	1	2	3	3	4	5	5	6
67	8261	8267	8274	8280	8287	8293	8299	8306	8312	8319	1	1	2	3	3	4	5	5	6
68	8325	8331	8338	8344	8351	8357	8363	8370	8376	8382	1	1	2	3	3	4	4	5	6
69	8388	8395	8401	8407	8414	8420	8426	8432	8439	8445	1	1	2	2	3	4	4	5	6
70	8451	8457	8463	8470	8476	8482	8488	8494	8500	8506	1	1	2	2	3	4	4	5	6
71	8513	8519	8525	8531	8537	8543	8549	8555	8561	8567	1	1	2	2	3	4	4	5	5
72	8573	8579	8585	8591	8597	8603	8609	8615	8621	8627	1	1	2	2	3	4	4	5	5
73	8633	8639	8645	8651	8657	8663	8669	8675	8681	8686	1	1	2	2	3	4	4	5	5
74	8692	8698	8704	8710	8716	8722	8727	8733	8739	8745	1	1	2	2	3	4	4	5	5
75	8751	8756	8762	8768	8774	8779	8785	8791	8797	8802	1	1	2	2	3	3	4	5	5
76	8808	8814	8820	8825	8831	8837	8842	8848	8854	8859	1	1	2	2	3	3	4	5	5
77	8865	8871	8876	8882	8887	8893	8899	8904	8910	8915	1	1	2	2	3	3	4	4	5
78	8921	8927	8932	8938	8943	8949	8954	8960	8965	8971	1	1	2	2	3	3	4	4	5
79	8976	8982	8987	8993	8998	9004	9009	9015	9020	9025	1	1	2	2	3	3	4	4	5
80	9031	9036	9042	9047	9053	9058	9063	9069	9074	9079	1	1	2	2	3	3	4	4	5
81	9085	9090	9096	9101	9106	9112	9117	9122	9128	9133	1	1	2	2	3	3	4	4	5
82	9138	9143	9149	9154	9159	9165	9170	9175	9180	9196	1	1	2	2	3	3	4	4	5
83	9191	9196	9201	9206	9212	9217	9222	9227	9232	9238	1	1	2	2	3	3	4	4	5
84	9243	9248	9253	9258	9263	9269	9274	9279	9284	9289	1	1	2	2	3	3	4	4	5
85	9294	9299	9304	9309	9315	9320	9325	9330	9335	9340	1	1	2	2	3	3	4	4	5
86	9345	9350	9355	9360	9365	9370	9375	9380	9385	9390	1	1	2	2	3	3	4	4	5
87	9395	9400	9405	9410	9415	9420	9425	9430	9435	9440	0	1	1	2	2	3	3	4	4
88	9445	9450	9455	9460	9465	9469	9474	9479	9484	9489	0	1	1	2	2	3	3	4	4
89	9494	9499	9504	9509	9513	9518	9523	9528	9533	9538	0	1	1	2	2	3	3	4	4
90	9542	9547	9552	9557	9562	9566	9571	9576	9581	9586	0	1	1	2	2	3	3	4	4
91	9590	9595	9600	9605	9609	9614	9619	9624	9628	9633	0	1	1	2	2	3	3	4	4
92	9638	9643	9647	9652	9657	9661	9666	9671	9675	9680	0	1	1	2	2	3	3	4	4
93	9685	9689	9694	9699	9703	9708	9713	9717	9722	9727	0	1	1	2	2	3	3	4	4
94	9731	9736	9741	9745	9750	9754	9759	9763	9768	9773	0	1	1	2	2	3	3	4	4
95	9777	9782	9786	9791	9795	9800	9805	9809	9814	9818	0	1	1	2	2	3	3	4	4
96	9823	9827	9832	9836	9841	9845	9850	9854	9859	9863	0	1	1	2	2	3	3	4	4
97	9868	9872	9877	9881	9886	9890	9894	9899	9903	9908	0	1	1	2	2	3	3	4	4
98	9912	9917	9921	9926	9930	9934	9939	9943	9948	9952	0	1	1	2	2	3	3	4	4
99	9956	9961	9965	9969	9974	9978	9983	9987	9991	9996	0	1	1	2	2	3	3	3	4

FOUR-FIGURE ANTILOGARITHMIC TABLES

ANTILOGARITHMS

	0	1	2	3	4	5	6	7	8	9	1	2	3	4	5	6	7	8	9
.00	1000	1002	1005	1007	1009	1012	1014	1016	1019	1021	0	0	1	1	1	1	2	2	2
.01	1023	1026	1028	1030	1033	1035	1038	1040	1042	1045	0	0	1	1	1	1	2	2	2
.02	1047	1050	1052	1054	1057	1059	1062	1064	1067	1069	0	0	1	1	1	1	2	2	2
.03	1072	1074	1076	1079	1081	1084	1086	1089	1091	1094	0	0	1	1	1	1	2	2	2
.04	1096	1099	1102	1104	1107	1109	1112	1114	1117	1119	0	1	1	1	1	2	2	2	2
.05	1122	1125	1127	1130	1132	1135	1138	1140	1143	1146	0	1	1	1	1	2	2	2	2
.06	1148	1151	1153	1156	1159	1161	1164	1167	1169	1172	0	1	1	1	1	2	2	2	2
.07	1175	1178	1180	1183	1186	1189	1191	1194	1197	1199	0	1	1	1	1	2	2	2	2
.08	1202	1205	1208	1211	1213	1216	1219	1222	1225	1227	0	1	1	1	1	2	2	2	3
.09	1230	1233	1236	1239	1242	1245	1247	1250	1253	1256	0	1	1	1	1	2	2	2	3
.10	1259	1262	1265	1268	1271	1274	1276	1279	1282	1285	0	1	1	1	1	2	2	2	3
.11	1288	1291	1294	1297	1300	1303	1306	1309	1312	1315	0	1	1	1	2	2	2	2	3
.12	1318	1321	1324	1327	1330	1334	1337	1340	1343	1346	0	1	1	1	2	2	2	2	3
.13	1349	1352	1355	1358	1361	1365	1368	1371	1374	1377	0	1	1	1	2	2	2	3	3
.14	1380	1384	1387	1390	1393	1396	1400	1403	1406	1409	0	1	1	1	2	2	2	3	3
.15	1413	1416	1419	1422	1426	1429	1432	1435	1439	1442	0	1	1	1	2	2	2	3	3
.16	1445	1449	1452	1455	1459	1462	1466	1469	1472	1476	0	1	1	1	2	2	2	3	3
.17	1479	1483	1486	1489	1493	1496	1500	1503	1507	1510	0	1	1	1	2	2	2	3	3
.18	1514	1517	1521	1524	1528	1531	1535	1538	1542	1545	0	1	1	1	2	2	2	3	3
.19	1549	1552	1556	1560	1563	1567	1570	1574	1578	1581	0	1	1	1	2	2	3	3	3
.20	1585	1589	1592	1596	1600	1603	1607	1611	1614	1618	0	1	1	1	2	2	3	3	3
.21	1622	1626	1629	1633	1637	1641	1644	1648	1652	1656	0	1	1	2	2	2	3	3	3
.22	1660	1663	1667	1671	1675	1679	1683	1687	1690	1694	0	1	1	2	2	2	3	3	3
.23	1698	1702	1706	1710	1714	1718	1722	1726	1730	1734	0	1	1	2	2	2	3	3	4
.24	1738	1742	1746	1750	1754	1758	1762	1766	1770	1774	0	1	1	2	2	2	3	3	4
.25	1778	1782	1786	1791	1795	1799	1803	1807	1811	1816	0	1	1	2	2	2	3	3	4
.26	1820	1824	1828	1832	1837	1841	1845	1849	1854	1858	0	1	1	2	2	3	3	3	4
.27	1862	1866	1871	1875	1879	1884	1888	1892	1897	1901	0	1	1	2	2	3	3	3	4
.28	1905	1910	1914	1919	1923	1928	1932	1936	1941	1945	0	1	1	2	2	3	3	4	4
.29	1950	1954	1959	1963	1968	1972	1977	1982	1986	1991	0	1	1	2	2	3	3	4	4
.30	1995	2000	2004	2009	2014	2018	2023	2028	2032	2037	0	1	1	2	2	3	3	4	4
.31	2042	2046	2051	2056	2061	2065	2070	2075	2080	2084	0	1	1	2	2	3	3	4	4
.32	2089	2094	2099	2104	2109	2113	2118	2123	2128	2133	0	1	1	2	2	3	3	4	4
.33	2138	2143	2148	2153	2158	2163	2168	2173	2178	2183	0	1	1	2	2	3	3	4	4
.34	2188	2193	2198	2203	2208	2213	2218	2223	2228	2234	1	1	2	2	3	3	4	4	5
.35	2239	2244	2249	2254	2259	2265	2270	2275	2280	2286	1	1	2	2	3	3	4	4	5
.36	2291	2296	2301	2307	2312	2317	2323	2328	2333	2339	1	1	2	2	3	3	4	4	5
.37	2344	2350	2355	2360	2366	2371	2377	2382	2388	2393	1	1	2	2	3	3	4	4	5
.38	2399	2404	2410	2415	2421	2427	2432	2438	2443	2449	1	1	2	2	3	3	4	4	5
.39	2455	2460	2466	2472	2477	2483	2489	2495	2500	2506	1	1	2	2	3	3	4	5	5
.40	2512	2518	2523	2529	2535	2541	2547	2553	2559	2564	1	1	2	2	3	4	4	5	5
.41	2570	2576	2582	2588	2594	2600	2606	2612	2618	2624	1	1	2	2	3	4	4	5	5
.42	2630	2636	2642	2649	2655	2661	2667	2673	2679	2685	1	1	2	2	3	4	4	5	6
.43	2692	2698	2704	2710	2716	2723	2729	2735	2742	2748	1	1	2	3	3	4	4	5	6
.44	2754	2761	2767	2773	2780	2786	2793	2799	2805	2812	1	1	2	3	3	4	4	5	6
.45	2818	2825	2831	2838	2844	2851	2858	2864	2871	2877	1	1	2	3	3	4	5	5	6
.46	2884	2891	2897	2904	2911	2917	2924	2931	2938	2944	1	1	2	3	3	4	5	5	6
.47	2951	2958	2965	2972	2979	2985	2992	2999	3006	3013	1	1	2	3	3	4	5	5	6
.48	3020	3027	3034	3041	3048	3055	3062	3069	3076	3083	1	1	2	3	4	4	5	6	6
.49	3090	3097	3105	3112	3119	3126	3133	3141	3148	3155	1	1	2	3	4	4	5	6	6

ANTILOGARITHMS

	0	1	2	3	4	5	6	7	8	9	1	2	3	4	5	6	7	8	9
.50	3162	3170	3177	3184	3192	3199	3206	3214	3221	3228	1	1	2	3	4	4	5	6	7
.51	3236	3243	3251	3258	3266	3273	3281	3289	3296	3304	1	2	2	3	4	5	5	6	7
.52	3311	3319	3327	3334	3342	3350	3357	3365	3373	3381	1	2	2	3	4	5	5	6	7
.53	3388	3396	3404	3412	3420	3428	3436	3443	3451	3459	1	2	2	3	4	5	6	6	7
.54	3467	3475	3483	3491	3499	3508	3516	3524	3532	3540	1	2	2	3	4	5	6	6	7
.55	3548	3556	3565	3573	3581	3589	3597	3606	3614	3622	1	2	2	3	4	5	6	7	7
.56	3631	3639	3648	3656	3664	3673	3681	3690	3698	3707	1	2	3	3	4	5	6	7	8
.57	3715	3724	3733	3741	3750	3758	3767	3776	3784	3793	1	2	3	3	4	5	6	7	8
.58	3802	3811	3819	3828	3837	3846	3855	3864	3873	3882	1	2	3	4	4	5	6	7	8
.59	3890	3899	3908	3917	3926	3936	3945	3954	3963	3972	1	2	3	4	5	5	6	7	8
.60	3981	3990	3999	4009	4018	4027	4036	4046	4055	4064	1	2	3	4	5	6	6	7	8
.61	4074	4083	4093	4102	4111	4121	4130	4140	4150	4159	1	2	3	4	5	6	7	8	9
.62	4169	4178	4188	4198	4207	4217	4227	4236	4246	4256	1	2	3	4	5	6	7	8	9
.63	4266	4276	4285	4295	4305	4315	4325	4335	4345	4355	1	2	3	4	5	6	7	8	9
.64	4365	4375	4385	4395	4406	4416	4426	4436	4446	4457	1	2	3	4	5	6	7	8	9
.65	4467	4477	4487	4498	4508	4519	4529	4539	4550	4560	1	2	3	4	5	6	7	8	9
.66	4571	4581	4592	4603	4613	4624	4634	4645	4656	4667	1	2	3	4	5	6	7	9	10
.67	4677	4688	4699	4710	4721	4732	4742	4753	4764	4775	1	2	3	4	5	7	8	9	10
.68	4786	4797	4808	4819	4831	4842	4853	4864	4875	4887	1	2	3	4	6	7	8	9	10
.69	4898	4909	4920	4932	4943	4955	4966	4977	4989	5000	1	2	3	5	6	7	8	9	10
.70	5012	5023	5035	5047	5058	5070	5082	5093	5105	5117	1	2	4	5	6	7	8	9	11
.71	5129	5140	5152	5164	5176	5188	5200	5212	5224	5236	1	2	4	5	6	7	8	10	11
.72	5248	5260	5272	5284	5297	5309	5321	5333	5346	5358	1	2	4	5	6	7	8	10	11
.73	5370	5383	5395	5408	5420	5433	5445	5458	5470	5483	1	3	4	5	6	8	9	10	11
.74	5495	5508	5521	5534	5546	5559	5572	5585	5598	5610	1	3	4	5	6	8	9	10	12
.75	5623	5636	5649	5662	5675	5689	5702	5715	5728	5741	1	3	4	5	7	8	9	10	12
.76	5754	5768	5781	5794	5808	5821	5834	5848	5861	5875	1	3	4	5	7	8	9	11	12
.77	5888	5902	5916	5929	5943	5957	5970	5984	5998	6012	1	3	4	5	7	8	10	11	12
.78	6026	6039	6053	6067	6081	6095	6109	6124	6138	6152	1	3	4	6	7	8	10	11	12
.79	6166	6180	6194	6209	6223	6237	6252	6266	6281	6295	1	3	4	6	7	9	10	11	13
.80	6310	6324	6339	6353	6368	6383	6397	6412	6427	6442	1	3	4	6	7	9	10	12	13
.81	6457	6471	6486	6501	6516	6531	6546	6561	6577	6592	2	3	5	6	8	9	11	12	14
.82	6607	6622	6637	6653	6668	6683	6699	6714	6730	6745	2	3	5	6	8	9	11	12	14
.83	6761	6776	6792	6808	6823	6839	6855	6871	6887	6902	2	3	5	6	8	9	11	13	14
.84	6918	6934	6950	6966	6982	6998	7015	7031	7047	7063	2	3	5	6	8	10	11	13	15
.85	7079	7096	7112	7129	7145	7161	7178	7194	7211	7228	2	3	5	7	8	10	12	13	15
.86	7244	7261	7278	7295	7311	7328	7345	7362	7379	7396	2	3	5	7	8	10	12	13	15
.87	7413	7430	7447	7464	7482	7499	7516	7534	7551	7568	2	3	5	7	9	10	12	14	16
.88	7586	7603	7621	7638	7656	7674	7691	7709	7727	7745	2	4	5	7	9	11	12	14	16
.89	7762	7780	7798	7816	7834	7852	7870	7889	7907	7925	2	4	5	7	9	11	13	14	16
.90	7943	7962	7980	7998	8017	8035	8054	8072	8091	8110	2	4	6	7	9	11	13	15	17
.91	8128	8147	8166	8185	8204	8222	8241	8260	8279	8299	2	4	6	8	9	11	13	15	17
.92	8318	8337	8356	8375	8395	8414	8433	8453	8472	8492	2	4	6	8	10	12	14	15	17
.93	8511	8531	8551	8570	8590	8610	8630	8650	8670	8690	2	4	6	8	10	12	14	16	18
.94	8710	8730	8750	8770	8790	8810	8831	8851	8872	8892	2	4	6	8	10	12	14	16	18
.95	8913	8933	8954	8974	8995	9016	9036	9057	9078	9099	2	4	6	8	10	12	15	17	19
.96	9120	9141	9162	9183	9204	9226	9247	9268	9290	9311	2	4	6	8	11	13	15	17	19
.97	9333	9354	9376	9397	9419	9441	9462	9484	9506	9528	2	4	7	9	11	13	15	17	20
.98	9550	9572	9594	9616	9638	9661	9683	9705	9727	9750	2	4	7	9	11	13	16	18	20
.99	9772	9795	9817	9840	9863	9886	9908	9931	9954	9977	2	5	7	9	11	14	16	18	20

Index

Some exciting gifts are up for grabs with this book!!

PHYSICAL CHEMISTRY PRACTICAL

Dr Saroj Kumar Maity and Dr Naba Kumar Ghosh

FREE Gifts Rs. 1000/-

DEAR READERS

By now you must have gone through this book in the entirety. So we invite your assessment on this title with absolute impartiality and candid observations. After all, we publishers blindly depend on your benchmarking. Friends, please fill up the open spaces below with CAPITAL LETTERS. You may add some extra sheets of paper if the space provided is not sufficient for you.

i. What are the important topics of your syllabus which are not adequately covered in this book?

ii. In which chapters and on what topics the treatment of the subject-matter by the author is not satisfactory and why?

iii. Have you come across the typos, factual inaccuracies, inconsistencies, and other forms of mistakes in the book? If yes, then please specify the mistakes and page numbers.

iv. Name top three books on the same subject, which, in your consideration, are better than this book. Could you provide some reasons?

Your 1st Preference: ....

Your 2nd Preference: ....

Your 3rd Preference: ....

v. Any suggestion to improve the quality of illustrations, diagrams, etc.?

vi. How did you like the cover design? Any suggestion on colours?

vii. Suggestions for further improvement of production-standard:

viii. Who recommended you this book:

☐ Teacher ☐ Friend ☐ Parents ☐ Bookseller ☐ Librarian

ix. Name of the person who has recommended:

Designation: ....

Address: ....

Phone No.: ....

e-mail: ....

x. Name and address of the bookseller from whom you have purchased this book:

....

....

....

xi. Name and address of the library where you would like to find the book:

....

....

xii. Name and address of your institution (Please mention the University or Board):

....

....

xiii. Your name, e-mail, and complete postal address:

....

....

....

xiv. A few other titles published by us that you would like to read:

....

....

The best assessment will be rewarded half-yearly. The award will be in the form of our publications, as decided by the Editorial and Marketing Operations Division, amounting to Rs. 1000/- (Rupees one thousand) only.

....

Please mail the filled up coupon at your earliest to :

Editorial & Marketing Operations Division

NEW CENTRAL BOOK AGENCY (P) Ltd

8/1 Chintamoni Das Lane, Kolkata 700 009

West Bengal, INDIA

Phone + Fax: 033 + 2241 8411, 2241 8573, 2241 9094

e-mail: ncbapvtltd@eth.net, ncba@cal2.vsnl.net.in

Website: www.centralgroup.com

